Evolution

Essays in honour of John Maynard Smith

EDITED BY

P. J. Greenwood
DEPARTMENT OF ADULT AND CONTINUING EDUCATION
UNIVERSITY OF DURHAM
DURHAM, UK

P. H. Harvey
DEPARTMENT OF ZOOLOGY
UNIVERSITY OF OXFORD
OXFORD UK

AND

M. Slatkin
DEPARTMENT OF ZOOLOGY
UNIVERSITY OF WASHINGTON
SEATTLE, WASHINGTON, USA

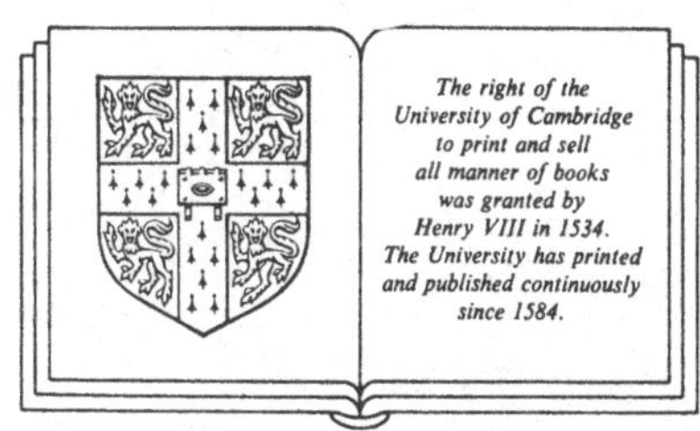

CAMBRIDGE UNIVERSITY PRESS
CAMBRIDGE
NEW YORK NEW ROCHELLE MELBOURNE SYDNEY

CAMBRIDGE UNIVERSITY PRESS
Cambridge, New York, Melbourne, Madrid, Cape Town, Singapore, São Paulo, Delhi

Cambridge University Press
The Edinburgh Building, Cambridge CB2 8RU, UK

Published in the United States of America by Cambridge University Press, New York

www.cambridge.org
Information on this title: www.cambridge.org/9780521348973

First published 1985
First paperback edition 1987
Re-issued in this digitally printed version 2009

A catalogue record for this publication is available from the British Library

Library of Congress Catalogue Card Number: 84-14221

ISBN 978-0-521-25734-3 hardback
ISBN 978-0-521-34897-3 paperback

Contents

The evolutionary ecology of sex

Contributors

B. O. BENGTSSON, Institute of Genetics, University of Lund, Sölvegatan 29, S-223 62 Sweden

J. W. BRADBURY, Department of Biology, University of California at San Diego, San Diego, La Jolla, CA 92093, USA

E. L. CHARNOV, Departments of Biology, Anthropology and Psychology, University of Utah, Salt Lake City, Utah 84112, USA

T. H. CLUTTON-BROCK, Large Animal Research Group, Department of Zoology, University of Cambridge, Cambridge, UK

D. CHARLESWORTH, School of Biological Sciences, University of Sussex, Falmer, Brighton BN1 9QG, UK

J. FELSENSTEIN, Department of Genetics SK-50, University of Washington, Seattle, Washington 98195, USA

R. GIBSON, Department of Biology, University of California at San Diego, San Diego, La Jolla, CA 92093, USA

J. L. GITTLEMAN, Department of Zoological Research, National Zoological Park, Smithsonian Institution, Washington, DC 20008, USA

P. J. GREENWOOD, Department of Adult and Continuing Education, University of Durham, Durham, UK

P. HAMMERSTEIN, Institute of Mathematical Economics, University of Bielefeld, FRG

P. H. HARVEY, Department of Zoology, University of Oxford, South Parks Road, Oxford OX1 3PS, UK

J. A. LEÓN, Instituto de Zoologia Tropical, Fac. Ciencias – UCV, Aptdo. 47058, Caracas 1041-A, Venezuela

R. C. LEWONTIN, Museum of Comparative Zoology, Harvard University, Cambridge, MA 02138, USA

R. M. MAY, Department of Biology, Princeton University, Princeton, New Jersey 08540, USA

R. E. MICHOD, Department of Ecology and Evolutionary Biology, University of Arizona, Tuscon, Arizona 85721, USA

C. PACKER, Department of Ecology and Behavioral Biology, University of Minnesota, 318 Church St SE, Minneapolis, Minnesota 55455, USA

G. A. PARKER, Department of Zoology, University of Liverpool, Liverpool, UK

A. PUSEY, Department of Ecology and Behavioral Biology, University of Minnesota, 318 Church St SE, Minneapolis, Minnesota 55455, USA

K. RALLS, Department of Zoological Research, National Zoological Park, Smithsonian Institution, Washington, DC 20008, USA

M. R. ROSE, Department of Biology, Dalhousie University, Halifax, Nova Scotia B3H 4J1, Canada

M. J. SANDERSON, Department of Ecology and Evolutionary Biology, University of Arizona, Tuscon, Arizona 85721, USA

J. SEGER, Department of Biology, Princeton University, Princeton, New Jersey 08540, USA

J. SILVERTOWN, Department of Biology, Open University, Walton Hall, Milton Keynes MK7 6AA, UK

M. SLATKIN, Department of Zoology, NJ-15, University of Washington, Seattle, Washington 98195, USA

N. C. STENSETH, Zoological Institute, University of Oslo, PO Box 1050, Blindern, Oslo 3, Norway

S. L. VEHRENCAMP, Department of Biology, University of California at San Diego, San Diego, La Jolla, CA 92093, USA

P. WHEELER, Department of Biology, Liverpool Polytechnic, Liverpool, UK

Editorial preface

If there is a single characteristic that has made John Maynard Smith's influence on evolutionary biology so great, it has been his gift for finding the right metaphor to describe succinctly an evolutionary problem. The terms he has introduced and his ways of viewing evolutionary phenomena have become so pervasive that it is easy to forget that he was the first to discuss such problems as the contrast between 'kin and group selection', to focus attention on the 'cost of meiosis', to ask about the 'hitch-hiking of genes', to elucidate 'Haldane's dilemma'. These are of course in addition to his many illuminating contributions to population genetic and ecological theory and his numerous efforts in the general area of theoretical biology. Like his mentor, J. B. S. Haldane, Maynard Smith has worked in so many fields that one might think there are several authors with the same name. The citations to Maynard-Smith and J. M. Smith have added to this impression.

Maynard Smith's influence in biology has been more than through his writings, however. With his boundless energy and his curiosity for all biological subjects, he has motivated and stimulated scientists wherever he has visited. And his personal influence on those of us fortunate to know him well has been profound. Whether over coffee or beer or on long walks, we have been treated to his narratives (some of which are remarkable for their evolutionary stability), advice, observations, and laws.* His stories are instructive at a variety of levels. Anyone who has heard about Maynard Smith being stranded in traffic in Haldane's automobile will forever be mindful of the precautions needed in the event of a fire.

* Three rules for scientific argument that have become known with sufficient retelling as 'Smith's Laws' are: (1) the bellman's theorem (which Maynard Smith attributes to Haldane), 'What I say three times must be true' (from Lewis Carroll); and either (2) Aunt Jabisco's theorem, 'It is a fact the whole world knows' (from Edward Lear), or (3) the third law, 'It is a truth universally acknowledged . . .' (from Jane Austen). Any scientific dispute that cannot be resolved on objective grounds (and some that can) can be settled by invoking these powerful theorems.

The authors who have contributed to this volume of essays feel a special debt to John Maynard Smith and we are pleased to dedicate this volume to him. The diversity of views represented and the diversity of approaches taken reflect at least partially the breadth of his own interests. We thank the authors, together with Martin Walters and Valerie Neal of Cambridge University Press, for their cooperation and patience throughout the preparation of this book.

Population genetics and evolution theory

1

Population genetics

R. C. LEWONTIN

Population genetics is one of the few biological sciences that has both a theoretical and an observational aspect. In this it resembles ecology, but it is unique in the degree to which the theoretical and the observational are tied to each other, at least in principle and in motivation, if not in useful results. There is no other biological science in which observations are so often directly motivated by formal theory and in which formal theory is so often constructed in an explicit attempt to make sense of the observations. The comparison with ecology is instructive. Theoretical population ecology is almost entirely the elaboration of a single underlying model, the logistic equation of population growth, for which there is virtually no empirical justification. At its most general, population ecological modelling does not take the logistic seriously, but supposes an unspecified multispecies interaction model which is then expanded in a Taylor's series, yielding, to the second term – the logistic model! Virtually all of observational ecology, on the other hand, is phenomenological. Do species interact? How? Can predation, competition, weather be shown to be causally efficacious or not in the determination of numbers of coexistence?

In contrast, population genetics begins with the undoubted facts of Mendelism, of chromosomal recombination, of mutation, of inbreeding, and builds a theoretical structure that is unassailable in its general outline. When called upon, it can even accommodate itself to non-Mendelian mechanisms of inheritance, to gene duplication and amplification, to any of the myriad genetic phenomena that are uncovered in the course of mechanical genetic investigation, as for example segregation distortion or gene conversion. At the same time, observational studies estimate the genetic variation that is the subject of the theoretical structure, or attempt to estimate the parameters of selection, migration, mutation, inbreeding, mating that are prescribed as relevant by the theoretical equations. The collaboration between Sewall Wright and Th. Dobzhansky in the observation and analysis of genic and chromosomal variation in natural populations has no parallel elsewhere in biology.* Nor was that union of theory and practice unique. The number of population geneticists, like John Maynard Smith, whose own research work has consisted of significant contributions to both theoretical and observation work, is quite remarkable.†

In view of the extraordinary interconnection between theory and observation in population genetics, we would expect to see a coherent field, well on its way to closure, having answered in an unambiguous fashion most or all of its leading questions. Instead, we see a field in disarray, with contending schools of explanation, apparently no closer to agreement on outstanding issues than they were 35 years ago when I first entered the professional study of the subject. At the moment, an appearance of calm pervades the population genetic arena, but a sharp ear will detect the panting of the out-of-breath contestants, as they try to regain some energy for yet another frustrating and indecisive round. What is truly remarkable is that, even when a question is decided, the result may have no perceptible effect on practice. It has now been 20 years since Moll, Lindsey and Robinson showed conclusively that overdominance is not the cause of heterosis in hybrid corn, a result with which no one apparently disagrees (Moll, Lindsey & Robinson, 1964), yet plant breeding continues to use a method designed for overdominance. How are we to explain the lack of real progress toward a consensus on the problems of population genetics, and what are the prospects for a better future?

In a previous work (Lewontin, 1974), I attempted to explain the unsatisfactory state of evolutionary genetics by the difficulty in measuring the actual quantities that appear in theoretical structures. The equations of population genetics deal with genotype frequencies, yet we have been obliged for most of the history of population genetics to study either phenotypes of unknown genotypic basis, or genotypes of no real interest in evolution. One would have thought that the last 20 years of observations on protein variation would have got around that epistemological paradox since protein variation usually has a well-defined genetic basis, and that variation is clearly something that lies at the base of a lot of non-trivial evolution. Yet struggles over what forces are operating on genetic variation in evolution still continue and, indeed, have been exacerbated by the recent interest in so-called punctuated equilibrium. Moreover, population genetics is not coextensive with evolutionary genetics, although people sometimes seem to think so. A large part of human population genetics,

* For details of the motivation, joint planning, and analysis of experiments by Dobzhansky and Wright, see Lewontin *et al.* (1981).

† In addition to Fisher, Wright, Haldane and H. J. Muller, I was able to list 19 in a quarter of an hour. A little more bibliographical effort would no doubt produce an equal number of others. Even a pure theoretician like M. Kimura is constantly motivated by and tied to experimental observations.

especially since the discrediting of eugenics in the 1950s, has concentrated on uncovering the causes of normal and abnormal variation in humans, in an attempt to be relevant to medicine and public health. Heritability studies of alcoholism and the 'genetic epidemiology' of feeble-mindedness use the apparatus of population genetics and are in at least as much contention as the problem of the evolution of enzymes. Agricultural population genetics, so called 'biometrical genetics', seems becalmed in Doldrums from which not even the combined huffing and puffing of statisticians and molecular geneticists has succeeded in moving it. Population genetics as a whole seems as unable as ever to solve the problems it has set for itself.

The difficulty is that in the development of population genetics there has been an inversion of the relationship of theory and observation. Population genetic theory begins as a deductive process. The phenomena of Mendelian segregation and chromosomal recombination, of mutation, migration, mating pattern, of the contingent development of the organism given the genotype and the environment, and of the differential survival and reproduction of genotypes, all subject to stochastic fluctuations, are built explicitly into the formal models of population genetic theory. All these phenomena are forces that are assumed to operate in a precisely defined way on the genotypic and phenotypic composition of the population. The straightforward problem of population genetics, then, is to compute the genetical state of the population at some future time from the present state and the forces. In formal terms, the state of the population at time t in the future is given by

$$x_t = g(x_0, \pi, t), \tag{1}$$

where π is the set of parameters of the process g. The process may also be computed backwards by solving the equation for x_0, given x_t. Adding a stochastic element changes (1) to a probabilistic statement,

$$\Pr\{x_t = x\} = h(x_0, \pi, t), \tag{2}$$

but otherwise the problem remains the same.

Given in this form, population genetics is a computational science. Tell me the genetics of the trait, the mutation rate, selective forces, population size, migration rates, etc., and I will tell you the trajectory (including the equilibrium) of the system. This computation can be carried out, however, only if the values of the parameters, π, are known. The straightforward approach would then be to devise independent methods of estimating the parameters, say of the amount of migration, by actually following individual organisms in their lifetimes, and to substitute the values found into (1) or (2). This deductive programme, however, has not been carried out for a number of reasons. First, it is very difficult to measure the actual rates of mutation and migration, the actual patterns of mating relations, actual $l(x)$ and $m(x)$ schedules of phenotypes, and the norms of reaction of the various genotypes, even if the genotypes could be identified. Second, the genotypes cannot always be recognized, especially for metric characters. Third, developmental patterns, probabilities of survival and reproduction, and behavioural phenomena like mating and migration, are all contingent on a variable environment, and they cannot be measured once for all. Their pattern of response to environment must be characterized, and then the environmental pattern must be recorded. Fourth, in the case of deducing the past from the present, we would need to know the past environments, an impossible task. In practice, if it were important enough, we might gather the necessary information for one trait, say sickle-cell anaemia in humans, where many of the parameters can be assumed to be constant and where the genetic basis of the trait is simple. But, as a practical programme for any reasonable number of evolutionary pathways, the computational path of deduction is clearly out of the question. Moreover, it is not clear that the computation question is really the one of interest. In 'genetic epidemiology', for example, the question is not 'what pattern of feeble-mindedness in the population would result from a given combination of genotypic and environmental variables' but, on the contrary, 'what are the developmental phenomena actually operating in families to produce the pattern we see?'. In evolutionary population genetics, we are generally more interested in questions like 'Is most of the amino acid substitution that has occurred in the evolution of the species a consequence of natural selection or of random forces?' than in predicting or retrodicting the evolutionary trajectory of a particular enzyme protein. That is, the questions are more often about the forces themselves than about their outcome, because the aesthetics of 'pure' science demands generality rather than specificity.

The consequence of the interest in the forces themselves, the parameters in Eqns (1) and (2), and of the immense practical difficulties of measuring these forces directly is that the theoretical formulations of population genetics have been inverted. Instead of computational devices to predict $x(t)$, they have become inferential structures to estimate π. This is, of course, the classical method of statistical inference: we deduce the observations that *would* be taken, given various hypotheses about the hidden universe, and then use the observations *actually* taken to choose the most likely universe.

Three examples of this method in population genetics will illustrate its operation. In his widely quoted but little-read paper, 'The covariance between relatives on the supposition of Mendelian inheritance', Fisher (1918) showed how the phenotypic variances and covariances of relatives of various degrees could be written in terms of the parameters for two loci with two alleles, each influencing the phenotype and, therefore, by implication, for any number of alleles at any number of loci. For two loci, there are two gene frequencies, two additive effects, two dominance effects, four epistatic parameters, and an environmental variance (a total of 11 parameters). This can be reduced to ten if one examines only crosses between inbred lines so that segregating gene frequencies will all be $p=q=\frac{1}{2}$. While there is no evolutionary dynamic here, it is still a computational problem analogous to Eqn (1), namely,

$$\sigma_{(R,R^1)_i}=g_i(\pi_1,\pi_2,\ldots,\pi_{11}). \tag{3}$$

That is, the expected phenotypic covariance between relatives R and R^1 of the ith degree can be computed by a specific function of the genetic and environmental parameters. But, of course, no one knows how to measure directly the additive, dominance, and epistatic effects of pairs of loci governing, say, yield in corn, not to speak of higher-order interactions for multiple loci. During the 1940s and 1950s, it became clear that rational choices of breeding methods could be made provided one could estimate these parameters, in particular the degree of dominance and epistasis in comparison with additive effects in a population. The result was a series of estimation equations of the type

$$\pi_i=h_i(\sigma(R,R^1)_1,\sigma(R,R^1)_2,\ldots), \tag{4}$$

a particular example of which is Comstock's estimate of the average degree of dominance over all loci:

$$\text{‘}a\text{’}=\frac{[2(\sigma_f^2-\sigma_m^2)]^{1/2}}{\sigma_m^2}, \tag{5}$$

where σ_f^2 and σ_m^2 are estimates of the variance among males and females within males in half-sib and full-sib families. The trouble with this estimate is that it is a fictitious ‘average dominance’ confounded by epistatic interactions and linked genes so that it estimated dominance only under certain simplifying assumptions. As a consequence, experiments of different structure gave different answers to the question of whether there was significant overdominance in corn, a question that was settled only after experiments specially designed to eliminate the effects of linkage were performed.

A less-happy outcome has characterized the second example, estimation of fitness in evolving populations. Beginning with a simple model of differential viability in an organism with discrete generations, we can predict successive generations. This relationship has then been inverted to make the observed ratios as the independent variables and the fitnesses as the dependent variables which can then be estimated from the inverse equations. However, Prout (1965) has shown that this procedure is completely invalid if there is differential fertility and will yield false frequency-dependent fitnesses, or even no evidence of selection when there is strong selection on fertility. Moreover, he has shown that *in principle* no procedure involving the genotypic frequencies in two successive generations will yield correct fitness estimates.

The third example illustrates a different sort of problem. Finite breeding size in a population results in a predictable probability of identity by descent of two alleles taken at random from a population. A special case of this identity is the frequency of allelism of lethals, which, roughly speaking, will be inversely proportional to the effective breeding size of the population. However, because the number of loci mutating to lethals is limited (about 400–600 per chromosome area in *Drosophila melanogaster*, for example), the observed frequency of allelism of lethals must be corrected for the allelism that arises from this limited number of lethal-bearing loci. That is, the allelism relevant to effective population size is

$$a_{\text{eff}} = \left(a_{\text{total}} - \frac{1}{n} \right),$$

where n is the number of loci mutating to lethals. In turn, the predicted a_{eff} is inversely proportional to effective population size N. That is,

$$a_{\text{eff}} \propto \frac{1}{2N},$$

so that inverting this relationship to estimate N gives us

$$N \propto \frac{1}{\left(a_{\text{total}} - \frac{1}{n} \right)}.$$

That is, the estimate of effective breeding size depends upon the reciprocal of the difference between two very small numbers, each of which has been derived from a very tedious experiment involving the test of allelism of a large number of lethals, and each with a considerable standard error. As a consequence the actual estimate of N may, with high probability, be negative or nearly infinite. An application of this technique to laboratory populations of known size had the result that it was not possible to distinguish between a population of 5000 from an infinite one (Prout, 1954).

These three examples illustrate different problems that arise when a causal relationship is inverted to attempt to estimate the causes. In the first case, the attempt to estimate the degree of dominance of genes suffers from the dependence of the covariances on a very large number of parameters of which the dominance is only one, so that inverting the deductive relationship can only be carried out by assuming values for the unknown parameters. In practice, these parameters can be estimated by yet other experiments or, as was the actual case for the dominance ratio in corn, experiments can be devised that reduce or remove the influence of the other factors such as linkage. In the case of corn, it was necessary to repeat the estimation experiment in various advanced generations of a random-mating population made from the original lines, an experiment that required a long-range plan and many years of experimental work.

In the second case, the possibility of inverting the causal relationship is structurally unstable to perturbations in the underlying model, and no experiments of this class will solve the problem. In the particular case of fitness estimation, this structural instability arises because meiosis and mating are randomizing events that actually destroy information created by the selection process, so that genotypic frequencies at two moments separated in the life cycle by meiosis and mating do not contain the necessary information to reconstruct the selection rules. On the other hand, two moments in the life cycle not separated by meiosis and mating do not contain any information about differential fertility, so the problem is insoluble by this approach.

The third case is the classic one of statistical estimation. Because the causes and effects are related as mathematical reciprocals, the sensitivity of dependent and independent variables to small perturbations are inverted. Large variation in N and n make only small absolute differences in the allelism outcomes, but reciprocally small differences in the observed allelism in an experiment result in immense variation in the estimate N, passing from the positive range through positive infinity into the negative half plane. When estimates are extremely sensitive to random variation in actual parameters or actual observations, the estimation procedure is useless. It is remarkable in population genetics how often a deductive relation has been inverted to create an estimation procedure without asking the question of its sensitivity to errors. For example, models of genetic epidemiology which depend upon path-coefficient analysis or related methods have seldom been subjected to extensive perturbation analysis to see how sensitive the estimates of genetic parameters are to different data sets. Monte Carlo sampling schemes or high-speed computers make this kind of perturbation analysis possible, although not trivial.*

The examples given are not exceptions, but the rule. Because of the large number of causal factors that enter into the determination of the genetic state of a population or an ensemble of populations and because some of these factors are randomizations that actually then destroy the products of information-creating causes, it will almost never be possible to invert the relation that predicts genetic structures from parameters of causal processes, in order to estimate the intensities of these processes. How, then, can population genetics do its business? Is it doomed to be nothing but perpetual number-juggling with no satisfactory closure? The answer lies in two directions, one being a proper understanding of what population

* The recent paper by Greenberg (1984) explicitly attacks this question. After a Monte Carlo simulation study to test how much power the methods have to distinguish between a one-locus and two-locus model with different degrees of dominance, about as simple a choice as one can imagine in genetics, the author notes: 'The studies reported here were extremely time consuming, both in human time and computer time. Yet such studies are an important way of testing the tools that geneticists use'. Right on!

genetics already provides and the second being a reorientation of population genetic research.

The first point is that population genetic theory is not designed to choose among competing hypotheses about causal forces. As I have explained, the attempt to use it for that purpose is destined, nearly always, to failure. Rather, population genetic theory is a descriptive theory that provides the mapping of causal processes as genetic outcomes. It says, '*if* mutation rates are such and such, *if* the mating pattern is such a one, *if* there are five genes affecting the character with the following norms of reaction, *then* the trajectory of the population in time, or the equilibrium state, or the steady state distribution of gene frequencies will be such and such'. This mapping serves two related functions. One is to provide qualitative prohibitions that enable us to exclude certain explanations of the observed genetic structure. So, for example, if we observe long-term stability and universal polymorphism for the Rh blood group in *Homo sapiens*, we cannot explain it by selection against the offspring of incompatible matings, because the mathematical analysis of such a selection process shows that it will lead to fixation of alleles. Or, if all the genetic variance for a trait is dominance variance, no argument about the relative advantage of increasing or decreasing the trait is relevant, because selection cannot be effective in the absence of additive variance. Nor are we allowed to explain a stable polymorphism as a consequence of random uncorrelated variation in selection coefficients.

More important, population genetic theory, precisely because of its weak inferential power, shows that claims for unambiguous explanations of observations are usually wrong. Claims that a certain pattern of genetic differentiation within and between populations are due to natural selection can almost always be shown to be too strong, because the same pattern can be produced by random drift and migration. So, for example, the claim of Prakash & Lewontin (1968) that the association between allozyme alleles and inversions in *Drosophila pseudo-obscura* and *D. persimilis* could only be the result of selection was contradicted in a stochastic analysis by Nei & Li (1975). In general, claims that differences in characters between species are unequivocal evidence that selection has been differential in the species are contradicted by the stochastic theory of population genetics of multiple locus traits. The theory of selection in linked multilocus systems which has become highly developed in the last 20 years does not have as its purpose the explanation of observed polymorphisms, but the addition of another mechanism which, under appropriate conditions of the parameters, must be taken into account in explanation. The theory of Evolutionarily Stable Strategies (ESS) must be seen in the same light. It is extremely unlikely that the necessary fitness parameters will be determined so as to show that particular behavioural repertoires are, in fact, evolutionarily stable. Rather,

ESS theory shows what is possible and what is impossible given various ranges of parameters. It tells us to be surprised at some outcomes but not surprised at others.

The delineation of the prohibited and the possible is the function of population genetic theory. The revelation of the actual is the task of population genetic experiments, a task that such experiments can accomplish provided they are freed of their strong dependence on the quantitative and statistical relations predicted by theoretical formulations and instead are constructed to provide unambiguous qualitative information. The possibility of the reorientation of experimental work, and the way such a change can give unambiguous evidence is shown by the history of studies of molecular polymorphism in natural populations.

Since 1966, when gel electrophoresis of proteins was introduced as a technique for studying genetic variation in natural populations, a very large portion of the work in experimental population genetics has been concerned with the study of enzyme polymorphisms. A general result has emerged that in a typical species about one third of structural gene loci studied show some polymorphism and that a typical individual is heterozygous at about 10% of its loci. These figures vary from species to species: mammals are somewhat less polymorphic than the average, insects rather more; an occasional species is reputed to be nearly totally without variation; one species of *Drosophila* is reported to have 85% of its loci polymorphic, but the modal figures are a reasonable characterization of vascular and non-vascular plants, bacteria, vertebrates and invertebrates of all sorts. The monotonous repeatability of this observation in so many different organisms has resulted in the major preoccupation of population genetics with an explanation of the polymorphism. Three approaches to explanation have been taken. One has been to compare gross statistics like the average heterozygosity or the total number of alleles segregating per locus, or the proportion of loci which are polymorphic, with theoretical predictions generated from selective and stochastic hypotheses. A second, more fine-grained, statistical approach has been to use the distribution of allelic frequencies within populations and the comparison of the distribution between populations to compare with predictions generated by competing hypotheses. A third, non-statistical, approach has been to attempt the direct measurement of selective differences or at least physiological differences between various genotypes segregating at structural gene loci. None of these attempts has been satisfactory. The statistical approaches have lacked the necessary discriminatory power for the reasons I have discussed above. The physiological approach has sometimes demonstrated selection but most often has failed because the necessary experimental power needed to detect the expected small selective differences between the genotypes has not been practical (Lewontin, 1974).

The situation has been made especially difficult by several uncertainties. If gel electrophoresis only detects charge substitutions as has been widely assumed on *a priori* grounds, then perhaps the apparently monomorphic loci are really polymorphic but do not happen to have amino acid substitutions that change the charge on the protein. Alternatively, perhaps the only polymorphism is, in fact, for charged amino acids since these residues occur predominantly on the outer surface of the folded polypeptide and are irrelevant to the function of the enzyme. This is an *a priori* hypothesis strongly favoured by the so-called neutralist school (Kimura & Ohta, 1971). In either case what is identified as an allele may be a composite of indistinguishable alleles so that neither the allelic distribution within populations nor the comparison of populations can be taken seriously. Attempts to take these ambiguities into account have only weakened the inferential power of the statistical approach by introducing yet more unknown parameters.

A second serious difficulty has been that no control values have existed against which one could compare supposedly selected loci. The comparison of gene frequency distribution and polymorphism always has been made against *theoretical* values without selection, but such theoretical values always are based upon unmeasured arbitrarily assigned parameters like effective population size, migration rates, mutation rates, etc. If, on the other hand, a large body of data existed on the polymorphism of genes known not to be selected (because they are not transcribed, for example) within a population, then this would serve as the control against which to test other loci without the necessity of making *a priori* guesses about parameters.

During the last 10 years, a series of experiments have appeared which speak directly to these issues and which, for the first time, give unambiguous qualitative information on polymorphism. One series of experiments using a variety of pH and gel concentrations (so-called sequential gel electrophoresis) has shown that, indeed, electromorphs may consist of a large number of different amino acid substitutions that had previously been lumped. Table 1 shows what happens when the usual allelomorphic classes of the xanthine dehydrogenase gene in *Drosophila pseudoobscura* are subjected to sequential gel electrophoresis (Singh, Lewontin & Felton, 1976). Eight alleles become 27, but one remains highly frequent while all other classes are much less frequent, many being unique in the sample. The same multiplication of alleles is seen for *Xdh* in *D. persimilis* (Coyne, 1976) and for esterase-5 in *D. pseudoobscura* (Keith, 1983; see also Table 2). On the other hand, when exactly the same procedure is applied to previously monomorphic loci (Coyne & Felton, 1977) or to loci with only two major polymorphic alleles (Kreitman, 1980), no breakup of allelic classes occurs. Sequential gel electrophoretic studies establish that monomorphic and

Table 1. *Frequency distribution of alleles of XDH from 12 populations of* D. pseudoobscura *after sequential gel electrophoresis*

Criteria used in sequence[a]				
(1)	(1+2)	(1+2+3)	(1+2+3+4)	Allelic code
1	1	1	1	*1000*
1	1	1	1	*2000*
5	5	5	5	*3000*
4	4	3	3	*4000*
		1	1	*4010*
19	9	1	1	*5000*
		1	8	*5010*
	10	1	1	*5100*
		7	2	*5110*
			5	*5111*
		2	2	*5120*
103	4	1	1	*6000*
		3	3	*6010*
	3	3	3	*6100*
	8	6	6	*6200*
		2	2	*6210*
	88	6	6	*6300*
		79	11	*6310*
			68	*6311*
		3	3	*6320*
7	6	1	1	*7000*
		1	1	*7010*
		4	1	*7020*
			3	*7021*
	1	1	1	*7100*
6	6	3	3	*8000*
		3	3	*8010*
No. of total alleles detected:				
8	13	24	27	
No. of new alleles added:				
—	5	11	3	

After Singh *et al.*, 1976.

[a] Criterion 1 = 5% gel, pH 8.9; No. alleles detected independently = 8.
Criterion 2 = 7% gel, pH 8.9; No. alleles detected independently = 13.
Criterion 3 = 5% gel, pH 7.1; No. alleles detected independently = 17.
Criterion 4 = 7% gel, pH 7.1; No. alleles detected independently = ?.

polymorphic loci are not simply the result of insufficient sensitivity of the electrophoretic method, but are genuinely different classes. Moreover, polymorphic loci fall into three categories. Some, like *Xdh*, have a single fairly common allele and a large number of rare alleles. Others, like *Adh*, have only two alleles forming a major polymorphism but are completely lacking the long list of rare alleles seen in Table 1. Finally, there are loci, like esterase-5, shown in Table 2, in which there is both a major polymorphism of two alleles and a long list of rare alleles.

Sequential gel electrophoresis experiments do not, in themselves, demonstrate that all, or nearly all, amino acid substitutions can be detected. The operating characteristics of gel electrophoresis remain unknown until an *a posteriori* experiment is carried out on proteins of known sequence. Such an experiment has been done using 32 variants of human haemoglobin of known amino acid sequence (Ramshaw, Coyne & Lewontin, 1979). This experiment showed that 85% of different amino acid substitutions in the molecule are detectably different from each other. More compelling, the *same* chemical substitutions but at different places in the molecule are distinguishable one from the other. For example, seven different substitutions of asparagine for aspartic acid at different positions in the α and β chains are all distinguishable electrophoretically from each other, although their nominal charge is the same. The possibility of distinguishing these substitutions arises because the degree of ionization of a charged group depends not only on the amino acid itself, but on its steric and chain neighbours as well. A polypeptide is an integral molecule, not a string of amino acids.

The ultimate possibility for making correct inferences about the forces operating on protein polymorphism does not lie with studies of the proteins themselves, however. What we must do is to take advantage of a very important property of the genes that specify the proteins: their internal heterogeneity. The stretch of DNA that we call the 'gene for ADH' is, in fact, a functionally complex array including non-transcribed flanking sequences, transcribed but non-translated leaders, transcribed but non-translated introns, exons that actually specify the amino acid sequence, and, within these, at every third base position, a redundant base whose identity can usually change without altering the codon. All of this functional heterogeneity lies within a stretch of the genome corresponding to about 0.01 centimorgans. Any differences we may observe in polymorphism between introns and exons, between third position bases and other bases, cannot be assigned to genetic drift or to chances of history or to differences in mutation rate, but must be referred directly to selection differentiating between functional classes of DNA. In this way, population genetics acquires a tool for inference that is different from its former weak dependence on the inversion of theoretical quantitative relations.

Table 2. *Frequency distribution of alleles of esterase-5 in two populations of* D. pseudoobscura *after five conditions of sequential gel electrophoresis*

Allele	No. of lines		
	James Reserve	Gundlach–Bundschu	Total
Null	1	0	1
0.85/1.00/1.00/1.00/1.00	1	2	3
0.89/1.00/1.00/1.00/1.00	1	0	1
0.89/1.02/1.00/1.00/1.02	3	0	3
0.89/1.02/1.02/1.02/1.02	2	0	2
0.95/1.00/1.00/1.00/1.00	2	6	8
0.95/1.00/1.00/1.02/1.00	1	0	1
0.95/1.00/0.96/0.95/1.00	0	1	1
0.95/1.00/0.97/0.97/1.04	0	1	1
0.98/0.98/0.98/0.98/0.98	1	0	1
0.98/1.00/1.00/1.00/1.00	0	1	1
0.98/1.00/1.07/1.07/1.04	0	1	1
0.98/1.00/1.08/1.09/1.00	1	2	3
0.98/1.02/1.08/1.09/1.03	1	0	1
0.98/1.02/1.08/1.09/1.04	2	0	2
0.98/M-D/1.07/M-D/1.00	2	0	2
0.98/1.22/1.07/1.13/1.00	0	1	1
0.98/1.24/1.07/1.15/1.00	1	2	3
1.00/1.00/0.96/0.96/0.96	0	1	1
1.00/1.00/0.98/0.98/1.00	5	5	10
1.00/1.00/1.00/1.00/1.00	40	44	84
1.00/1.00/1.01/1.00/1.00	1	6	7
1.00/1.00/1.00/1.00/1.02	1	3	4
1.04/0.98/1.01/1.01/0.97	1	0	1
1.06/0.99/0.94/0.96/1.00	1	0	1
1.06/0.99/0.96/0.98/1.00	2	0	2
1.06/0.99/1.00/0.98/1.00	1	1	2
1.06/1.00/0.98/0.98/1.00	4	3	7
1.06/1.00/1.00/1.00/1.00	24	26	50
1.06/1.16/1.00/1.32/1.02	0	2	2
1.06/M-D/0.99/M-D/1.02	2	0	2
1.07/1.00/1.00/1.00/1.00	1	0	1
1.09/1.00/1.00/1.00/1.00	3	4	7
1.09/1.00/1.02/1.02/1.02	1	0	1
1.12/0.99/1.02/1.00/0.98	1	1	2
1.12/0.99/1.02/1.00/0.99	0	1	1
1.12/0.99/1.02/1.02/0.99	1	0	1
1.12/1.00/1.00/1.00/1.00	1	0	1
1.12/1.00/1.02/1.02/1.00	5	7	12
1.16/1.00/1.00/1.00/1.00	1	0	1
1.16/1.00/1.00/1.02/1.02	1	0	1
Total	33 alleles (19 unique) 116 genes	22 alleles (8 unique) 121 genes	41 alleles 237 genes

After Keith, 1983.

The first experiment on DNA sequence polymorphism has been carried out by Kreitman (1983) with striking results. The ADH locus in *D. melanogaster* codes for a small polypeptide of 255 amino acids. The coding region is interrupted by two small introns of 65 and 70 bases, respectively. There is also a 5′ adult intron, not transcribed in larvae. In addition, Kreitman sequenced about 1000 bases of 3′ flanking sequence. In the gene proper, Kreitman found a 6% polymorphism of bases for the introns and a 7% polymorphism of bases in exons when only redundant (mostly third position) base positions are considered. In contrast, there was absolute monomorphism for first and second position bases in the exons, with the exception of the single major electrophoretic polymorphisms already known. Thus there is no hidden amino acid polymorphism undetected by sequential gel electrophoresis, confirming the accuracy of the technique for assessing amino acid polymorphism. Most important, the complete lack of amino acid substitutions, despite a 7% polymorphism of bases, is exactly the reverse of what is to be expected from random mutations. For ADH, three-quarters of all random base substitutions will result in amino acid replacements. Thus, selection has removed all substitutions with a remarkable consistency and fidelity. There appear to be no 'neutral' amino acid position, despite the fact that the protein is 30% isoleucine and valine. Selection need not be very strong to oppose the force of mutation, but it must be *very* discriminatory.

A second feature of the DNA sequence studies is the ability to follow the phylogeny of genes because of the individuality of the total DNA sequence. Each ADH gene carries its sequence 'signature', a haplotype specified by a stretch of 2000 bases. Kreitman's data show crossovers that have occurred between fast and slow electrophoretic alleles, and that fast alleles are much less variable at the level of bases than are slow alleles, despite the recombination. Thus, fast alleles must be a relatively recent derivative from slow. Most important from the standpoint of population genetic theory, it is now possible to distinguish in practice what has only been an abstract distinction in theory up to this point: the distinction between *identity by kind* and *identity by descent*. This difference lies at the heart of explanation in population genetics, because it distinguishes selective homozygosity from drift homozygosity. So, in the future, we will be able to distinguish between monomorphic loci (at the protein level) that are also totally homozygous by descent and those that have considerable DNA heterogeneity and so are monomorphic because of purifying selection. That is, the selection-drift controversy, which depends for its continued existence on the ambiguity of observations, will be resolved.

The observations of sequential gel electrophoresis, of calibration experiments, and of DNA sequencing have already resolved ambiguities produced by previous observations. First, it is now clear that 'heterozygosity' and 'polymorphism' are inadequate descriptions of variation because they are heterogeneous. At the amino acid level there are two sorts of heterozygosity revealed by these experiments. There is major polymorphism of two or three alleles in roughly equally high frequency, as for example the fast and slow alleles of ADH, and there is the vast array of minor polymorphisms producing a long list of rare alleles which altogether contribute a significant heterozygosity, as in the esterase-5 locus of *D. pseudoobscura*. Some loci are totally monomorphic, lacking both kinds of heterozygosity. Some, like ADH, have only the first sort of polymorphism. Others, like xanthine dehydrogenase, have only the long list of minor variants with one common allele, while yet other loci, like esterase-5, possess both sorts of polymorphism. To confound these quite different phenomena by a calculation of heterozygosity confounds different causal pathways. Moreover, in the future, we will be able to distinguish the two classes of monomorphic loci, again distinguishing selection from accidental forces.

Population genetics has been in difficulty because a large array of causal forces interacting with each other map onto what has been, up to now, a simple set of observations. In that mapping, information is irretrievably lost so that the process of inferring the causes from the observations leads to great ambiguities. The solution to this dilemma does not appear to lie in making yet further deductions about the statistics of gene frequency distributions in the hope that a combination of inferences will sort out the truth. The union of all the inferences appears to contain little more information than any smaller set of them. Rather, it appears that qualitatively new observations are demanded, observations that give a finer structure of genetic detail corresponding to the detail that is built into the assumption of the theory. It seems that with the help of molecular genetics, we may be on our way to constructing the detailed knowledge of the genetic structure of populations.

References

Coyne, J. A. (1976). Lack of genic similarity between two sibling species of *Drosophila* as revealed by varied techniques. *Genetics*, **84**, 593–607.

Coyne, J. A. & Felton, A. A. (1977). Genic heterogeneity at two alcohol dehydrogenase loci in *D. pseudoobscura* and *D. persimilis*. *Genetics*, **87**, 235–304.

Fisher, R. A. (1918). The correlation between relatives on the supposition of Mendelian inheritance. *Transactions Royal Society Edinburgh*, **52**, 399–433.

Greenberg, D. A. (1984). Simulation studies of segregation analysis: Application to two-locus models. *American Journal of Human Genetics*, **36**, 167–76.

Keith, T. P. (1983). Frequency distribution of esterase-5 alleles in two populations of *D. pseudoobscura*. *Genetics*, **105**, 135–55.

Kimura, M. & Ohta, T. (1971). *Theoretical Aspects of Population Genetics.* Princeton: Princeton University Press.

Kreitman, M. (1980). Assessment of variability within electromorphs of ADH in *D. melanogaster. Genetics*, **95**, 467–75.

Kreitman, M. (1983). Nucleotide polymorphism at the alcohol dehydrogenase locus of *D. melanogaster. Nature*, **304**, 412–17.

Lewontin, R. C. (1974). *The Genetic Basis of Evolutionary Change.* New York: Columbia University Press.

Lewontin, R. C., Moore, J. A., Provine, W. B. & Wallace, B. (1981). *Dobzhansky's Genetics of Natural Populations*, vol. I–XLIII. New York: Columbia University Press.

Moll, R. H., Lindsey, M. F. & Robinson, H. F. (1964). Estimates of genetic variances and level of dominance in maize. *Genetics*, **49**, 411–23.

Nei, M. & Li, W. H. (1975). Probability of identical monomorphism in related species. *Genetic Research*, **26**, 31–43.

Prakash, S. & Lewontin, R. C. (1968). A molecular approach to the study of genic heterozygosity. III. Direct evidence of co-adaptation in gene arrangements of *Drosophila. Proceedings of the National Academy of Sciences, USA*, **59**, 398–405.

Prout, T. (1954). Genetic drift in irradiated experimental populations of *Drosophila melanogaster. Genetics*, **39**, 529–45.

Prout, T. (1965). The estimation of fitness from genotype frequencies. *Evolution*, **19**, 546–51.

Ramshaw, J. A. M., Coyne, J. A. and Lewontin, R. C. (1979). The sensitivity of gel electrophoresis as a detector of genetic variation. *Genetics*, **93**, 1019–37.

Singh, R. S., Lewontin, R. C. & Felton, A. A. (1976). Genetic heterogeneity within electrophoretic 'alleles' of xanthine dehydrogenase in *D. pseudoobscura. Genetics*, **84**, 609–29.

2

Somatic mutations as an evolutionary force

M. SLATKIN

The effectiveness of natural selection depends on genetic variation among individuals in those traits exposed to selection. Any force that promotes genetic variation will intensify the response to selection. Much of population genetic theory is devoted to models of the different forces and their roles in maintaining genetic variability among individuals in the face of directional selection and genetic drift, both of which reduce variability. Most models are of species made up individuals that are genetically homogeneous, as is appropriate for many higher animals. There are, however, many species in which the germ line is not separate until later in life and which have the possibility, at least, of several different parts of an individual producing gametes (Buss, 1983). In such species, there can be selection among somatic cells before gametes are produced. Assuming that an individual develops from a single zygote and assuming that there is no transfer of genetic material among individuals during development, any genetic variation among somatic cells must be due to mutations occurring during development. In this context, mutation is used in the broadest sense as meaning any change in the genome of a cell that can be transmitted to gametes descended from that cell.

The effectiveness of somatic mutations in promoting genetic evolution, particularly in higher plants, has been suggested by Whitham & Slobodchikoff (1981) and D. E. Gill (personal communication). They argue that the potentially large number of different flowers and the late separation of the germ line allow the possibility of genetic variation within a plant, with that variation being immediately exposed to natural selection. Plants could then evolve rapidly in the face of new conditions with some of the change occurring within a generation. The potential importance of this process depends on several factors. To assess the roles of those factors, I will develop some simple models of somatic mutations. The problem is very similar to that of group selection (Maynard Smith, 1964) with each individual being analogous to a group.

In this chapter, I will not be concerned with another issue raised by

Whitham & Slobodchikoff (1981) and Whitham (1982), that heterogeneity within individuals is itself adaptive. As these authors argue, it is possible that heterogeneity, caused by genetic or non-genetic processes, could lead to greater resistance of plants to various herbivores and increase the plants' viability. This possibility would fall in the realm of conventional population genetic models with the fitnesses of individuals determined by their phenotypic properties and their interactions with herbivores.

The individual

To develop any model, we need a fairly explicit description of how somatic mutations can occur and be transmitted to gametes. A zygote has a known genotype and, in the absence of somatic mutations, that genotype and Mendel's laws will determine what gametes are produced. If somatic mutations occur, their effect will be to change the distributions of the genotypes from the expectations under Mendelian inheritance. Those changes can be due both to the mutations themselves and to any selection acting among somatic cells before gamete formation.

For illustrative purposes we will model a population of trees, with each tree assumed to have the same number, B, of branches, each of which produces hermaphroditic flowers. There is no variation among flowers on a single branch but there may be among branches. We assume that somatic mutations can occur at random times during development. The number of branches that will contain a mutant depends on the time of occurrence. If it occurs early enough, all branches would be affected. If it occurs after all branches are distinct, then only one would be affected. For the purposes of this model, the development of the tree can be summarized by the probability distribution of the number of branches affected by a typical somatic mutation, ϕ_i. That distribution can be derived from a model of branch formation. I will also assume that selection, if it occurs, is only among the branches after they are formed.

To illustrate how ϕ_i is derived from a model of tree growth, consider the following, highly idealized, example. Assume that the tree grows at a uniform rate and that, at k uniformly spaced times, each stem divides to produce two stems. This produces what mathematicians call a bifurcating tree that will have $B=2^k$ branches. If a mutation can occur with equal probability at any time during the growth of the tree, then we can compute ϕ_i, the probability that the mutation occurs in exactly i of the branches. There are $k+1$ time intervals during which the mutation can occur and it is equally likely that it will occur in any one. If it occurs during the first interval, it will affect all B branches since there have been no divisions yet. If it occurs in the second interval, it will affect $B/2$ or 2^{k-1} branches. If it occurs in the third interval, it will affect $B/4$ or 2^{k-2} branches and so on until, in the last interval, it will affect only one branch. Therefore, $\phi_i = 1/(k+1)$ for $i = 1, 2, 4, \ldots, 2^k$ and $\phi_i = 0$ otherwise, assuming that only one mutation can occur.

Hard and soft selection

Assume that a population of N trees has discrete non-overlapping generations with zygotes for each generation being the result of random combinations of gametes from a gamete pool produced by the previous generation. It is easiest to model the frequencies of alleles in that gamete pool.

I will use two assumptions about how trees contribute to the gamete pool. One is that the contribution of each branch depends only on its genotype, an assumption that corresponds to what Wallace (1975) calls 'hard' selection, and the other is that the contribution of each tree depends on the genotype of the zygote producing the tree, which corresponds to what Wallace calls 'soft' selection. Under soft selection, branches in the same tree are in competition with each other for the opportunity to make up the tree's contribution to the gamete pool. The distinction between hard and soft selection was made by Dempster (1955) and Maynard Smith (1962, 1970*a*) in discussing the maintenance of genetic variability in subdivided populations. In this context, the assumption of soft selection is the same as the assumption made by Levene (1953). It is difficult to tell which model is more realistic for plants. We could imagine that flowers on each branch have the same chance of being pollinated and setting seed (hard selection) or we would imagine that each tree could be visited by the same number of pollinators and seed dispersers regardless of the number of branches (soft selection).

The probability of fixation of a single somatic mutation

The first problem is to find the probability of fixation of a single advantageous mutation. To do so we can use the result of Kimura (1962) for the probability of fixation of mutations in a finite population. Kimura's result is for a population of genetically homogeneous, diploid individuals of effective size N. For a diallelic locus with alleles A and a and relative fitnesses given by

$$AA{:}\ 1+s, \quad Aa{:}\ 1+hs, \quad aa{:}\ 1; \tag{1}$$

the probability of fixation of A is

$$u = \frac{\int_0^p \mathrm{e}^{-2NsDx(1-x)-2Nsx}\,\mathrm{d}x}{\int_0^1 \mathrm{e}^{-2NsDx(1-x)-2Nsx}\,\mathrm{d}x}, \tag{2}$$

where p is the initial frequency of A in the pool of gametes and $D = 2h-1$.

Eqn (2) is derived using a diffusion approximation that is valid when s is small in magnitude and N is large. If p is also small, as will be the case in considering the fixation probability of a newly arising mutation, then the numerator in Eqn (2) can be approximated and we can write

$$u=\frac{p}{2I}\,(1-e^{-2NsDp(1-p)-2Nsp}), \tag{3}$$

where I is the denominator in Eqn (2). This formula will be useful for comparing probabilities of fixation under different assumptions.

For finding the probability of fixation of a mutation that initially appears in a somatic cell, the distinction between the frequency among adults and the frequency in the gamete pool becomes important. Consider a typical somatic mutation. On the average, it will be present in $\bar{i}$ of the branches where

$$\bar{i}=\sum_{j=1}^{B} j\phi_j. \tag{4}$$

Under the assumption of hard selection, each of the B branches contributes to the gamete pool with the contribution depending on its relative fitness (that is, on whether or not it carries the new mutation). The contribution of each tree depends on the average fitness of its branches. If each heterozygous branch contributes a fraction $(1+hs)$ relative to those not containing the mutation, then, p, the initial frequency of A in the gamete pool is given by

$$p=\frac{\bar{i}(1+hs)}{2(NB+\bar{i}hs)}. \tag{5}$$

In subsequent generations, the probability of fixation of the mutation is given by Eqn (3). After the first generation, all individuals are assumed to be genetically homogeneous so Kimura's result (Eqn 2) can be used.

Under the assumption of soft selection, the result is slightly different. The initial frequency in the gamete pool is now

$$p=\frac{\bar{i}(1+hs)}{2N(B+\bar{i}hs)}, \tag{6}$$

which is always less than the value given by Eqn (5) if $N>1$. The reason for this difference is that the assumption of soft selection ensures that a somatic mutation in one tree can never have an initial frequency in the gamete pool greater than $1/2N$. If there is very strong selection in favour of the somatic mutation, as might be expected if the mutation conferred considerable resistance to a common herbivore, then the difference between these models is substantial. In the case of hard selection, s can be made sufficiently large that the somatic mutation is almost certain to be fixed. In contrast, the probability of fixation under soft selection can never exceed the probability for a single mutation that appears in a zygote, for which $p = 1/2N$.

To compare these results with those from models of mutations that occur at the gametic stage, we must be careful about describing when the mutation is counted as having occurred. A mutation carried by a single gamete has a very low probability of appearing even in a single zygote because its frequency in the gamete pool is minute. On the other hand, if we count mutations of this type as having occurred only if they appear in a zygote, then the initial frequency is $1/2N$. In examining models of recurrent mutation, this difference is not important because the fraction of zygotes that carry mutations is the same as the fraction of the gametes.

Whitham & Slobodchikoff (1981) argue that somatic mutations are more likely to be fixed than gametic mutations. Implicit in their argument is an assumption about when mutations are to be counted. If a gametic mutation is counted only if it appears in a zygote and if it confers the same advantage as a somatic mutation, then its fixation probability is larger than a comparable somatic mutation, because it is equivalent to a somatic mutation that occurs in all branches. If, however, the mutation is counted in the gamete pool, then its probability of fixation is much smaller than that of a comparable somatic mutation because there are, by assumption, many more gametes. In that case, Whitham & Slobodchikoff are correct.

We can consider the relative rates of evolution due to continued mutational pressure of both types, using arguments analogous to those of Kimura & Ohta (1971). For gametic mutations, the expected number of such mutations appearing in the zygotes in a generation is $2N\mu_g$ where μ_g is the gametic mutation rate. The initial frequency of each is $1/2N$ so the chance that each mutation is fixed is given by Eqn (3). Therefore, the rate of substitution of such mutations is

$$\frac{\mu_g}{2I}\,(1 - e^{-sD(1-1/2N)-s}), \qquad (7)$$

where I is the denominator in Eqn (2).

Let μ_s be the rate of occurrence of somatic mutations per tree per generation. That is, a fraction μ_s of the alleles will mutate at some time between zygote formation and gamete production. The probability of fixation of each of these mutants is given by Eqn (3) with p given by (5) or (6), depending on the model of selection.

The rate of substitution of somatic mutations is

$$\frac{2N\mu_s p}{2I}\,(1 - e^{-2NsDp(1-p)-2Nsp}). \qquad (8)$$

We can compare these rates by assuming s is small and N is large. Under that assumption (8) will exceed (7) when, approximately,

$$(2Np)^2\mu_s > \mu_g. \tag{9}$$

Note that this condition does not depend on D or s. It is difficult to use this result to argue whether somatic mutations are likely to be important evolutionary forces or not because there seem to be so few relevant data. For alleles that confer the same advantage, the rate of somatic mutations must be somewhat larger than the gametic mutation rate because the relative fixation probability is smaller. Somatic mutation rates could indeed be much larger than gametic mutation rates because of the potentially large number of mitotic cell divisions occurring in many plants before gamete formation.

Genetic diversity

The potential role of somatic mutations is only partly revealed by examining the fate of single mutations, although that discussion does point out the importance of the overall population structure. As discussed by Whitham & Slobodchikoff (1981), the process of somatic mutation may well lead to a greater genetic variation in a population, at least in part because of the greater number of opportunities for a mutation to occur. To investigate this process, I will develop a simple model of the maintenance of genetic variability at a selectively neutral locus to find the contribution of somatic mutations. While the neutral mutations of the kind modelled are not the basis of adaptive evolution, new environmental conditions could cause some neutral or nearly neutral mutations to become advantageous. If that is possible, the more genetic variability that is maintained, the greater will be the opportunities for adapting to new conditions.

The model will be developed in terms of probabilities of identities by descent of pairs of alleles. This method was introduced by Malécot (1948) and has been used to analyse subdivided populations by Maruyama (1970) and Maynard Smith (1970*b*) among others. The approach I will take here is similar to that of a model related to group selection (Slatkin, 1981). This model will be used to answer the question of the extent of genetic variability maintained within and among individuals.

Consider a single genetic locus and assume that mutations can occur both during mitotic cell division and during meiosis. Let μ_s be the rate at which mutations occur between zygote formation and gamete formation. The rate μ_s is the rate per locus per individual and incorporates the number of mitotic cell divisions. As in the previous section, a mutation that occurs during development has a probability ϕ_i of being present in i branches. Let μ_g be the mutation rate during gamete formation. A fraction μ_g of the gametes produced by each branch in each generation will carry a mutation at the locus of interest. Assume that each mutation is to a new, neutral allele – the 'infinite alleles' model of Kimura & Crow (1964).

Let f_0 be the probability of identity by descent of two alleles chosen with replacement from gametes produced by the same branch. The assumption of sampling with replacement is equivalent to the assumption that there is an infinite number of gametes produced by each branch. Let f_1 be the probability of identity by descent of alleles from two gametes chosen from different branches on the same tree and f_2 be the same probability but for gametes from different trees. If we assume discrete non-overlapping generations, we can derive recursion equations for these three quantities.

In the gamete pool, the average probability of identity by descent of two alleles from two gametes chosen at random (with replacement) from the pool is

$$\bar{f}=(1-\mu_g)^2\left[\frac{1}{NB}f_0+\frac{1}{N}\left(1-\frac{1}{B}\right)f_1+\left(1-\frac{1}{N}\right)f_2\right], \tag{10}$$

where, as before, B is the number of branches per tree and N is the number of trees. Since zygotes are formed by choosing two gametes at random, $\bar{f}$ is also the probability of identity by descent of the two alleles in a zygote. To find f_0, we account first for the possibility that the tree we choose is unaffected by a somatic mutation, an event which occurs with probability $1-\mu_s$. In that case, the probability of identity by descent of two alleles chosen from the same branch is $(1+\bar{f})/2$. If the tree has been affected by a somatic mutation, the chance that a randomly chosen branch carries the mutation is $\bar{i}/B$ and the chance that it does not is $(1-\bar{i}/B)$. In the former case, the probability of identity by descent is 1/2 and in the latter case it is $(1+\bar{f})/2$. Therefore

$$f'_0=(1-\mu_s)\left(\frac{1+\bar{f}}{2}\right)+\mu_s\left[\frac{\bar{i}}{2B}+\left(1-\frac{\bar{i}}{B}\right)\left(\frac{1+\bar{f}}{2}\right)\right], \tag{11}$$

which can be simplified to

$$f'_0=\frac{1+\bar{f}}{2}-\frac{\mu_s\bar{i}\bar{f}}{2B}. \tag{12}$$

To compute f_1, we first note that there are three possibilities that must be accounted for. For two different branches on the same tree, the probability that neither branch contains a mutation that is present in exactly i branches is $(B-i)(B-i-1)/B(B-1)$, the probability that one carries the mutation and the other does not is $2i(B-i)/B(B-1)$, and the probability that both carry the mutation is $i(i-1)/B(B-1)$. The net probabilities are found by averaging over all i.

To compute f'_1, the probability in the next generation of identity by descent of gametes chosen from different branches on the same tree, we account for the three possibilities. If neither branch carries the mutation, the probability of identity by descent of alleles in two gametes chosen at random is $(1+\bar{f})/2$ because there is an equal chance that the same or different alleles can be chosen. If only one branch carries the mutation the relevant probability is $(1+\bar{f})/4$ because, if the mutant allele is chosen, it cannot be identical by descent with any other allele. If both branches carry the mutation, the relevant probability is 1/2. Summing these components and multiplying by the probability that such a mutation has occurred, we obtain

$$f'_1=(1-\mu_s)\left(\frac{1+\bar{f}}{2}\right)+\frac{\mu_s}{2B(B-1)}$$
$$\times\{\bar{f}[B(B-1)-\bar{i}(B-1)]+B(B-1)-\bar{i}B+\bar{i}^2\}, \qquad (13)$$

where the first term accounts for the possibility that no mutation occurs. Eqn (13) can be simplified to yield

$$f'_1=\frac{1+\bar{f}}{2}-\mu_s[\bar{i}\bar{f}(B-1)+\bar{i}B-\overline{i^2}], \qquad (14)$$

where $\overline{i^2}$ is the average of i^2.

To calculate f'_2, we must account for the possibility that neither of the branches chosen from different trees carries a somatic mutation, that one does, or that both do. The probability that neither carries a new somatic mutation is $(1-\mu_s\bar{i}/B)^2$; the probability that one does is $2(\mu_s\bar{i}/B)(1-\mu_s\bar{i}/B)$; and the probability that both do is $(\mu_s\bar{i}/B)^2$. In the first case, the probability of identity by descent of two alleles is just $\bar{f}$; in the second case the probability is $\bar{f}/2$; and in the last case it is $\bar{f}/4$. Therefore,

$$f'_2=\left(1-\frac{\mu_s\bar{i}}{B}\right)^2\bar{f}+2\left(\frac{\mu_s\bar{i}}{B}\right)\left(1-\frac{\mu_s\bar{i}}{B}\right)\frac{\bar{f}}{2}+\left(\frac{\mu_s\bar{i}}{B}\right)^2\frac{\bar{f}}{4}.$$
$$=\bar{f}-\frac{\mu_s\bar{i}}{B}\bar{f}+\frac{1}{4}\left(\frac{\mu_s\bar{i}}{B}\right)^2\bar{f}. \qquad (15)$$

From these three equations, we can obtain a single, albeit large, recursion equation for $\bar{f}$, the average identity by descent in the gamete pool, as a function of its value in the preceding generation, of the two mutation rates, μ_s and μ_g and of the parameters describing the population, B and N. Once the value of $\bar{f}$ is known in each generation, the relative amounts of genetic variability with and among trees can be found by comparing f_1 and f_2. Here, I will find the equilibrium values of the various quantities. To simplify the analysis, I will assume that both μ_s and μ_g are sufficiently small that terms containing the squares or higher powers of those quantities can be ignored.

The recursion equation for $\bar{f}$ is

$$\bar{f}(t+1)=(1-2\mu_g)\left\{\frac{1}{NB}\left[\frac{1+\bar{f}}{2}-\frac{\mu_s\bar{i}\bar{f}}{2B}\right]+\frac{1}{N}\left(1-\frac{1}{B}\right)\right.$$
$$\left.\left[\frac{1+\bar{f}}{2}-\mu_s(\bar{i}\bar{f}(B-1)+\bar{i}B-\overline{i^2})\right]+\left(1-\frac{1}{N}\right)\left[\bar{f}-\frac{\mu_s\bar{i}\bar{f}}{B}\right]\right\}, \qquad (16)$$

where $\bar{f}$ on the right-hand side is the value in generation t. This equation has the approximate solution at equilibrium:

$$\bar{f}=\frac{1+2\mu_s\left(1-\frac{1}{B}\right)(\bar{i}B-\overline{i^2})}{1+4N\mu_g+2N\mu_sC}, \tag{17}$$

where

$$C=\frac{\bar{i}}{2B^2N}+\frac{1}{N}\left(1-\frac{1}{B}\right)(\overline{i(B-1)})+\left(1-\frac{1}{N}\right)\frac{\bar{i}}{B}. \tag{18}$$

Unless the number of branches is very large, so that the second term in the numerator of Eqn (17) becomes of order 1 in magnitude, the principal effect of the somatic mutations is to augment the gametic mutation rate. If we assume the numerator is approximately 1, we can rewrite (17) as

$$\bar{f}=\frac{1}{1+4N\mu'} \tag{19}$$

where μ' is an effective mutation rate defined as

$$\mu'=\mu_g+\frac{C\mu_s}{2}. \tag{20}$$

Somatic mutations will increase the overall mutation rate in the population and reduce the average probability of identity by descent in the gamete pool.

As a means of augmenting the genetic diversity among trees, somatic mutations could be very important. Because there are potentially many mitotic cell divisions, there may be many opportunities for somatic mutations to occur. In Eqn (20), it is easy to imagine means by which μ_s is an order of magnitude or more greater than μ_g, making somatic mutations the principal cause of variation among trees. That does not imply that there will be very much standing variation within each tree on which natural selection could act at any time. In this context, the value of f_1 is not a good indicator of variation within a tree because gametes from different branches may differ because the tree is a heterozygote. Every branch, however, will be genotypically the same except for somatic mutations which occur in that generation. The order of magnitude of variation among branches in genotype is that of the somatic mutation rate, μ_s. That will be much smaller than the genotypic variation among trees unless both μ_s and μ_g are very small, small enough that effectively each mutation passes through the population before another appears.

Discussion

The models developed in the preceding sections cannot tell us whether or not somatic mutations are an important evolutionary force. What they can do is suggest some aspects of the biology of organisms that might have bearing on the question. For example, we have found that the probability of fixation of a somatic mutation depends in part on how population numbers are determined in each generation. It is certain that the simple distinction made between hard and soft selection does not exactly pertain to any real population, but it seems likely that the essential difference between those two types of selection is important. If different parts of the same organism are primarily in competition with each other for reproductive opportunities, either vegetative or sexual, then the model of soft selection suggests that newly arising, advantageous somatic mutations would have a much lower chance of being fixed than a comparable gametic mutation. On the other hand, if parts of an individual organism are competing as much with parts of other organisms as with themselves, then the model of hard selection suggests that strong selection in favour of a somatic mutation could lead to its fixation more readily than it would for a gametic mutation.

The second model analysed shows that somatic mutations can substantially increase the genetic diversity in a population. But most of the diversity will be among different individuals. Somatic mutations cannot maintain much variation within individuals because the variation is lost in each generation. Only in very long-lived organisms which reproduce primarily by asexual modes could the variation within individuals be comparable to the variation among individuals.

The main conclusion is that the potential evolutionary importance of somatic mutations depends largely on the relative rates of somatic and gametic mutations. As Whitham & Slobodchikoff (1981) point out, long-lived organisms with numerous reproductive parts might have many opportunities for somatic mutations to produce genetic variation within individuals but few opportunities for gametic mutations to become established because few zygotes will produce new individuals. If somatic mutations are important for the genetic evolution of such species, then we could expect there to be genetic variation among different parts of an individual. That suggests the measurement of genetic variation within individuals would be essential to finding the evolutionary role of somatic mutations.

Conclusion

Organisms which do not sequester their germ cells early in development and which have multiple reproductive parts have the potential for possessing genetic variation among different parts of the same individuals through the accumulation of somatic mutations during development. Several authors have argued for the evolutionary importance of selection acting among different parts of an individual in allowing for adaptation to new conditions within the lifetime of an individual. Two models of the population genetics of somatic mutations are introduced and discussed. Both models depend on the number of different parts of the same individual that are affected by a typical somatic mutation. One model is of the rate of substitution of advantageous somatic and gametic mutations and the other is of the extent of genetic variability within and among individuals due to somatic and gametic mutations. The potential importance of somatic mutations in both models depends strongly on the relative somatic and gametic mutation rates. Only if the somatic mutation rates are much larger than the gametic mutation rates will somatic mutations be of more importance. Even then, the primary role of somatic mutations is to increase the effectiveness of selection among individuals. Somatic mutations are confined to a single individual for too short a time to have their evolutionary dynamics determined by the processes acting within each individual unless selection is very strong.

This research has been supported in part by the National Science Foundation (USA), Grant No. DEB-81-20580. A. Harper, M. Kirkpatrick, S. Palumbi, S. Rudolph, and T. G. Whitham have made several useful suggestions on an earlier version of this paper.

References

Buss, L. W. (1983). Evolution, development and the units of selection. *Proceedings of the National Academy of Sciences, USA*, **80**, 1387–91.

Dempster, E. R. (1955). Maintenance of genetic heterogeneity. *Cold Spring Harbor Symposia on Quantitative Biology*, **20**, 25–32.

Kimura, M. (1962). On the probability of fixation of mutant genes in a population. *Genetics*, **47**, 713–19.

Kimura, M. & Crow, J. F. (1964). The number of alleles that can be maintained in a finite population. *Genetics*, **49**, 725–38.

Kimura, M. & Ohta, T. (1971). *Theoretical Aspects of Population Genetics*. Princeton, N.J.: Princeton University Press.

Levene, H. (1953). Genetic equilibrium when more than one ecological niche is available. *American Naturalist*, **87**, 331–3.

Malécot, G. (1948). *Les Mathématiques de l'Hérédité*. Paris: Masson & Cie.

Maruyama, T. (1970). Effective number of alleles in a subdivided population. *Theoretical Population Biology*, **1**, 273–306.

Maynard Smith, J. (1962). Disruptive selection, polymorphism and sympatric speciation. *Nature*, **195**, 60–2.

Maynard Smith, J. (1964). Group selection and kin selection. *Nature*, **201**, 1145–7.

Maynard Smith, J. (1970*a*). Genetic polymorphism in a varied environment. *American Naturalist*, **104**, 487–90.

Maynard Smith, J. (1970*b*). Population size, polymorphism, and the rate of non-Darwinian evolution. *American Naturalist*, **104**, 231–7.

Slatkin, M. (1981). Populational heritability. *Evolution*, **35**, 859–71.

Wallace, B. (1975). Hard and soft selection revisited. *Evolution*, **29**, 465–73.

Whitham, T. G. (1982). Individual trees as heterogeneous environments: Adaptation to herbivory or epigenetic noise? In *Species and Life History Patterns: Geographic and Habitat Variations*, ed. R. F. Denno & H. Dingle, pp. 9–27. New York: Springer.

Whitham, T. G. & Slobodchikoff, C. N. (1981). Evolution by individuals, plant–herbivore interactions, and mosaics of genetic variability: The adaptive significance of somatic mutations in plants. *Oecologia*, **49**, 287–92.

3

The flow of genes through a genetic barrier

B. O. BENGTSSON

Gene flow is a problem for evolution by descent with modification. An exchange of genetic material between two forms prevents their independent modification or, at least, makes it more difficult to come about. Evolutionary biologists, discussing why two species have been formed, have therefore often erected mountain ridges or (for aquatic organisms) built land bridges through the range of an ancestral species, to explain why gene flow between the incipient species did not prevent their differentiation.

A less dramatic way to obtain a decrease in genetic contact is to assume that there can be barriers to the flow of genes in the genetic material itself. Two forms may – during only a brief period of isolation – become genetically different for a small number of factors, but different in such a way that other genes are hindered in their flow from one form to the other.

A simple example of this process would be the following. In a local population, a chromosome translocation has spread to fixation, making all organisms in the population homozygous for the new chromosomal type. All genes coming in with migrants from other parts of the species will then, in the first generation hybrids, occur in a chromosomally heterozygous genetic background. Such heterozygotes have a lower fertility than other organisms due to meiotic problems, and there is therefore a possibility that an incoming gene will already be lost at this stage and not transmitted any further. In the next hybrid generation, translocation heterozygotes will continue to occur and add to the probability that newly introduced genes are lost from the population; and the situation will be the same in all the following generations. The final effect is that fewer of the incoming genes will ultimately become incorporated in the local genetic background than could be expected from the number of incoming migrants. The 'effective migration' will thus be less than the physical migration of organisms (or gametes) would lead us to believe.

This is, of course, the idea underlying the mode of speciation called 'stasipatric' (White, 1968, 1978; see also Futuyma & Mayer, 1980, for a recent discussion of this and other non-allopatric speciation models). But also other situations, not necessarily involving chromosome mutations, can be imagined in which a single or only a small number of genetic differences act as barriers to gene flow. The purpose of this article is to describe what strength such genetic barriers will have. In particular I will show (i) that it is difficult to build efficient genetic barriers with only a few factors, and (ii) that the strength of a barrier, given a certain loss of fitness for the first generation hybrids, depends more on the interactions between the barrier-building factors than on their number.

The effect of a single translocation

Let us return to the example where a local population has become homozygous for a translocation, and let us calculate the probability that a gene coming in with a migrant from elsewhere in the species will become incorporated into the local genetic background. We assume that the marker gene is unlinked to the translocation and that it has no fitness effects of its own. The relative fitness of the translocation heterozygotes is assumed to be $1-s$, compared with a fitness of unity for other types in the population. A balance between migration and selection will develop for the different chromosomal types, and the chromosome configuration typical for the rest of the species will be rare in the local population if the incoming migration rate is small and the selective disadvantage of the chromosomal heterozygotes is noticeable. We will make these assumptions and can then to a fair degree of accuracy assume that the chromosomal heterozygotes will exclusively mate with organisms homozygous for the local chromosome type.

The marker gene will always exist in a chromosomally heterozygous background in the first hybrid generation. The probability that the gene is transmitted to the next generation and there occurs in the local chromosomal background is $(1-s)/2$, which is also the probability that it is transmitted but occurs in the same genetic background as before. One more generation increases the probability of incorporation of the gene into the local genetic background to $(1-s)/2+(1-s)^2/4$; and it is easy to generalize and find that the probability of ultimate inclusion of the gene is $\sum_{i=1}^{\infty}(1-s)^i/2^i$, which is the same as $(1-s)/(1+s)$ (Bengtsson, 1974).

This probability of inclusion of an incoming gene will be called the 'gene flow factor', gff, for the genetic barrier in question. Thus, for a single translocation

$$\text{gff}=(1-s)/(1+s). \tag{1}$$

The gene flow factor measures the strength of the genetic barrier and may, when migration rates are small, be used to calculate the 'effective migration' for a population with a genetic barrier:

$$m_e=\text{gff}\cdot m. \tag{2}$$

The effective migration rate, m_e, is that rate of migration which would have the same evolutionary effect in a population with no genetic barrier as the actual migration rate (m) now has in the population with a barrier.

The probability of ultimate inclusion has been calculated under a number of assumptions. That the assumption about hybrids mating exclusively with the local type does not introduce any considerable error in the estimate is shown by comparing the present result with the results obtained by Spirito, Rossi & Rizzoni (1983). They have investigated the increase over time of a marker gene in a population when the population exchanges migrants with another, karyotypically different population. Their method of analysis is more accurate than the one used here in that it includes all possible mating types, but is also more complicated, in particular if other types of genetic barriers are to be studied. A nice property of the gene flow factor as defined here is that it can easily be calculated for most types of genetic barriers with the help of a simple matrix method. The method is described in the appendix at the end of the article.

The assumption that has a strong influence on the value the gene flow factor takes is the degree of linkage between the marker gene and the translocation break-point. If the marker gene is closely linked to the translocation, then the transfer of the marker gene into the local genetic background becomes more difficult, and the gene flow factor goes down. The formula for the gene flow factor for a gene r recombination units away from one of the translocation break-points can be shown to be (see the appendix):

$$\text{gff} = r(1-s)/[1-(1-r)(1-s)]. \tag{3}$$

Values of the gene flow factor for two different strengths of linkage are given in Table 1, and the case of an unlinked marker gene is illustrated by the continuous line in Fig. 1. It is clear that close linkage of the marker gene to the translocation decreases the effective gene flow considerably. However, the main impression given by this table and this figure is that a translocation is rather a poor barrier to gene flow for the great majority of genes that are unlinked to the translocation. A Robertsonian translocation (a centric fusion or fission), for example, with a fitness effect of $s = 0.10$ will only decrease the effective migration rate by about 18%. This, at the same time as the selection acting on the translocation itself, is very strong and will lead to a strict differentiation between the local and the standard karyotypes with only a few translocation heterozygotes to be found anywhere.

That single chromosome changes are relatively ineffective as barriers to gene flow is a conclusion also reached by Barton (1979) and Spirito *et al.* (1983). Barton has made a detailed investigation of the delay a chromosomal cline will cause in the spread of an allele and has found that, in

Table 1. *The gene flow factor, i.e. the probability of ultimate inclusion in the local genetic background, for a gene, linked or unlinked to a translocation with disadvantage s in heterozygote form*

	Gene flow factor	
s	$r=0.1$	$r=0.5$
0.0	1.00	1.00
0.1	0.47	0.82
0.2	0.29	0.67
0.3	0.19	0.54
0.4	0.13	0.43
0.5	0.09	0.33
0.6	0.06	0.25
0.7	0.04	0.18
0.8	0.02	0.11
0.9	0.01	0.05
1.0	0.00	0.00

Fig. 1. The gene flow factor, gff, as a function of s, the decrease of fitness of the first-generation hybrids, when the hybrid disadvantage is caused by a single factor (continuous line) and a large number of independently acting and unlinked factors (broken line).

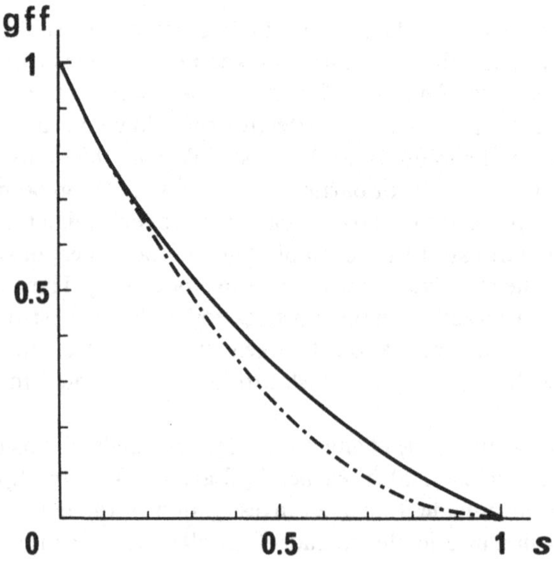

particular, selectively favoured alleles will not be noticeably hindered by a geographic difference in karyotypes. The importance of this finding for our understanding of the relationship between chromosome mutations and species formation is described and discussed in an interesting article by Barton & Hewitt (1981).

The effect of two or more independent factors

What happens if the local population differs from the standard karyotype of the species by two translocations (or inversions or other factors causing heterozygote disadvantage)? Let us assume that one of the translocations is associated with heterozygote disadvantage s_1 and that the corresponding value for the other translocation is s_2. It can then be shown with the matrix method that the gene flow factor for a gene unlinked to any of the translocations is

$$\text{gff} = (1-s_1)(1-s_2)/(1+s_1)(1+s_2), \tag{4}$$

which means that the gene flow factor for the barrier is the product of the gene flow factors for each of the two translocations. An example: The gene flow factor produced by a reciprocal translocation with heterozygote fitness 0.67 is 0.50. If this translocation occurs together with the Robertsonian translocation discussed in the preceding section, their joint effect will be $0.50 \cdot 0.82 = 0.41$, i.e. the effective migration rate will be decreased to slightly below one half of the demographic value.

To see what can happen when the number of factors involved in the genetic barrier increases, let us consider the situation where the barrier is built by n independent and unlinked factors. The fitness of the first generation hybrids is $\Pi(1-s_i)$, and the corresponding gene flow factor is

$$\text{gff} = \Pi(1-s_i)/(1+s_i). \tag{5}$$

If the number of factors is large and they all have equal effects, then the gene flow factor (5) can be approximated by

$$\text{gff} = (1-s)^2, \tag{5a}$$

where s (with no subscript) now stands for the loss of fitness suffered by the first generation hybrids.

We have thus been able to find a description of the strength of the barrier expressed as a function of the fitness of the first generation hybrids (which in many situations can be measured), and not in terms of the fitness effects associated with the individual factors building the barrier (which are much more difficult to study). The gene flow factor for this case is plotted against the selective disadvantage of the first-generation hybrids in Fig. 1 (broken line), together with the case when only one factor is involved.

The interesting observation to make from the figure is the close similarity of the two curves. The gene flow factor is about the same, irrespective of

Table 2. *The fitness effects assumed in the Dobzhansky model*[a]

	A_1A_1	A_1A_2	A_2A_2
B_1B_1	1	1	1
B_1B_2	1	$1-s$	$1-v$
B_2B_2	1	$1-t$	$1-w$

[a] The v and w values play no part in the present considerations.

whether the genetic barrier is built by one factor or many factors with correspondingly smaller effects, at least as long as the fitness of the first-generation hybrids is not very strongly reduced.

The calculations in this section have all been made under the assumption that the marker gene is unlinked to any of the factors building the genetic barrier. This is, of course, unrealistic when the number of factors becomes very large, but the purpose of the analysis has been to show that an increase in the number of factors building the genetic barrier does not – by itself – particularly influence the gene flow factor, as long as the strength of the factor is measured relative to the fitness of the first-generation hybrids.

The Dobzhansky model

The genetic barriers discussed so far are all associated with a problem. If they are to be effective as barriers, the disadvantage of being a heterozygote must be considerable – but then why did the factors causing heterozygote disadvantage spread in the local population in the first place? This question, which relates to the general problem of karyotype evolution, has been much discussed during the last decade and will not be reviewed here (see the article by Walsh, 1982, and the references given by him); instead, we will turn to an alternative way of building genetic barriers that was first described by Dobzhansky (1937).

Dobzhansky envisaged a two-locus situation with complementary dominant alleles with deleterious effects as in Table 2. An ancestral population of genotype $A_1A_1B_1B_1$ is split into two parts. Allele B_2 spreads to substitute for B_1 in the first population, while allele A_2 substitutes for A_1 in the rest of the species. These gene substitutions may come about by 'almost neutral evolution', since there are no strong fitness differences at any of the loci as long as the other locus is close to fixation for the original allele.

If migrants now enter the first population from the rest of the species, then hybrids with the double heterozygous genotype $A_1A_2B_1B_2$ will be formed. These hybrids are associated with a loss of fitness, according to the assumptions made, as are also one of the single heterozygote types that will be produced by backcrosses between the first-generation hybrids and organisms with the local genotype. A genetic barrier has thus been created between the population and the rest of the species, and without an evolutionary 'jump' having been made over a region of decreased population mean fitness. This model of genetic differentiation has been further analysed by, in particular, Nei, Maruyama & Wu (1983) and Bengtsson & Christiansen (1983).

The gene flow factor for a Dobzhansky type of genetic barrier can be calculated with the same method as used before. It is

$$\text{gff} = (1-s)[1+t+r(1-t)]/(1+t)[2-(1-r)(1-s)], \tag{6}$$

where s and t are as defined in Table 2, and r is the recombination fraction between the two loci building the genetic barrier. The factor is calculated for a gene that is unlinked to any of the two loci, and is valid as long as the local population retains $A_1A_1B_2B_2$ as its standard genotype.

If the two loci building the barrier are unlinked, the gene flow factor simplifies to

$$\text{gff} = (1-s)(3+t)/(3+s)(1+t). \tag{6a}$$

And if the two interacting alleles act exactly as true complementary dominants, i.e. the fitness effects s and t are equal, then – irrespective of the linkage value – the gene flow factor becomes

$$\text{gff} = (1-s)/(1+s). \tag{6b}$$

Thus, the gene flow factor is the same for this, the original Dobzhansky model, and the situation where the barrier is built by a single factor alone (see Eqn (1)).

A model of general hybrid effects

We shall finally consider a more general case where the genetic barrier is described via the fitness effects associated with different kinds of backcross hybrids. Assume that it is alleles from exactly two loci that give rise to fitness effects when they occur in the local genetic background. Heterozygotes for both of the loci have fitness $1-s_2$, while single locus heterozygotes have fitness $1-s_1$, irrespective for which of the loci the organism is a heterozygote. We can if we wish let s_2 take a negative value and thereby allow heterosis in our model. One should then, however, assume that $-s_2$ is smaller than s_1; otherwise the heterotic effect is stronger than the hybrid disadvantage and the two alleles will increase in frequency in the local population.

The gene flow factor for this general two-locus barrier is

$$\text{gff} = (1-s_2)[2(1-rs_1)-(1-r)(1-s_1)]/(1+s_1)[2-(1-r)(1-s_2)], \quad (7)$$

where r, as before, is the recombination fraction between the two loci giving rise to the barrier, and the gene flow factor is calculated for genes unlinked to these loci.

If the two loci are unlinked, the expression simplifies to

$$\text{gff} = (3-s_1)(1-s_2)/(1+s_1)(3+s_2). \quad (7a)$$

We shall here only consider two special cases, describing the strongest and weakest possibilities of hybrid disadvantage. The strongest instance occurs when all heterozygotes have the same fitness disadvantage, i.e. $s_1 = s_2 = s$. The gene flow factor is then

$$\text{gff} = (3-4s+s^2)/(3+4s+s^2). \quad (7b)$$

The weakest case occurs when the two loci interact in such a way that only the double heterozygotes have a fitness disadvantage ($s_2 = s$ and $s_1 = 0$), and it has the gene flow factor

$$\text{gff} = 3(1-s)/(3+s). \quad (7c)$$

The gene flow factors for these two cases are illustrated in Fig. 2, where it is seen that the difference in the fitness of the single heterozygotes can lead to quite a substantial difference in effective gene flow. But one can also see that

Fig. 2. As in Fig. 1, but the hybrid disadvantage is now caused by two interacting loci. The continuous line describes the situation when the single heterozygotes are perfectly fit, while the broken line describes the case when the single heterozygotes have the same disadvantage as the double heterozygotes.

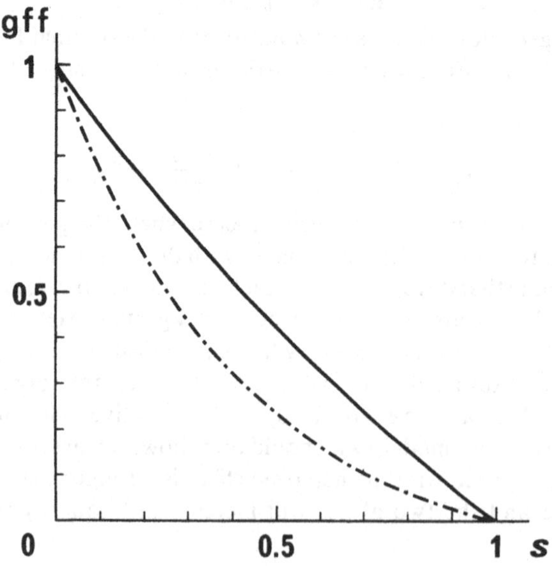

Table 3. *The loss of fitness the first-generation hybrids must show for the genetic barrier to decrease the gene flow by a half* (*i.e.* $s_{\frac{1}{2}}$ *is the value for which the relevant gene flow factor is 0.5*)

Model	$s_{\frac{1}{2}}$
2 loci: single and double heterozygotes equal	0.26
Dobzhansky model with $t=2s$	0.28
Many independent factors	0.29
One translocation, and the Dobzhansky model with $t=s$	0.33
Dobzhansky model with $t=s/2$	0.37
2 loci: single heterozygotes perfectly fit	0.43

not even the strongest case of hybrid disadvantage with two loci produces a particularly efficient genetic barrier.

The method outlined here for studying general hybrid disadvantage can easily be extended to situations with more than two loci involved in the barrier. The analysis will not be taken any further in this chapter, but it is clear already from the results obtained for two loci that the gene flow factor will go towards 0 for the strongest case and towards $1-s$ for the weakest case, when the number of loci causing hybrid disadvantage increases.

Conclusion

The results obtained here about the gene flow factor seem to me to indicate that it is difficult to build efficient genetic barriers from a small number of Mendelian factors. Table 3 gives the fitness disadvantage the first-generation hybrids must have if the effective migration is to be reduced to one half of the demographic value. It can be seen that very strong selective effects are necessary to obtain this rather restricted decrease of gene flow.

Increasing the number of factors in the barrier does not *automatically* increase the strength of the barrier if the fitness of the first-generation hybrids is kept constant, as seen in Fig. 1. However, if the number of factors building the barrier becomes large, then the majority of the genome will be linked to at least one such factor, with a concomitant decrease in the effective gene flow. It is also possible to get a strong genetic barrier, if a fair number of factors are involved in causing the loss of fitness and if these factors interact epistatically (as in the strongest case of hybrid disadvantage, discussed above).

The results we have reached imply that genetic barriers built by one or a few independent factors only rarely – if ever – can play a role in speciation processes by *initiating* a genetic differentiation that later goes to completion in full or partial sympatry. This does not, however, make such barriers evolutionarily unimportant. A population, which during a period of isolation has developed both ecological and reproductive differences, will only be able to coexist as a separate form with the originally identical mother-species if the exchange of genes between the two forms is below a critical value (see, for example, the discussions in Wallace, 1959; Bengtsson, 1982). A genetic barrier of the type discussed here can in such situations increase the probability that the new form will not become extinct or that its ecological innovations will be taken up as a polymorphism within the old species. This evolutionary role may, however, at best be called secondary, hardly primary.

Appendix

Call the genetic backgrounds in which an incoming gene may occur g_i ($i=1, \ldots, n$), where g_1 is the first-generation hybrid's and g_n is the normal genetic background in the local population. Let the probability that the marker gene occurs in background g_j in the next generation, when it in the present generation occurs in background g_i, be denoted by b_{ij}. This value includes both the effect of the fitness of genotype g_i and the probability, as given by the genetic assumptions, that a mating of type $g_i \times g_n$ produces an offspring with genotype g_j.

If the order of the genotypes is chosen in a suitable way, then the b_{ij} values form a non-negative, triangular $n \times n$ matrix, which we will call B. The probability of ultimate inclusion of the incoming gene in the local genetic background, gff, can with B be written as the solution of the following equation

$$\lim_{t\to\infty} (1, 0, 0, \ldots, 0, 0)B^t = (0, 0, 0, \ldots, 0, \text{gff}). \tag{A1}$$

The matrix B has its eigenvalues on the diagonal and the largest is always 1, which occurs as b_{nn}. By using standard spectral theory it is easy to show that the solution to Eqn ($A1$) can be written

$$\text{gff} = v_1/v_n, \tag{A2}$$

where v_1 and v_n are the first and last elements in a column eigenvector to B, corresponding to the eigenvalue 1.

This method of finding the gene flow factor will be illustrated by an example. Consider the situation where the marker gene is linked to one of the translocation break-points. There are only two genetic backgrounds that have to be taken into consideration: g_1 which is the chromosomally heterozygous background, and g_2 which is the standard genetic background in the local population. The elements in the matrix B are then

$$b_{11} = (1-s)(1-r)$$
$$b_{12} = (1-s)r$$
$$b_{21} = 0$$
$$b_{22} = 1,$$

where s is the disadvantage for the chromosomal heterozygotes and r is the recombination fraction between the marker gene and the translocation break-point. A column vector with elements v_1 and v_2 is an eigenvector to B with eigenvalue 1 if

$$(1-s)(1-r)v_1 + (1-s)rv_2 = v_1;$$

and, thus, it follows that the gene flow factor in this case is

$$\mathrm{gff} = v_1/v_2 = r(1-s)/[1-(1-r)(1-s)],$$

as given by (3).

I would like to thank Walter Bodmer, Freddy Christiansen and Nick Barton for interesting discussions on the topic of genetic barriers.

This work has been supported, in parts, by the Swedish Natural Science Research Council.

References

Barton, N. H. (1979). Gene flow past a cline. *Heredity*, **43**, 333–9.

Barton, N. H. & Hewitt, G. M. (1981). Hybrid zones and speciation. In *Essays on Evolution and Speciation in Honor of M. J. D. White*, ed. W. R. Atchley & D. S. Woodruff, pp. 109–45. Cambridge: Cambridge University Press.

Bengtsson, B. O. (1974). Karyotype evolution *in vivo* and *in vitro*. PhD thesis, Oxford University.

Bengtsson, B. O. (1982). The effect of gene flow and competition on the coexistence of two related forms. In *Evolution and Genetics of Populations*, ed. S. Jayakar & L. Zonta, pp. 19–29. Supplement to *Atti Associazione Genetica Italiana*, vol. 29.

Bengtsson, B. O. & Christiansen, F. B. (1983). A two-locus mutation-selection model and some of its evolutionary implications. *Theoretical Population Biology*, **24**, 59–77.

Dobzhansky, Th. (1937). *Genetics and the Origin of Species*. New York: Columbia University Press.

Futuyama, D. J. & Mayer, G. C. (1980). Non-allopatric speciation in animals. *Systematic Zoology*, **29**, 254–71.

Nei, M., Maruyama, T. & Wu, C.-I. (1983). Models of evolution of reproductive isolation. *Genetics*, **103**, 557–79.

Spirito, F., Rossi, C. & Rizzoni, M. (1983). Reduction of gene flow due to partial sterility of heterozygotes for a chromosome mutation. I. Studies on a 'neutral' gene not linked to the chromosome mutation in a two population model. *Evolution*, **37**, 785–97.

Wallace, B. (1959). The influence of genetic systems on geographical distribution. *Cold Spring Harbor Symposia on Quantitative Biology*, **24**, 193–204.

Walsh, J. B. (1982). Rate of accumulation of reproductive isolation by chromosome rearrangements. *American Naturalist*, **120**, 510–32.
White, M. J. D. (1968). Models of speciation. *Science*, **159**, 1065–70.
White, M. J. D. (1978). *Modes of Speciation*. San Francisco: Freeman.

4

Intraspecific resource competition as a cause of sympatric speciation

J. SEGER

Rosenzweig (1978), Bengtsson (1979), and Gibbons (1979) argue that intraspecific resource competition might cause sympatric speciation. Bengtsson emphasizes that the environment need not be spatially or temporally heterogeneous in any way.

> [In most models of speciation], the fitness value of an animal is determined by the genotype it has, and the habitat in which it lives. In a more realistic model it should also depend on how much necessary resource is available for the animal and the competition from other genotypes for this resource.
>
> Models can be constructed which take into account such competition between genotypes. In a special case one can find the exact conditions for the stable coexistence of two incipient species, which have some degree of gene flow between them but also slightly different resource utilization distributions.
>
> An interesting property of this class of models is that they show how sympatric speciation can occur in a species where all animals live in the same habitat and under the same fitness regime, but differ, due to their genetic constitution, in their resource utilization.

Here I elaborate this proposal and describe a simple genetic model in which intraspecific resource competition leads to speciation. In the discussion, I point out that models of this kind are more closely related than they may at first appear to be, to some previously studied models of sympatric speciation.

Phenotype-dependent resource competition

Consider a species in which individuals differ with respect to a trait that enables them to exploit different regions of a resource spectrum. To fix ideas, think of a quantitative character (say, beak size) that tends to scale with overall body size, and let the corresponding resource be a food (say, seeds) that occurs in a more or less continuous distribution of sizes. For the sake of simplicity, suppose that each individual will consume only those food items that fall within a well-defined range of sizes, the particular range being determined by the individual's phenotype. Thus individuals with large beaks tend to be large and to take large seeds, while those with small beaks tend to be small and to take small seeds. Given these assumptions, it follows almost inevitably that individuals of a given size will stand more directly in competition with each other than they will with individuals of sizes different from their own. Other things being equal, the expected fitness of the members of a given size class will be negatively frequency-dependent if there is any competition for seeds.

Let beak size be influenced by genes at many polymorphic loci and also by environmental variation. Then if the species mates randomly, there will be an approximately normal distribution of beak sizes. If the distribution of available seed sizes is also normal (after any necessary corrections for allometry), then the population can easily evolve to an equilibrium distribution of gene frequencies such that the resulting distribution of beak sizes 'matches' the distribution of seed sizes. At this equilibrium all phenotypes (and therefore all genotypes) have equal expected fitnesses (Slatkin, 1979).

But what happens if the distribution of available seed sizes is *not* normal? For example, suppose that over the range of seed sizes used by the species the distribution is uniform. Then individuals in the tails of the beak-size distribution will enjoy higher fitnesses than will those in the centre, and the species will be subject to disruptive selection on beak size. Given our assumptions of polygenic control, moderate heritability, random mating, and phenotype-dependent resource utilization, there is no way the species can exhibit a distribution of phenotypes that matches a uniform or other non-normal distribution of available resources.

How might a better match be realized between the distributions of phenotype and of resource? Each of the assumptions just mentioned suggests a different way in which this might be done. For example, the development of beak size could be uncoupled from direct genotypic control, such that a rectangular distribution of beak sizes resulted. Or the preferences of individuals of a given size for seeds of a given size might somehow be modified. In principle, each of these scenarios is equally plausible. But it is easy to imagine constraints of various kinds (mechanical, developmental, or merely phylogenetic) that could prevent such changes from occurring, at least to the extent needed to bring about a good fit between the distribution of phenotypes and that of resources.

The remaining possibility is that panmixia might give way to a system of positively assortative mating. This would flatten the distribution of phenotypes, without requiring that the genetics or development of the trait be modified in any way. It seems intuitively clear that assortative mating could be favoured under these conditions. Compared with parents who mated randomly, those who mated assortatively would have a greater number of offspring with extreme (and therefore relatively fit) phenotypes.

This intuition is supported by the many explicit models in which assortative mating evolves under regimes of disruptive selection (e.g. Maynard Smith, 1966; Dickinson & Antonovics, 1973; Udovic, 1980; Felsenstein, 1981). I have used a simple deterministic simulation to study the conditions under which different schemes of assortative mating may lead to speciation, under the assumptions discussed above.

Assortative mating with resource competition

All versions of the model share the following features. The species is haploid and has a resource-utilization phenotype that is influenced by two independently segregating loci, A and B. Alleles A_1 and B_1 each contribute a genotypic value of '1', and alleles A_2 and B_2 each contribute '2'. Thus the four haploid genotypes map onto three (two-locus) genotypic values. But genotypic values do not absolutely determine phenotypic values. There is assumed to be a substantial and unavoidable component of environmental variation, such that each genotypic value gives rise to a phenotypic value immediately below or above itself with probability $\frac{1}{4}$. Thus there are five distinct phenotypic values, and the four genotypes give rise to them according to the following scheme.

Genotype	Genotypic value	Phenotypic values				
		1	2	3	4	5
A_1B_1	2	$\frac{1}{4}$	$\frac{1}{2}$	$\frac{1}{4}$		
A_1B_2	3		$\frac{1}{4}$	$\frac{1}{2}$	$\frac{1}{4}$	
A_2B_1	3		$\frac{1}{4}$	$\frac{1}{2}$	$\frac{1}{4}$	
A_2B_2	4			$\frac{1}{4}$	$\frac{1}{2}$	$\frac{1}{4}$

The phenotypic values correspond to intervals along the resource axis. Individuals of phenotype '1' take *only* resources of type '1' (e.g. small birds take small seeds), and so on for the other phenotypes and resource classes. Thus direct resource competition takes place only between members of a given phenotype class, but three or even all four of the genotypes may be present within a given phenotype class.

All members of a phenotype class have the same expected fitness, which is directly proportional to the resource abundance, and inversely proportional to the number of individuals competing for that resource. Letting i index phenotype and resource classes, $W_i = R_i/F_i$, where W is fitness, R is resource density, and F is population density. Thus the summed fitness of all of the members of phenotype class i must be equal to R_i. Individual fitnesses are treated formally as viabilities (that is, as probabilities of entering the mating pool), but they could just as well be thought of as relative fecundities (for females) and as relative probabilities of mating, given a particular pattern of female choice (for males).

Generations are discrete and non-overlapping. After selection, mating takes place in a single large pool. The implicit primary sex ratio is 1:1, and there are no sex differences in viability, so male and female genotype frequencies are always equal. In some versions of the model there is a locus governing female choice, with alleles C_1 (choosing) and C_2 (random mating). Females carrying the C_1 allele select mates according to one or another of the schemes described below. The A, B, and C loci are all unlinked. In some versions of the model there is another unlinked locus (D) that determines an arbitrary, selectively neutral phenotypic marker.

A fully deterministic algorithm was used to advance the population from one generation to the next, according to the rules and assumptions outlined above. This algorithm is equivalent to a system of recurrence equations in the genotype frequencies.

Four different resource distributions were employed, as set out in the following table.

	Class				
Distribution	1	2	3	4	5
Binomial	0.0625	0.25	0.375	0.25	0.0625
Uniform	0.2	0.2	0.2	0.2	0.2
Overdispersed	0.075	0.25	0.35	0.25	0.075
Underdispersed	0.05	0.25	0.4	0.25	0.05

Note that all four distributions are symmetrical. Thus the equilibrium gene frequencies at the A and B loci are equal to $\frac{1}{2}$, at least under random mating. The 'binomial' resource distribution is the null case, corresponding to the phenotype distribution that occurs under random mating. The 'uniform' distribution is extremely overdispersed. The 'overdispersed' and 'underdispersed' distributions are only mildly so, relative to the binomial distribution.

Five different schemes of female choice were considered. These schemes are described below, together with the results. A summary is given in Table 1.

(1) *Genotypic assortative mating (A and B loci).* Under this scheme of mating, a C_1 female mates only with males whose genotypes at the A and B loci are identical to her own. For example, a female of genotype $A_1B_2C_1$ will

Table 1. *Results of five schemes of female choice under three distributions of resource*

Basis of assortative mating	Resource distribution		
	Binomial	Overdispersed (incl. uniform)	Underdispersed
(1) Genotype (A/B) Choice polymorphic	Choice is neutral	Choice is favoured and speciation occurs	Choice is favoured and speciation occurs
(2) Phenotype (A/B) Choice polymorphic	Choice is eliminated	Choice is favoured but speciation does not occur	Choice is eliminated
(3) Phenotype (A/B) and markers (D) Choice fixed	Choice would not be favoured	Character-displaced species of equal abundance form around the two markers	Choice would not be favoured
(4) Markers (D) Choice fixed	Markers are neutral, do not change in frequency	Character-displaced species of equal abundance form around the two markers	A low-variance species (mainly A_1B_2 or A_2B_1) forms around one marker, and a high-variance species (all A/B genotypes) forms around the other
(5) Markers (D) Choice polymorphic	Not considered	Choice is favoured only very weakly until it is at high frequency; then character-displaced species form, as above	Not considered

mate either with $A_1B_2C_1$ or with $A_1B_2C_2$, the relative frequencies of the two possible matings being equal to the relative frequencies of the two kinds of acceptable males. This is a highly artificial system of choice, because it allows females to distinguish between male genotypes that express the same phenotypic value. In the case of A_1B_2 and A_2B_1, females can even choose between genotypes with the same genotypic value.

Under *any* resource distribution other than the null (binomial) distribution, the choice gene (C_1) goes quickly to fixation. At fixation for C_1 there are four 'species', each of which is genetically homogeneous at the A and B loci. Choice is neutral under the binomial resource distribution because a binomial distribution of phenotypes can be maintained with any amount of assortative mating, when the frequencies of A_1 and B_1 are equal to $\frac{1}{2}$.

(2) *Phenotypic assortative mating* (*A and B loci*). Under this scheme of mating, females can distinguish only between the phenotypes of potential mates, not between genotypes or even genotypic values. (Of course, males of phenotypes '1' and '5' must be A_1B_1 and A_2B_2, respectively.) The choice gene is eliminated under the binomial and underdispersed resource distributions, because phenotypic choice inevitably flattens the phenotype distribution (as genotypic choice does not). If the resource distribution is highly overdispersed (uniform), then the choice gene goes quickly to fixation. But speciation does not occur, because all possible matings continue to take place at appreciable frequencies. (Note that all four A–B genotypes are present in the central phenotypic class.) Speciation could occur under this resource distribution and scheme of choice only if there were no environmentally induced phenotypic variation.

The mildly overdispersed resource distribution would be matched exactly with A-locus and B-locus gene frequencies of $\frac{1}{2}$, and a moderate amount of assortative mating. Thus, under this distribution, the two alleles at the choice locus are expected to find a stable interior equilibrium. They do, but the equilibrium frequency of C_1 (the choice allele) is *higher* than expected, giving rise to an equilibrium phenotype distribution (0.076, 0.25, 0.348, 0.25, 0.076) that is even flatter than the resource distribution.

Why is there *more* assortative mating at equilibrium than is needed to equalize individual fitnesses? The reason seems to be that female choice creates a weak inclusive fitness effect (Hamilton, 1964). At equilibrium, the frequency of C_1 is 0.645 in the terminal phenotypic classes (1 and 5), 0.620 in the subterminal classes (2 and 4), and 0.609 in the central class (3). Thus members of the same phenotype class are positively related at the C locus. On average, a female who chooses will thereby increase the reproductive success of a male who is more likely to be carrying the C_1 allele than is a male taken at random from the population. The coefficient of relatedness (Michod & Hamilton, 1980; Seger, 1981) is very small ($R \simeq 0.0004$), but so is the expected reduction of fitness in the offspring produced by assortative matings.

This is not an instance of runaway selection (Fisher, 1930; O'Donald, 1980; Lande, 1981; Kirkpatrick, 1982), at least not in the usual sense, because the *average* female preference is for an *average* male phenotype. Perhaps it could be viewed as a case of simultaneous runaway selection in opposite directions. In any event, it suggests that directional runaway selection might usefully be analysed in terms of inclusive fitness effects at the choice locus (or loci).

(3) *Phenotypic assortative mating* (*A and B loci*) *with genotypic assortative mating* (*D locus*). Under this scheme the population is assumed to be fixed for allele C_1, so that all females are mating assortatively with respect to phenotypes, as described under scheme (2), above. This assumption is logical only if the resource distribution is overdispersed, because only in that case would choice be favoured. The C_1 allele is now assumed to confer on its bearers a *generalized* preference for mates that resemble them in any detectable way. Alleles D_1 and D_2 are introduced, controlling some completely arbitrary, selectively neutral, visible phenotypic difference (say, bright versus dull plastic leg bands). Thus females mate only with males whose functional phenotypes *and* marker phenotypes are the same as their own.

If the D alleles are introduced at a low level of phase disequilibrium with the alleles at A and B, the initial phase disequilibria grow rapidly, and the frequencies of D_1 and D_2 converge on $\frac{1}{2}$. Depending on the signs and relative magnitudes of the initial phase disequilibria, A_1 and B_1 are drawn into association with one of the D alleles, while A_2 and B_2 are drawn into association with the other D allele. Under the uniform resource distribution this process goes to completion, leaving in the end only two genotypes, $A_1B_1D_i$ and $A_2B_2D_j$, at equal frequencies.

It is of no interest that 'speciation' takes place under this scheme. The assumption that females will mate assortatively with respect to a number of different traits (including the arbitrary markers) means that from the beginning the markers necessarily define two reproductively isolated mating pools. The interesting result is that, as expected (Slatkin, 1980), these mating pools become genetically and phenotypically differentiated.

(4) *Genotypic assortative mating* (*D locus*). This scheme is the same as scheme (3), above, except that females choose *only* with respect to the arbitrary marker, not with respect to the functional phenotypes of their potential mates. As with the previous scheme, two character-displaced species form around the two markers if the resource distribution is overdispersed. Under the null (binomial) resource distribution the markers are neutral, because choice with respect to an arbitrary marker *need not*

lead to phenotypic assortment and *does not* do so unless such assortment is favoured for other reasons.

A strange pattern of association builds up if the resource distribution is underdispersed. A low-variance species consisting mainly of A_1B_2 or A_2B_1 forms around one of the markers, and a high-variance species consisting of all four genotypes forms around the other. The two species have nearly identical average phenotypes, so they are not character-displaced in the usual sense. Instead, the low-variance species specializes on the centre of the resource distribution while the high-variance species specializes on the tails. Given the highly artificial genetic constraints built into the model, this is the most efficient way to divide the resource distribution. Presumably the high-variance species would itself speciate into a small (A_1B_1) and a large (A_2B_2) species, each one exploiting a single tail of the resource distribution, if another phenotypic marker were introduced.

(5) *Genotypic assortative mating* (*C and D loci*). This final mating scheme involves all four loci. It is just like scheme (4) above, except that the population is *not* initially fixed for the choice gene. Only those females who actually carry allele C_1 choose mates who are like them at the D locus. The rest mate randomly with all of the available males.

Will choice itself be favoured, under an overdispersed resource distribution? The outcomes of schemes (1) and (2) clearly suggest that it should be, if D_1 and D_2 become associated with A_1B_1 and A_2B_2 (thereby signalling the presence of low and of high genotypic values). And the outcome of scheme (4) suggests that choice will tighten these very associations. Thus the entire process ought to lift itself up by its own bootstraps.

Unfortunately, this does not usually happen. In all of the cases studied the resource distribution was made uniform, providing very strong selection in favour of assortative mating. All runs were started with gene frequencies of $\frac{1}{2}$ at loci A, B and D, and with the largest possible degree of phase disequilibrium (all $A_1B_1D_1$ and $A_2B_2D_2$). If C_1 is introduced at a frequency of 0.25, it quickly rises to a frequency of 0.28, but then its progress comes to a virtual halt as the disequilibria between D and A–B decay to very small values. (Selection in favour of extreme phenotypes maintains the phase disequilibrium between A and B at a moderately high level.) If C_1 is introduced at a frequency of 0.5 it goes to effective fixation within 240 generations, but it does so by the dynamically strange route illustrated in Fig. 1. There is a sharp rise (generations 1–20) owing to the initial, artificially high phase disequilibria. Then there is a long plateau (generations 20–100) during which the frequency of C_1 very slowly increases while the phase disequilibria hold fairly steady at moderate values. Then there is a second, accelerating rise to fixation (generations 100–240), during which the phase disequilibria increase. At the end there is complete disequilibrium and speciation.

Although a strong positive feedback between choice and the conditions favouring choice does appear during this final period, it seems to require that the choice gene already be at high frequency, and it therefore fails to explain how choice with respect to an arbitrary marker could spread (at least at an appreciable rate) when rare.

Discussion

Taken together, the models described here clearly support Bengtsson's claim that 'sympatric speciation can occur in a species where all animals live in the same habitat and under the same fitness regime, but differ, due to their genetic constitution, in their resource utilization'. It should be stressed that speciation occurs not merely because individuals *differ* genetically in their resource utilization, but also, and in particular, because there are *constraints* on the possible *distributions* of their phenotypic differences.

Fig. 1. Gene-frequency and phase disequilibrium trajectories in the four-locus model (scheme 5). The illustrated run was started with equal frequencies of genotypes $A_1B_1C_1D_1$, $A_1B_1C_2D_1$, $A_2B_2C_1D_2$, and $A_2B_2C_2D_2$. (*a*) Gene frequencies. The choice gene (C_1) goes to fixation, as described in the text. (*b*) Standardized pairwise phase disequilibria for loci *A–B* (dotted line), and for loci *A–D* and *B–D* (solid line). The phase disequilibria decay to intermediate values from their initially maximal values, and then increase to maximal values as the choice gene goes to fixation and speciation occurs.

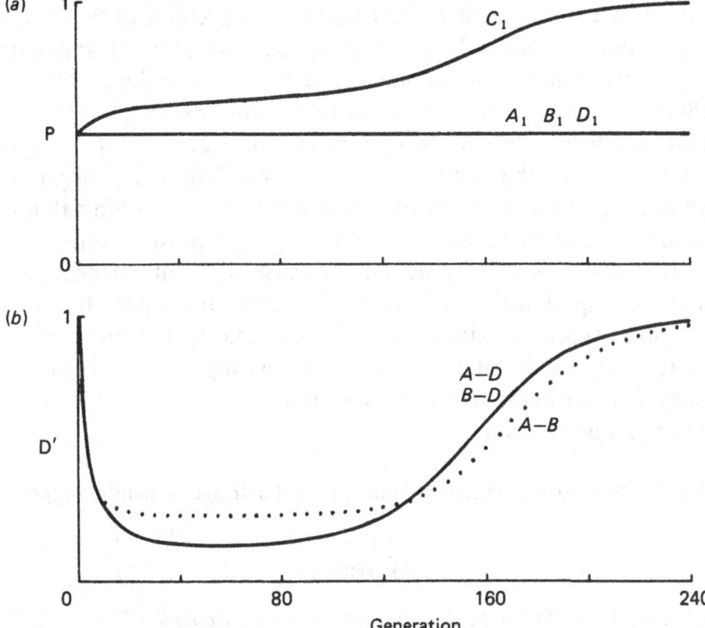

These constraints arise from fundamental properties of the species' genetic and developmental systems. In the context of the models these properties are taken to be unalterable, but in reality it would be necessary only that the genetic and developmental constraints be less readily alterable than the mating system. Thus the models suggest that sympatric speciation could provide, figuratively speaking, an easy way out of a difficult ecological and developmental bind.

But if there is environmentally induced variation of the functional phenotype so that genotypes cannot be distinguished unambiguously, then speciation will not occur unless mating is assortative with respect to some other, genetically unambiguous phenotypic difference, one that can be used as a proxy for the functional genotypes that are under disruptive selection. The four-locus model (scheme 5) suggests that such a system will not necessarily evolve spontaneously, even though it would be advantageous and stable once evolved. This would seem to pose a serious difficulty for this mechanism of sympatric speciation, unless *generalized* systems of female preference (along the lines of scheme 3) tend to evolve, preadapting the species for incorporation of an arbitrary marker into its system of assortative mating.

Many recent models of sympatric speciation trace back directly to Maynard Smith's (1966) development of Levene's (1953) model. The model developed here is no exception. It is, in fact, a Levene model with habitat selection (the process Maynard Smith considered most likely to cause speciation). But the form of habitat selection is peculiar in two respects. First, it is only *partial* habitat selection, because there is unavoidable environmental variation of the phenotype. Second, individuals 'return' to the different 'habitats' (= resource classes) in proportions that are not free to evolve independently. The genetic system constrains the phenotypes to occur in proportions that will, in general, *not* match the carrying capacities of the habitats, even if there is perfect assortative mating within habitats (as in scheme 2, phenotypic assortative mating). In most Levene models, positively assortative mating evolves because different phenotypes can specialize on qualitative differences between the habitats. In this one the phenotypes are already fully specialized, to the point of being *limited* to different 'habitats'. Assortative mating evolves because the frequency distribution of 'habitats' does not match that of phenotypes, if mating occurs at random.

I thank B. O. Bengtsson, J. Maynard Smith and L. Partridge for helpful suggestions.

References

Bengtsson, B. (1979). Theoretical models of speciation. *Zoologica Scripta*, **8**, 303–4.

Dickinson, H. & Antonovics, J. (1973). Theoretical considerations of sympatric divergence. *American Naturalist*, **107**, 256–74.

Felsenstein, J. (1981). Skepticism towards Santa Rosalia, or why are there so few kinds of animals? *Evolution*, **35**, 124–38.

Fisher, R. A. (1930). *The Genetical Theory of Natural Selection.* Oxford: Clarendon Press.

Gibbons, J. R. H. (1979). A model for sympatric speciation in *Megarhyssa* (Hymenoptera: Ichneumonidae): competitive speciation. *American Naturalist*, **114**, 719–41.

Hamilton, W. D. (1964). The genetical evolution of social behaviour. I. *Journal of Theoretical Biology*, **7**, 1–16.

Kirkpatrick, M. (1982). Sexual selection and the evolution of female choice. *Evolution*, **36**, 1–12.

Lande, R. (1981). Models of speciation by sexual selection on polygenic traits. *Proceedings of the National Academy of Sciences, USA*, **78**, 3721–5.

Levene, H. (1953). Genetic equilibrium when more than one ecological niche is available. *American Naturalist*, **87**, 331–3.

Maynard Smith, J. (1966). Sympatric speciation. *American Naturalist*, **100**, 637–50.

Michod, R. E. & Hamilton, W. D. (1980). Coefficients of relatedness in sociobiology. *Nature*, **288**, 694–7.

O'Donald, P. (1980). *Genetic Models of Sexual Selection.* Cambridge University Press.

Rosenzweig, M. L. (1978). Competitive speciation. *Biological Journal of the Linnean Society*, **10**, 275–89.

Seger, J. (1981). Kinship and covariance. *Journal of Theoretical Biology*, **91**, 191–213.

Slatkin, M. (1979). Frequency- and density-dependent selection on a quantitative character. *Genetics*, **93**, 755–71.

Slatkin, M. (1980). Ecological character displacement. *Ecology*, **61**, 163–77.

Udovic, D. (1980). Frequency-dependent selection, disruptive selection, and the evolution of reproductive isolation. *American Naturalist*, **116**, 621–41.

5

Darwinian evolution in ecosystems: the Red Queen view

N. C. STENSETH

Introduction

> I have been watching it [the Darwinian theory of evolution] slowly unravel as a universal description of evolution [...]. I have been reluctant to admit it [...] but [...] that theory, as a general proposition, is effectively dead, despite its persistence as a text-book orthodoxy.
>
> [Gould, 1980]

Is a new paradigm of stasis-plus-punctuation (see Eldredge & Gould, 1972) about to replace the gradualistic Darwinian theory of evolution? Several paleontologists (e.g. Gould, 1980, 1982; Williamson, 1981*a*,*b*; Stanley, 1982) argue strongly in favour of this. But is their attack justified? Several evolutionary biologists (e.g. Stebbins & Ayala, 1981; Charlesworth, Lande & Slatkin, 1982; Hoffman, 1982; Maynard Smith, 1982*a*, 1983; Schopf, 1982; Schopf & Hoffman, 1983) do not think so.

Darwinists have two reasons for focussing on gradual changes (Maynard Smith, 1983): a large change is unlikely to improve adaptation since a finely graded variation is needed to produce a detailed adaptive fit to current conditions, and since existing organisms are mostly presumed to be close to their adaptive peak (Fisher, 1930). In neither case is the occurrence of 'hopeful monsters' (individuals carrying a genetic mutant with large phenotypic effect; Goldschmidt, 1940) excluded, but in both cases they will need fine tuning by selection of mutants with *small* phenotypic effects before detailed adaptation to the current environment is achieved.

The punctuationists, on the other hand, claim that the fossil record reveals long periods of little or no evolutionary change (i.e. stasis), punctuated by spells of rapid change, usually associated with the division of an ancestral species into two daughter species. Given this evolutionary pattern, the punctuationists promulgate that macroevolution must be 'decoupled' from microevolution and that an alternative to the Darwinian theory is needed (Stanley, 1979, 1982). However, as long as we have no fully developed theory for the long-term behaviour of ecosystems – nor for the behaviour of the fossil record – such a claim is unjustified. But, being unable at present to predict the various patterns seen in the fossil record is *not* the same as saying that the Darwinian theory is incompatible with the fossil record. Nevertheless, that is essentially what the punctuationists like Gould do (e.g. Maynard Smith, 1982*a*).

Evolutionary biology ought to be – but often is not – the child of a lasting marriage between genetics and ecology; Lewontin (1979), for instance, bemoaned the fact that ecology and evolutionary genetics 'remain essentially separate disciplines, travelling separate paths while politely nodding to each other as they pass'. In this essay I review how these fields may be integrated; I focus on some recent attempts in formulating a Darwinian theory for evolution in ecosystems.

What kind of theory do we need?

According to Maynard Smith (1982*a*), we

> need a theory which says something about selection, and hence about the environment. Since the major component of the environment of most species consists of other species in the ecosystem, it follows that we need a theory of ecosystems in which the component species are evolving by natural selection.

John Maynard Smith and I have recently presented some fragments of such a Darwinian theory for evolution in ecosystems (Stenseth & Maynard Smith, 1984). This model is based on Van Valen's (1973) Red Queen hypothesis which asserts that any evolutionary change in any species is experienced by coexisting species as a change of their environmental conditions. Hence, if a species is to continue its existence, it must constantly evolve as rapidly as possible. The Red Queen explained to Alice in Wonderland that

> . . . it takes all the running *you* can do, to keep in the same place. If you want to get somewhere else, you must run at least twice as fast as that!
>
> (L. Carroll, *Through the Looking-glass*)

This view of evolution leads us all the way back to Darwin (Van Valen, 1973, 1977):

> [as] the most important of all causes of organic change is one which is almost independent of [. . .] altered physical conditions, namely, the mutual relation of organism to organism [. . .] if some of these many species become modified and improved, others will have to be improved in a corresponding degree or they will be exterminated.
>
> (Darwin, *The Origin of Species*)

Maynard Smith (1969) pointed out that what we need

> is first a theory of ecological permanence, and then a theory of evolutionary ecology. The former would tell us what must be the relationships between the species

composing an ecosystem if it is to be 'permanent', that is, if all species are to survive, either in a static equilibrium or in a limit cycle. In such a theory, the effect of each species on its own reproduction and on that of other species would be represented by a constant or constants [...]. In evolutionary ecology these constants become variables, but with a relaxation time large compared to the ecological time scale. Each species would evolve so as to maximize the fitness of its members.

Such a theory must allow for changes in the number of species in the ecosystem. It is hard to formulate; the data needed to test it would be those of paleontology and ecology.

The Darwinian theory of evolution

According to Darwinism, evolution of adaptations takes place as a result of natural selection (e.g. Maynard Smith, 1969; Dawkins, 1982). By neo-Darwinism we mean this idea, but add a theory of heredity; this is what I call the Darwinian theory of evolution. In its most general form, the theory of heredity goes back to Weismann (1886), who claimed that acquired characters could not be inherited: Weismann 'separated' the germ line from the soma line. This is consistent with the central dogma of molecular genetics which states that protein may not be translated back to DNA, and the central dogma of embryology which states that bodily form and behaviour may not be translated back to proteins (Dawkins, 1982, p. 168 & pp. 173–6). If Weismann was right (and we still have no reason to believe he was not), Darwin's theory is greatly strengthened, since natural selection then becomes the *only* process leading to adaptations rather than just a possible one; this is so because Weismann's assumption dictates that there is *no* relation between the need for a new variety (i.e. a mutation) and its occurrence. Weismann's view also makes it *possible* to understand evolution without understanding development (Maynard Smith, 1982*b*); all we need, in order to make the Darwinian theory work, is that genetic differences contribute significantly to the phenotypic variation within a population – and this is indeed a fact. This insight is of great assistance in the study of evolution in ecosystems.

The Darwinian theory assumes that populations consist of individuals which produce progeny like themselves and that coexisting individuals vary. Given these properties, it follows that the variety(-ies) which, in the current environment, produces more offspring than its conspecifics in the same population, and whose offspring survive better until maturity, will come to dominate the population through the process of natural selection; this is *not* a hypothesis – it is a logical consequence of our assumptions of how life is organized.

The Darwinian theory may be expressed more precisely: let $N_{j,t}$ be the number of a *particular* variety or phenotype, j, in a monomorphic population at time t. Let B_j be the *expected* number of progeny produced by each of them, $s_{y,j}$ the *expected* survival of young, and $s_{a,j}$ the *expected* survival of adults until the next generation. The triplet $(B_j, s_{a,j}, s_{y,j})$ defines the jth variety's ecological strategy and is determined by the variety's physiology, morphology (e.g. birds' bills) and behaviour (e.g. flocking and altruistic helping of conspecifics) influencing an individual's ability to utilize available resources and to avoid environmental hazards. (Obviously, there are some constraints on how B_j, $s_{a,j}$ and $s_{y,j}$ may vary for any given species; 'If there were no constraints on what is possible, the best phenotype would live for ever, would be impregnable by predators, would lay eggs at an infinite rate, and so on' (Maynard Smith, 1978*a*); i.e. $s_{a,j} = s_{y,j} = 1$ and $B_j = \infty$.) For a discretely reproducing population, we then have

$$N_{j,t+1} = B_j(E)s_{y,j}(E)N_{j,t} + s_{a,j}(E)N_{j,t}, \tag{1}$$

where E defines the contemporary environment determined by abiotic as well as biotic factors (including conspecific individuals). The mean, or expected (inclusive) fitness, $\bar{W}_i(E)$, of the ith species in the contemporary environment is defined by

$$\bar{W}_i(E) = \sum_j p_j \cdot [B_j(E) \cdot s_{y,j}(E) + s_{a,j}(E)], \tag{2}$$

where p_j is the relative frequency of the jth strategy – or variety – in the current population. Natural selection favours those strategies – B_j, $s_{a,j}$, $s_{y,j}$ – where the resulting population (mono- or polymorphic) cannot be invaded by any mutant strategy under prevailing conditions; this is an Evolutionarily Stable Strategy (or ESS; Maynard Smith, 1982*b*; see p. 17 & Appendix D for specifications). In the current environment, E, this will – for a variety of biological situations – correspond to maximizing $\bar{W}_i(E)$ over available strategies. From (2), it then follows that if either the physical or the biotic component of the environment (E) changes, some evolutionary change in the ith species will – in general – necessarily occur. That is, the Red Queen hypothesis is at the heart of the Darwinian theory (Stenseth, 1984*a*). Indeed, from Eqn (2), it appears that evolution cannot be understood without first understanding ecology.

The Red Queen and the lag load

According to the Red Queen hypothesis, most organisms are most of the time *un*likely to be *at* their adaptive peak since the Wrightian adaptive landscape (Wright, 1982) changes as a result of evolution in coexisting species as well as a result of altered physical conditions. (Notice that Wright's figure 7*c* depicts a typical Red Queen situation; he then gives the conditions under which it is likely to occur.) Following the lead of Sir Ronald Fisher (1930), Maynard Smith (1976*a*) formalized the Red Queen idea further by introducing his lag load concepts. The *lag* – or *evolutionary – load* for the ith population (or species) is defined as

$$L_i = L_i(E) = [\hat{W}_i(E) - \bar{W}_i(E)]/\hat{W}_i(E), \tag{3}$$

where $\bar{W}_i(E)$ – defined by (2) – is the mean fitness of the current ith species or population and $\hat{W}_i(E)$ is the corresponding fitness of the theoretically optimal genotype for the contemporary environment, E, including all favourable mutations, whether or not they are yet segregating in the ith population. The lag, L_i, may decrease by increases in $\bar{W}_i(E)$ resulting from evolution through natural selection in species i. Both $\bar{W}_i(E)$ and $\hat{W}_i(E)$ may change as a result of altered environmental conditions, E, either through evolution in coexisting species or through altered physical conditions. Obviously, evolution in sympatric species may improve *as well as* deteriorate the ith species' conditions for existence; however, as most organisms are usually close to their adaptive peak, a deterioration is most likely.

A generalized version of Fisher's (1930) fundamental theorem of natural selection implies that the rate of evolution – or the rate of adaptive change – in the ith species will increase with increasing L_i.

Ecology and evolution

Three time scales

Distinguish between three time scales:

(i) The *ecological time scale* in which the abundances and distributions of species are the variables. This is the time scale to which ecological models – and theories – refer (e.g. Maynard Smith, 1974; Pimm, 1982). The number of species, as well as the nature and strength of the interactions between coexisting species (i.e. the food web) are constants.

(ii) The *gene-frequency time scale* in which the constants defining the strength of the connections in the ecological food web become variables; these will change through the process of natural selection operating on each of the coexisting species. The number of species as well as the structure of the food web are, however, fixed. The parameters in the ecological community models become evolutionary variables – the number of equations in these models is, however, fixed. This is the time scale to which population genetics models – and theories – refer. The ESS concept commonly refers to this time scale.

(iii) The *speciation–extinction time scale* in which the number of coexisting species becomes a variable; it changes as a result of invasion (succession as well as speciation) and extinction. Here the structure of the food web is no longer fixed. Neither is there any fixed number of equations in the ecological community models. Theories for the behaviour of the fossil record refer to this time scale.

Distinguishing between these time scales simplifies the analysis of evolution in ecosystems (Stenseth & Maynard Smith, 1984) since we then may assume that processes operating at the faster time scales have reached their equilibrium when considering the slower time scales. Collectively, (ii) and (iii) may be called the *evolutionary time scale*; however, distinguishing between time scales (ii) and (iii) does not necessarily imply that different evolutionary processes (such as individual and species selection) need be involved – contrary to what the punctuationists claim (e.g. Stanley, 1979, 1982; Gould, 1982). The models discussed below assume only the operation of individual selection.

Ecological load and evolutionary load

The evolutionary load differs from Darwin's '*conditions of existence*'; the former refers to time scale (ii) whereas the latter refers to time scale (i). In order to distinguish between these, Van Valen (1973, 1976*a*) introduced the terms 'environmental load' and 'amount of competition'. A general definition of the latter is what I call the *ecological load*, C_i, of the *i*th species under prevailing environmental conditions:

$$C_i = C_i(E) = [\bar{W}_i(E_o) - \bar{W}_i(E)]/\bar{W}_i(E_o), \tag{4}$$

where E_o is the theoretically optimal environment for the current form(s) of the *i*th species. For instance, for a competitive community in a physically optimal environment, we would obtain E_o by putting all competing species' densities equal to zero.

The average fitness or specific net growth rate, $\bar{W}$, enters the definitions of both load concepts: the *evolutionary load* is developed for comparing the degree of adaptation of various strategies – or combinations of strategies – *to a particular* environment, E; i.e. $\bar{W}_i'(E)$ and $\bar{W}_i''(E)$ – representing fitness of two different strategies – are to be substituted for $\bar{W}_i(E)$ in (3). The *ecological load*, on the other hand, is developed for comparing the degree of adaptation *of a particular strategy* to various environmental conditions; i.e. we compare $\bar{W}_i(E')$ and $\bar{W}_i(E'')$ for $E' \neq E''$. The difference between the ecological load, C, and the evolutionary load, L, is illustrated in Fig. 1 (see Stenseth & Maynard Smith, 1984). The evolutionary load, L, for species II (presumed to be optimally placed between species I and III so that $\bar{W}_{II}(E) = \hat{W}_{II}(E)$) will increase regardless of whether a species IV is added, or species III is deleted. The ecological load, C, for species II, will, on the other hand, increase in the first case, but decrease in the latter; in all three cases, E_o will be the same and defined as that with no competitors. Deleting species III will make it ecologically easier for species II; however, species II could do it *even better* by evolutionarily changing its niche position.

In evolutionary time, L will – for a fixed E – be minimized through changes in the phenotypic composition of the population (Maynard Smith, 1976*a*). In ecological time, C will be minimized through optimal food and habitat selection.

The magnitude of C_i for a given species i determines its probability of extinction: the larger C_i is, the larger will be the probability of species i becoming extinct. *That is, it is the species' 'conditions of life' which determine whether or not it will persist in the ecosystem.* Essentially, this is what Gause's (1934) principle and MacArthur & Levins' (1967) concept of 'limiting similarity' are all about.

With increasing average evolutionary load, $\bar{L}$, for the coexisting species, the average rate of the species' microevolution in the ecosystem will increase. In that case, the average ecological load will, due to changes in E, frequently be too large to be compatible with continued existence; several species will fall too far behind in the evolutionary 'race for life' (to quote *The Origin*). Subsequently, such extinction will leave room for invading species; i.e. species turnover may increase with increasing L.

Evolution in ecosystems – a general approach

The ecosystem analogue of Maynard Smith's ESS concept is defined as *a community being such that neither a mutant strategy of an existing species nor any new species can invade* (Allan, 1976; Lawlor & Maynard Smith, 1976; Roughgarden, 1979; Reed & Stenseth, 1984). If such a non-invadable

Fig. 1. Each species' utilization function – or niche – is assumed to be composed of the niches of many morphs (corresponding to each genotype) as discussed by, for instance, Roughgarden (1979). The initial state is shown in the top diagram with three species uniformly spaced along a resource axis; species II is *at* its adaptive peak. This state can be altered in two ways: (a) a fourth species may be added, or (b) a species may be removed. Change (a) would deprive species II of resources, and change (b) would provide species II with additional resources, yet both changes would result in a similar increase in the evolutionary load, L_{II}, of species II. Modified from Stenseth & Maynard Smith (1984).

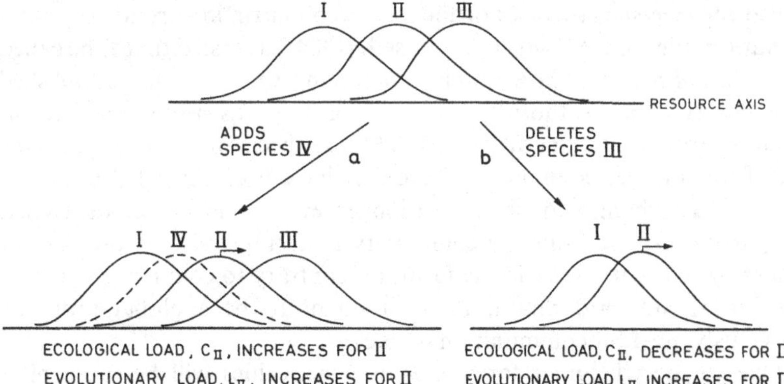

ecosystem exists, evolution will eventually cease in a physically stable environment. In order to study this ESS property of ecosystems, let $X=(X_1, X_2,\ldots, X_n)$ denote the density-vector of the populations in the current ecosystem. The dynamics of the coexisting species are given by

$$\frac{\mathrm{d}X}{\mathrm{d}t}=f(X), \tag{5}$$

which either has a locally stable equilibrium at $X=X^*$ or some stable limit cycle. Hence, it is resistant towards *ecological* perturbations. Nevertheless, it may not be resistant towards *evolutionary* perturbations. To study the latter property, we must extend Eqn (5) so as to include varieties of the already existing species, or entirely new species attempting to invade the current community. Let $Y=(Y_1, Y_2,\ldots, Y_m)$ be the density-vector of those trying to invade. The extended model is defined as

$$\begin{aligned}\frac{\mathrm{d}X}{\mathrm{d}t}&=F(X, Y)\\ \frac{\mathrm{d}Y}{\mathrm{d}t}&=G(X, Y),\end{aligned} \tag{6}$$

where $F(X,0)=f(X)$ and $G(X,0)=0$. If (6) is to describe evolution, G must be constrained by some phenotype sets (Maynard Smith, 1978*a*) which, of course, are hard to formulate. Assuming that all invaders are rare to begin with, the *evolutionary stability* of the original ecosystem (5) is determined by the *ecological stability* of the equilibrium $X=X^*$, $Y=0$ of (6) for *any* possible Y-vector or G. From the assumptions regarding (6), it follows that its Jacobian matrix, P, evaluated at the equilibrium $(X^*,0)$ is given by

$$P=\begin{bmatrix}\dfrac{\partial\widehat{F}}{\partial X} & \dfrac{\partial\widehat{F}}{\partial Y}\\ 0 & \dfrac{\partial\widehat{G}}{\partial Y}\end{bmatrix}=\begin{bmatrix}Q & U\\ 0 & R\end{bmatrix}, \tag{7}$$

where $Q=\partial\widehat{F}/\partial X=\partial F(X^*,0)/\partial X=\partial F(X^*)/\partial X$, etc. The Q-matrix is called the *community-matrix* or the food web matrix in current ecological literature (Pimm, 1982). Reed & Stenseth (1984) suggested that R be called the *invasion matrix*. Those species attempting to invade the established community – that is, those species defining R – consist of the new forms resulting from mutations at the site and those species immigrating into the area from another community where they have been subjected to micro-evolutionary changes for shorter or longer (evolutionary) periods. Hence, for Q to correspond to an ESS community, it must be resistant towards any – slightly or greatly – deviating form; i.e. contrary to common usage (e.g. Lawlor & Maynard Smith, 1976), it ought to be a global property. Obviously, no ESS community has to exist.

Due to the triangular form of P, its eigenvalues will be those of Q

(evaluated at $X = X^*$) *plus* those of R (evaluated at $X = X^*$, $Y = 0$). Since (5) is ecologically stable to begin with, the *evolutionary stability of (5) is defined by the eigenvalues of the R-matrix only*; obviously, this simplifies greatly. It follows that $\bar{L} > 0$ corresponds to a Q-matrix being such that an R-matrix for which at least one eigenvalue having positive real part could be found. Evolution will cease if the ecosystem is ecologically stable and resistant towards any invading form. Specifically, on time scale (ii), evolution will cease in a physically stable environment if mean fitness, $\bar{W}_i$, is maximized for each of the coexisting species (see Roughgarden, 1979).

This approach emphasizes that evolution cannot be understood without first understanding the coexisting species' population dynamics, and the ecological competition between established strategies *and* those strategies attempting to invade the community. Certainly, this approach is a feasible one for studying the long-term behaviour of ecosystems, and hence the behaviour of fossil records.

A macroscopic approach to evolution in ecosystems

The variables in (5) and (6) are the densities of each species; these may be called 'microscopic' variables. Alternatively we may formulate models in terms of some 'macroscopic' variables. Unfortunately, it is unclear what are appropriate 'macroscopic' variables in evolutionary ecology (e.g. Maynard Smith, 1974). It could, for instance, be the species' biomasses or densities. Having such macroscopic variables, we could derive 'laws' connecting them in terms of the dynamics of the interacting species. One such attempt is the statistical mechanics of evolutionary ecology (Kerner, 1957, 1959; Leigh, 1965, 1968; for review see Maynard Smith, 1974; Stenseth, 1979). It is uncertain – at least to me just now – whether this latter approach may improve our understanding of evolution in ecosystems. But the general philosophy is important! Something *like* this is indeed needed; my hunch is that progress will be made by analysing models like (5) and (6) from this perspective.

Joint work together with Maynard Smith has suggested another 'macroscopic' formulation (Stenseth & Maynard Smith, 1983). Our model – referring to time scales (ii) and (iii) – is expressed in terms of the average evolutionary load, $\bar{L}$, and the number of coexisting species, S. I will briefly review – and extend – this model.

The Red Queen in a physically stable environment

On the gene-frequency time scale (ii), a change in the evolutionary load (L_i) of species i will – according to the Red Queen hypothesis – be given as (Maynard Smith, 1976*a*)

$$\delta L_i = \delta_c L_i - \delta_g L_i, \quad i = 1, 2, 3, \ldots, S, \tag{8}$$

where $\delta_c L_i$ is the increase in the evolutionary load of the ith species caused by microevolution in coexisting species, and $\delta_g L_i$ is the reduction in the evolutionary load of the ith species caused by its own microevolution. Hence,

$$\delta L_i = \sum_{j=1}^{S} \beta_{ij} \cdot \delta_g L_j - \delta_g L_i, \quad i = 1, 2, 3, \ldots, S, \tag{9}$$

where β_{ij} ($\beta_{ii} = 0$) is the increase in L_i due to unit change in L_j. Assuming – as did Fisher (1930) – that most species are usually close to their adaptive peak, any evolutionary change in any species is most likely experienced as a deterioration by all other species; i.e. $\beta_{ij} > 0$. If $\delta_g L_i = k_i \cdot L_i$ (compare Fisher's fundamental theorem; see p. 59) we can – referring to time scale (ii) – rewrite (9) as

$$\frac{dL_i}{dt} = \sum_{j=1}^{S} \beta_{ij} \cdot k_j \cdot L_j - k_i \cdot L_i, \quad i = 1, 2, 3, \ldots, S. \tag{10}$$

Taking the average of both sides in (10), we obtain after minor rearrangement

$$\frac{d\bar{L}}{dt} = \frac{k}{S} \cdot \sum_j (D_j - 1) \cdot L_j. \tag{11}$$

where D_j – the 'distributed decrement' (Maynard Smith, 1976*a*) – is defined as $D_j = \sum_i \beta_{ij}$ and $k_i = k = \text{const}$. Some function $d(\beta, L)$ – where $\beta = (\beta_{ij})$ and $L = (L_i)$ – may be found so that

$$\frac{d\bar{L}}{dt} = d(\beta, L) \cdot \bar{L}, \tag{12}$$

where $\bar{L}$ is the average load. The β_{ij} depend on S and are constrained by the coexisting species' phenotypes sets. The β_{ij} determine the equilibrium, $\bar{L}^*$, in (11) and hence whether or not the currently coexisting species will cease to evolve.

On the gene-frequency time scale (ii), the R-matrix (p. 62) is the ecological equivalent to the population geneticists' β_{ij}-matrix. Both define whether or not invasion will occur and describe how *coexisting species interact (micro-)evolutionarily*. The β_{ij}-matrix and the R-matrix are the evolutionary analogues of the ecologists' community matrix (Q; p. 62) which describes how *coexisting species interact ecologically*.

On the speciation-extinction time scale (iii), S is no longer a constant; the β_{ij}-values may be functions of S (Stenseth & Maynard Smith, 1984). Because Maynard Smith (1976*a*) treated the β_{ij}-values as constants, he was unable to make sense out of the Red Queen hypothesis (Stenseth, 1979). However, incorporating the fact that the β_{ij} may change and approximating d by a linear function of S and $\bar{L}$, the following Darwinian model for the long-term behaviour of ecosystems results:

$$\frac{dS}{dt} = e + m \cdot \bar{L} + n \cdot S$$
$$\frac{d\bar{L}}{dt} = (a + b \cdot \bar{L} + c \cdot S) \cdot \bar{L}, \tag{13}$$

where $e>0$, $n<0$ and $b<0$ whereas a, c and m are of uncertain sign (Stenseth & Maynard Smith, 1984). The dS/dt equation is based on island biogeographic arguments (Rosenzweig, 1975; Stenseth, 1979) together with arguments relating species turnover to $\bar{L}$ (p. 61). The $d\bar{L}/dt$ equation is found by Taylor-expanding $d(\beta, L)$.

The dynamic behaviour of the ecosystem described by (13) is discussed thoroughly by Stenseth & Maynard Smith (1984). Two of the resulting patterns are more interesting in the present context than the others (Fig. 2); for these the extinction rate increases more than the speciation rate increases with increases in $\bar{L}$; i.e. $m<0$. If $\bar{L}^*=0$, *stasis* is expected to be common in physically stable environments; if $\bar{L}^*>0$, continued evolution is expected to occur in physically stable environments. That is, the Darwinian theory of evolution predicts the views of both schools of thought *depending on how the variables are related to each other.* Evolution in a *physically constant* environment may continue for ever at a constant non-zero rate in what is presumably *a species-rich ecosystem* (Fig. 2*a*) or *cease in a species-poor ecosystem* (Fig. 2*b*); Van Valen's (1973) 'law of extinction' follows only in the former case. These results are indeed very important, as they demonstrate that Darwinian evolution *does not* necessarily cease in a physically stable environment: the Red Queen view of evolution is indeed

Fig. 2. Phase diagram depicting the dynamic behaviour of the evolutionary models (13) – assuming a physically stable environment – for two biologically plausible cases. A depicts what most probably corresponds to a species-rich ecosystem being so that d(β, L) increases with increasing values of S. B depicts what most probably corresponds to a species-poor ecosystem being so that d(β, L) is independent of S. See the main text for further discussion.

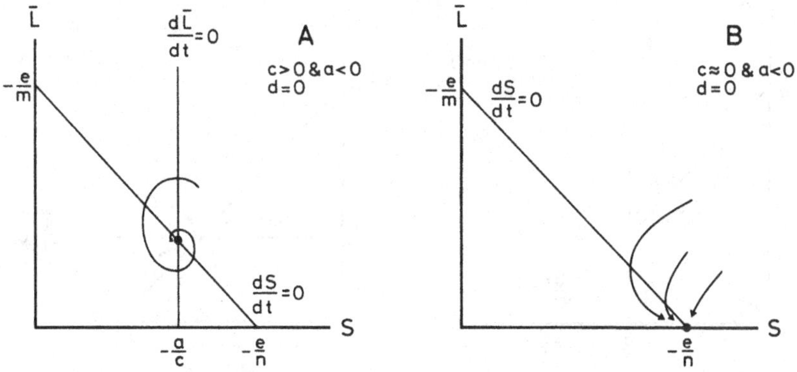

plausible! Major physical disturbances of the ecosystem or the crossing of some evolutionary threshold may change the long-term behaviour of the ecosystem from one mode to another (i.e. from a to b in Fig. 2); such major changes are presumably rare phenomena. However, they could change the system so that a new equilibrium number of species, S^*, emerges.

The Red Queen in a physically unstable environment

Real ecosystems are not physically stable. Hence, some background level of physical disturbance must be incorporated. This amounts to adding a constant $p>0$ to the $d\bar{L}/dt$ equation in (13). The resulting long-term dynamic behaviour is depicted in Fig. 3. Continued evolution is expected in a species-poor as well as in a species-rich ecosystem (i.e. $\bar{L}^*>0$); the local stability properties differ, however, in the two cases. Hence, the fossil record may still be used to distinguish between the two modes: in an ecosystem exhibiting the *Red Queen type of gradual evolution*, long time-spans are needed to restore long-term stability after a major perturbation. In an ecosystem exhibiting *stasis-plus-punctuation type of evolution* a relatively short time-span is needed before stability is restored. Notice, that *true stasis is never expected* in any real ecosystem. Van Valen's (1973) 'law of constant extinction' is now predicted in species-poor *as well as* in species-rich ecosystems.

Conclusion

The Red Queen helps us understand the world around us

Van Valen (1973) proposed the Red Queen hypothesis in order to *explain* the empirical 'law of constant extinction'. His paper has been criticized for

Fig. 3. Same as Fig. 2 but for a physically *un*stable environment; p is the rate of increase in L due to random changes in the physical environment. See the main text for discussion.

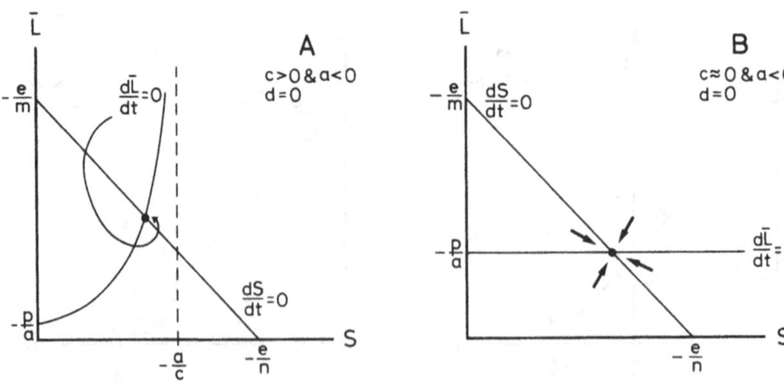

two reasons. First, the 'law of constant extinction' may only be an artefact of improper analysis of data (e.g. Hallam, 1976; see also Van Valen, 1976*b*). Second, it was not obvious that a constant rate of extinction necessarily follows from the Red Queen hypothesis (e.g. Maynard Smith, 1975, 1976*a*; see also Van Valen, 1977). Among the available models for the long-term behaviour of ecosystems (for review, see Stenseth, 1979), the one originally suggested by MacArthur (1969; his Fig. 4) and further developed by Rosenzweig (1975), is – I now think – the most promising one; much of this essay has been a review and further development of a model combining the approaches suggested by Rosenzweig (1975) and Maynard Smith (1976*a*). Based on this model, it appears that ecosystems are in fact expected to exhibit a long-term stable number of species. Further, *depending upon the nature of the ecological interactions between the coexisting species*, the ecosystem may exhibit a continued non-zero rate of microevolution, with constant rates of speciation and extinction, or the ecosystem may settle down in a fixed state so that each of the coexisting species is at its adaptive peak (i.e. stasis) and with no species turnover. However, in a physically stable ecosystem with a fixed *set* of species, stasis is always expected. The Red Queen also helps us understand other aspects of the living world (see also Dawkins, 1982).

Evolutionists (e.g. Medawar, 1967) repeatedly return to the question of whether there is any biological law which, in Maynard Smith's (1970) words,

> [. . .] might enable us to put an arrow on time in evolutionary processes, as the second law of thermodynamics enables us to put an arrow on physical processes.

At that time, Maynard Smith could not find a good basis for the common-sense feeling that there is a steady evolutionary increase in complexity and improved adaptedness. These difficulties seem to result from ignoring the biotic component of the environment. With the Red Queen view of evolution, such a continued increase in complexity is indeed expected; if other evolving species are the major component of a species' environment, an evolutionary 'race for life' seems unavoidable. One evolutionary improvement achieved by one species must necessarily be neutralized by an even better adaptation by the coexisting species if they are to continue to exist. This may, for instance, explain – without assuming any non-Darwinian selection process – long-term evolutionary trends like Cope's rule (i.e. the steady increase in average body size within lineages; e.g. Simpson, 1953). It would indeed be interesting to know whether the lineages obeying Cope's rule are those for which large body size confers intra- and interspecific advantages (see, e.g. Grant, 1972).

The Red Queen view also suggests one out of many possible explanations for why organisms may not be perfectly adapted to their current 'conditions of existence': as soon as the organism evolves to meet some particular environmental conditions, some other conditions will prevail. Darwin claimed that the 'Unity of Type', although important, is subordinate to the 'conditions of existence' because the 'Type' *was* once an organism which evolved to meet some particular conditions. This is why evolutionary biologists reject 'typological thinking' (Mayr, 1963). Notice that saying that stasis is a real phenomenon implies 'typological thinking' (Stenseth, 1984*b*). Hence, stasis is not expected – as a general phenomenon – on the basis of the Darwinian theory (see also Stebbins & Ayala, 1981; Maynard Smith, 1983). The Red Queen also provides a Darwinian solution to St George Mivart's (1871) dilemma; he could not see how evolution would proceed by gradual accumulation of small changes, since most intermediate forms seemed maladaptive. However, since the 'conditions of existence' change continuously, intermediate forms may in fact have been adaptive. The punctuationist, on the other hand, 'solves' Mivart's dilemma by arguing that the intermediate forms never existed.

It is interesting that Maynard Smith (1976*b*) once remarked, when discussing 'why sex?' – the queen of problems in evolutionary biology – that

[one] is left with the feeling that some essential feature of the situation is being overlooked.

The Red Queen view implies that evolutionists erroneously omitted the effects of continuously changing ecological conditions: an organism must necessarily reproduce sexually – and hence maintain a high evolutionary potential – if it strongly interacts ecologically with other species; if not, sex is not so essential (Maynard Smith, 1978*b*). Thus, it is pleasing to know that sexual reproduction particularly occurs in habitats with high species diversity (Maynard Smith, 1978*b*).

'Gradualism or punctualism' – or – 'gradualism and punctualism'

The most important (empirical) disagreement between gradualists and punctuationists, is over the existence of temporal stasis rather than punctuation (e.g. Maynard Smith, 1983). It is difficult to establish (or dismiss) the reality of gradual transition from one species to another; what appears to be gradual to a population biologist (ecologist or geneticist) may appear as a punctuation to a paleontologist (e.g. Stebbins & Ayala, 1981; see also p. 59). Nor do we have any way of being sure that what appears to be a punctuation in the fossil record is not simply the result of a gradual change of a species in another ecosystem followed by reimmigration into the ecosystem from which the fossil record originates. Recall what Darwin wrote:

Local varieties will not spread into other and distant regions until they are considerably modified and improved; and when they do spread, if discovered in a geological formation, they will appear as if suddenly created there, and will simply be classified as new species.

(Darwin, *The Origin of Species*)

I cannot resist a reminder of Gause's principle, or MacArthur & Levins' concept of limiting similarity. If a species attempting to invade is to succeed, it cannot be *too* similar to any of the species already in the ecosystem. Hence, there is a methodological snag with the paleontologists' analyses 'establishing' temporal stasis. Relative to the variation *between* species, the variation *within* species will always be small (Stenseth, 1984*b*). This is not to say that I dismiss the existence of a few 'living fossils'; however, since evolutionary theory – as the gas theory – only makes statements about classes of species (or molecules), singular observations of 'living fossils' are irrelevant in the present context. What is under discussion is whether this is a common pattern.

The Lake Turkana mollusc sequence has been claimed to be one of the better – maybe even the best – example documenting the reality of the punctualistic view (Williamson, 1981*b*). *Most* observed speciation events in this sequence seemed to be synchronized and to occur at times of easily detectable physical stress – a major lowering of the water level in the lake, a major volcanic eruption producing heavy ashfall onto the lake, or both: Fig. 2*b* – and the punctualistic view seems to apply. However, this need not be correct since these easily detectable environmental changes necessarily will overshadow the Red Queen type of gradual evolutionary changes (i.e. as in Fig. 2*a* or Fig. 3*a*); neither can we – on the basis of the data (compare above comments) – reject the possibility that the observed extensive evolutionary changes associated with the physical stress of the environment indeed were gradual. Only if *all* lineage splitting had occurred simultaneously – and coinciding with the physical disturbances – would the Turkana sequence exemplify the punctualistic view.

Punctuationists repeatedly claim that the (presumed) *temporal* stasis is mirrored by what might be called *spatial* stasis (e.g. Stanley, 1979; Williamson, 1981*a*). Many data suggest, however, that most widely distributed species do vary considerably – often gradually – over their geographic ranges; this is observed for electrophoretically detected protein polymorphism, colour, body-size (Bergmann's rule), size of appendages (e.g. Allan's rule), litter size, etc. (see any modern ecology text); certainly, it is difficult to reconcile these observations with the *claimed* commonness of temporal stasis. An interesting example is Kurtén's (1958) fossil record – which covers several glacial periods – of the cave bear (*Ursus spelaeus*) demonstrating that it was large during glacial periods and small during warm inter-glacial periods. This is indeed a nice temporal analogue to Bergmann's spatial rule. Another important example demonstrating how one kind blends into another is provided by the distribution of various subspecies of the herring gull (*Larus argentatus*) and the lesser black-headed gull (*L. fuscus*) around the North Pole (Mayr, 1963). The various subspecies form a *continuous* chain around the pole where each subspecies can breed with both neighbouring subspecies except in Europe; here the continuous spatial changes have produced sufficient differences to 'prohibit' interbreeding. Finally, in many widely distributed species, the most extreme forms often differ as much – or more – than do those of sympatric species. For instance, the most extreme forms of the weasel (*Mustela erminea*) differ in weight by a factor of more than 5; coexisting *Mustela* species differ, on the other hand, only by a factor of 1.2–2.3 (1.7 on the average) (for documentation, see Stenseth, 1984*b*). Paleontologists might indeed classify the extreme forms of *M. erminea* as different species!

What rings through is this: evolution is – except for rare instances – expected never to cease; we rather expect a *continuum* of evolutionary rates which, for some groups of species in some ecosystems, might be almost constant for geologically long time-spans. It seems impossible to sort out which species do what, where and when unless we study paleontology *as well as* ecology. Much remains to be done, but thus far we have no reason to dismiss the validity of the orthodox Darwinian theory. Certainly we must remember that

> [it] is the nature of science that once a position becomes orthodox it should be subjected to criticism [... but it] does not follow that, because a position is orthodox, it is wrong ...
>
> (Maynard Smith, 1976*c*)

Supported by NFR (Sweden) and the Nansen-foundation (Norway).

References

Allan, P. M. (1976). Evolution, population dynamics, and stability. *Proceedings of the National Academy of Sciences, USA*, **73**, 665–8.

Charlesworth, B., Lande, R. & Slatkin, M. (1982). A neo-Darwinian commentary on macroevolution. *Evolution*, **36**, 474–98.

Dawkins, R. (1982). *The Extended Phenotype*. San Francisco: W. H. Freeman.

Eldredge, N. & Gould, S. J. (1972). Punctuated equilibria: an alternative to phyletic gradualism. In *Models in Paleobiology*, ed. T. J. M. Schopf, pp. 82–115. San Francisco: W. H. Freeman.

Fisher, R. A. (1930). *The Genetical Theory of Natural Selection*. Oxford: Clarendon Press.

Gause, G. F. (1934). *The Struggle of Existence*. New York: Hafner Press.

Goldschmidt, R. B. (1940). *The Material Basis of Evolution*. New Haven: Yale University Press.

Gould, S. J. (1980). Is a new and general theory of evolution emerging? *Paleobiology*, **6**, 119–30.

Gould, S. J. (1982). Darwinism and the expansion of evolutionary theory. *Science*, **216**, 380–7.

Grant, P. R. (1972). Interspecific competition between rodents. *Annual Review of Ecology and Systematics*, **3**, 79–106.

Hallam, A. A. (1976). The Red Queen dethroned. *Nature*, **259**, 12–13.

Hoffman, A. (1982). Punctuated versus gradual mode of evolution: A reconsideration. *Evolutionary Biology*, **15**, 411–36.

Kerner, E. H. (1957). A statistical mechanistics of interacting biological species. *Bulletin of Mathematical Biophysics*, **19**, 121–46.

Kerner, E. H. (1959). Further considerations of the statistical mechanistics of biological associations. *Bulletin of Mathematical Biophysics*, **21**, 217–35.

Kurtén, B. (1958). Life and death of Pleistocene cave bears. *Acta zoologica fennica*, **95**, 1–59.

Lawlor, L. R. & Maynard Smith, J. (1976). The coevolution and stability of competing species. *American Naturalist*, **110**, 79–99.

Leigh, E. G. (1965). On the relation between the productivity, biomass, diversity and stability of a community. *Proceedings of the National Academy of Sciences, USA*, **53**, 777–83.

Leigh, E. G. (1968). The ecological role of Volterra's equations. In *Some Mathematical Problems in Biology*, ed. M. Gerstenhaber, pp. 1–66. Providence: American Mathematical Society.

Lewontin, R. C. (1979). Fitness, survival and optimality. In *Analysis of Ecological Systems*, ed. D. J. Horn, G. R. Stairs & R. D. Mitchell, pp. 3–21. Columbus: Ohio State University Press.

MacArthur, R. H. (1969). Patterns of communities in the tropics. *Biological Journal of the Linnean Society*, **1**, 19–30.

MacArthur, R. H. & Levins, R. (1967). The limiting similarity, convergence and divergence of coexisting species. *American Naturalist*, **101**, 377–85.

Maynard Smith, J. (1969). The status of neo-Darwinism. In *Towards a Theoretical Biology, 2: Sketches*, ed. C. H. Waddington, pp. 82–9. Edinburgh: Edinburgh University Press.

Maynard Smith, J. (1970). Time in the evolutionary process. *Studium Generale*, **23**, 266–72.

Maynard Smith, J. (1974). *Models in Ecology*. Cambridge University Press.

Maynard Smith, J. (1975). *The Theory of Evolution* (3rd ed.). Harmondsworth: Penguin Books.

Maynard Smith, J. (1976*a*). A comment on the Red Queen. *American Naturalist*, **110**, 331–8.

Maynard Smith, J. (1976*b*). A short term advantage for sex and recombination through sib-competition. *Journal of Theoretical Biology*, **63**, 245–58.

Maynard Smith, J. (1976*c*). Group selection. *Quarterly Review of Biology*, **51**, 277–83.

Maynard Smith, J. (1978*a*). Optimization theory in evolution. *Annual Review of Ecology and Systematics*, **9**, 31–56.

Maynard Smith, J. (1978*b*). *The Evolution of Sex*. Cambridge University Press.

Maynard Smith, J. (1982*a*). Evolution – sudden or gradual? In *Evolution Now*, ed. J. Maynard Smith, pp. 125–8. London: Macmillan.

Maynard Smith, J. (1982*b*). *Evolution and the Theory of Games*. Cambridge University Press.

Maynard Smith, J. (1983). Current controversies in evolutionary biology. In *Dimensions of Darwinism: Themes and Counterthemes in Twentieth Century Evolution Theory*, ed. M. Greene, *in press*. Cambridge University Press.
Mayr, E. (1963). *Animal Species and Evolution*. Cambridge: Harvard University Press.
Medawar, P. B. (1967). *The Art of the Soluble*. London: Methuen.
Mivart, St G. J. (1871). *On the Genesis of Species*. New York: Macmillan.
Pimm, S. L. (1982). *Food Webs*. London: Chapman & Hall.
Reed, J. & Stenseth, N. C. (1984). On evolutionarily stable strategies. *Journal of Theoretical Biology*, in press.
Roughgarden, J. (1979). *Theory of Population Genetics and Evolutionary Ecology: An Introduction*. New York: Macmillan.
Rosenzweig, M. L. (1975). On continental steady states of species diversity. In *Ecology and Evolution of Communities*, ed. M. L. Cody & J. M. Diamond, pp. 121–40. Cambridge: Harvard University Press.
Schopf, T. J. M. (1982). A critical assessment of punctuated equilibria. I. Duration of taxa. *Evolution*, **36**, 1144–57.
Schopf, T. J. M. & Hoffman, A. (1983). Punctuated equilibria in the fossil record. *Science*, **219**, 438–9.
Simpson, G. G. (1953). *The Major Features of Evolution*. New York: Columbia University Press.
Stanley, S. M. (1979). *Macroevolution: Pattern and Process*. San Francisco: W. H. Freeman.
Stanley, S. M. (1982). Macroevolution and the fossil record. *Evolution*, **36**, 460–74.
Stebbins, G. L. & Ayala, F. J. (1981). Is a new evolutionary synthesis necessary? *Science*, **213**, 967–71.
Stenseth, N. C. (1979). Where have all the species gone? On the nature of extinction and the Red Queen Hypothesis. *Oikos*, **33**, 196–227.
Stenseth, N. C. (1984*a*). What constitutes an organism's environment? *Evolutionary Theory*, in press.
Stenseth, N. C. (1984*b*). Are geographically widely distributed recent species usually phenotypically uniform? *Paleobiology*, **38**, 870–80.
Stenseth, N. C. & Maynard Smith, J. (1984). Coevolution in ecosystems: Red Queen evolution or stasis? *Evolution*, in press.
Van Valen, L. (1973). A new evolutionary law. *Evolutionary Theory*, **1**, 1–30.
Van Valen, L. (1976*a*). Energy and evolution. *Evolutionary Theory*, **1**, 179–229.
Van Valen, L. (1976*b*). The Red Queen lives. *Nature*, **260**, 575.
Van Valen, L. (1977). The Red Queen. *American Naturalist*, **111**, 809–10.
Weismann, A. (1886). *Essays Upon Heredity and Kindred Biological Problems*. Oxford: Clarendon Press.
Williamson, P. G. (1981*a*). Morphological stasis and developmental constraints: real problems for neo-Darwinism. *Nature*, **294**, 214–15.
Williamson, P. G. (1981*b*). Paleontological documentation of speciation in Cenozoic molluscs from Turkana Basin. *Nature*, **293**, 437–43.
Wright, S. (1982). Character change, speciation, and the higher taxa. *Evolution*, **36**, 427–43.

6

Game theory and animal behaviour

G. A. PARKER AND P. HAMMERSTEIN

Introduction

The study of animal behaviour has developed from two very disparate ancestries. The American 'behaviourists' (stemming from Thorndike and Watson) and the 'reflexologists' (from Bekhterev and Pavlov) concentrated almost exclusively on learning phenomena, culminating in the Skinnerian view (1953) that virtually all observable behaviour is the product of conditioning and experience. In contrast, the early naturalists, and later Darwin, viewed the behaviour they observed in nature as the product of inborn skills or instincts. Following this tradition, the pioneers of 'ethology' (Lorenz, 1965; Tinbergen, 1951; von Frisch, 1965) concentrated almost exclusively on the 'function' (= adaptive value) of behaviour patterns. The 'nature versus nurture' debate is now generally accepted as being sterile, but in practical terms there appears to be remarkably little dialogue between modern sociobiologists and operant psychologists.

Regrettably, ethology suffered alongside other branches of biology in that the functions ascribed to many phenomena often relied on group selection, a feature that has largely disappeared since the rise of sociobiology, which could be termed the 'selfish gene' approach to ethology (Dawkins, 1976).

Sociobiology has developed predominantly as a result of Hamilton's (1964) now legendary insights into the evolution of sociality through kin selection. Their effect has been so overwhelming that many (particularly outside the discipline) have tended to equate sociobiology with kin selection and altruism. However, much (and perhaps most) of behavioural interaction involves conflict in its various forms. The sociobiology of conflict and its resolution has attracted attention more recently. The formal approach to the theory of animal conflict suggested by John Maynard Smith (1972; Maynard Smith & Price, 1973) provides sociobiology with a major new conceptual tool, and one which is certainly no less important for most other branches of evolutionary theory.

Maynard Smith's idea was that, in cases of conflict, one should seek for an 'evolutionarily stable strategy' or ESS. A strategy is simply one of a set of possible alternative behavioural 'programmes' that could be 'played' by individuals in a population. An ESS is a strategy which, when played by most individuals, cannot be beaten by any rare, alternative strategy. In more rigorous terms, a strategy is an ESS if, when fixed in the population, no rare mutant strategy can invade and spread. To deduce what will be an ESS, Maynard Smith used techniques that are modified from the mathematics of game theory (von Neumann & Morgenstern, 1944).

Why do we need the ESS concept? Classically, sociobiologists have drawn graphs of the costs and the benefits of increasing expenditure in a given action, and argued that selection will favour maximizing net benefit (benefit minus cost). There is essentially nothing wrong with this procedure if the animal's strategy is played against a fixed environment. However, the approach is unsatisfactory for most behavioural interactions since costs and benefits usually depend on the strategies currently being played by other members of the population. In short, we need ESS when fitness is frequency dependent. A full treatment of the current state of evolutionary game theory is now available (Maynard Smith, 1982; see also the shorter reviews of Parker, 1984; Riechert & Hammerstein, 1983).

Several authors had applied what was in essence an ESS approach, before Maynard Smith's general development of the concept (see Maynard Smith, 1982). Fisher's (1930) cryptic verbal treatment of sex-ratio evolution may be the first example, but Shaw & Mohler's (1953) modelling of the same problem is perhaps the first (and rather remarkable) use of the ESS analytical technique (see Charnov, 1982). Also notable is Hamilton's (1967) search for an 'unbeatable' sex ratio. Lewontin (1961) appears to have been the first explicitly to use game theory in relation to evolution. Perhaps the greatest debt is owed to the extensive studies of frequency-dependent selection made by population geneticists (e.g. Clarke & O'Donald, 1964). But the dramatic step was made by John Maynard Smith, who has developed the framework of evolutionary game theory and applied it to so many fundamental problems that had been previously poorly understood because of a lack of adequate modes of analysis.

Our main aim in this paper is to draw objective parallels between ESS philosophy and game theory. We appreciate that mathematicians will regard ESS theory simply as an adjunct to the theory of games, and that biologists will see game theory as just another mathematical technique by which they can make inferences about function. We nevertheless hope that scholars of both disciplines may benefit by learning more about the links they share with each other.

Why is ESS theory so important for understanding behaviour?

Payoffs in behavioural interactions almost always depend on the 'play' of an animal's opponent. Thus if two animals are in dispute over some resource, the best strategy for one depends critically on the strategy that will be played by the other. Such two-player games are common in behaviour, and are generally termed contests or pairwise interactions (Maynard Smith, 1982). However, many (perhaps even most) behaviour games are *n*-person games where an individual is characteristically 'playing the field' (Maynard Smith, 1982; termed 'scrambles' by Parker, 1984). An obvious example, first discussed by Pulliam, Pyke & Caraco (1982) is the 'vigilance game' that is analysed in more detail below. Birds in flocks are commonly concerned with two major fitness-related activities: foraging for food, and looking out for predators. The ideal solution for an individual bird is to be able to forage continuously whilst the others perform the vigilance, though this is hardly likely to suit the birds that are continuously vigilant. Once again, payoffs to an individual depend both on its own strategy, and on the strategy of the other players in the game.

It is becoming clear that most social behaviour requires interpretation in terms of ESS. Contest behaviour in disputes over territories, females, food items, nest sites, or other fitness-related resources can usually be analysed as pairwise interactions. Dominance behaviours are also essentially pairwise but have the property that they are repeated interactions, a feature which can greatly affect the ESS solutions (Selten & Hammerstein, 1984). Analyses of the 'scramble' or 'playing the field' type have been applied to a wide variety of problems such as habitat selection and distribution patterns (Fretwell, 1972; Pulliam & Caraco, 1984), mate searching (e.g. Parker, 1978) and alternative mating strategies in sexual selection (e.g. Rubenstein, 1980; Dunbar, 1982; Parker, 1982), sexual advertisement (Andersson, 1982; Parker, 1982), parent–offspring conflict (Parker & Macnair, 1979), and vigilance in flocks (Pulliam *et al.*, 1982).

ESS theory is not simply a technique for analysing conflict, as an alternative to using kin selection to analyse cooperation and altruism. Evolutionary game theory can equally well help to interpret cooperative interactions, in which case we may need to use kin selection and relatedness in order to calculate payoffs (Grafen, 1979; Hines & Maynard Smith, 1979). ESS theory indicates which strategies will be stable out of a set of alternatives; its use is by no means restricted to conflict behaviour.

Analogies between classical and evolutionary game theory

Classical game theory

Game theory originated in the social sciences as an attempt to produce a theory of rational behaviour in interpersonal human conflicts (von Neumann & Morgenstern, 1944). What exactly is meant by the term 'rational behaviour'? This proves to be a crucial issue; rather surprisingly, the concept of rationality has changed markedly during the development of game theory. We will now outline briefly the history and philosophy of these changes.

An early idea of rational behaviour was that the players (i.e. the interacting opponents) act according to the so-called minimax (or maximin) solution. This means that they should choose their action in the conflict as follows: 'for every possible action calculate the worst possible outcome (i.e. that with minimum benefit); now choose the action for which the worst possible outcome is least bad'.

However, there is a logical inconsistency with the minimax principle as outlined in Fig. 1*a*, which shows the hypothetical payoffs that two individuals (player 1, 2) would obtain in a game in which each player has just two possible strategies. It is easy to see that player 1's minimax strategy must be strategy *B*. On the other hand, player 2's minimax strategy is strategy *C*. But if player 2 can be expected to choose this strategy, then it pays player 1 to deviate from the minimax solution by playing strategy *A*. Hence the minimax principle cannot be a general rule for rational behaviour, because the expectation that the rule will be followed involves an incentive to deviate from it.

This kind of paradox prompted Nash (1951) to propose an alternative rule for rational behaviour. The clear difficulty with minimax is that it can pay to deviate if one's opponent plays minimax. Nash chose to define rational behaviour in such a way as specifically to exclude this problem. A *Nash equilibrium point* is defined as a combination of strategies, one for each player, such that no player could improve his payoff by deviating his strategy unilaterally from the combination. In Fig. 1*a*, the Nash equilibrium is for player 1 to choose strategy *A* and for player 2 to choose *C*, which incidentally happens to yield the highest payoffs to each player in our hypothetical game.

This leads to another early concept of game theory, that of Pareto optimality. The outcome of a game is said to be Pareto-optimal if no player could receive a higher payoff without anyone else being worse off. This is best illustrated by the payoff matrix in Fig. 1*b*, which represents the well-known 'Prisoner's Dilemma' (for an interpretation of this game see Riechert & Hammerstein, 1983). There are three outcomes in the Prisoner's

Fig. 1. (*a*) A hypothetical two-person game in which player 1 has two strategies *A*, *B*, and player 2 has two strategies *C*, *D*. The game is asymmetric. Player 1's minimax strategy is *B*, since the worst possible outcome is 0 for *A*, which is less than 1 for *B*. Player 2's minimax strategy is *C*, since 20 > 10. The only Nash equilibrium point for this game is (*A*, *C*). If a player expects his opponent to play this solution, there is no incentive to deviate. By deviating, player 1 would get 50 instead of 100 units; the same applies to player 2. (*b*) The Prisoner's Dilemma. Two players can either cooperate or not. At first glance, it appears that both players should play the Pareto optimum in which both cooperate. However, there is an incentive to deviate from this solution, since 10 > 9. The only Nash equilibrium point is for both not to cooperate. This leads to much lower payoffs than the cooperative optimum.

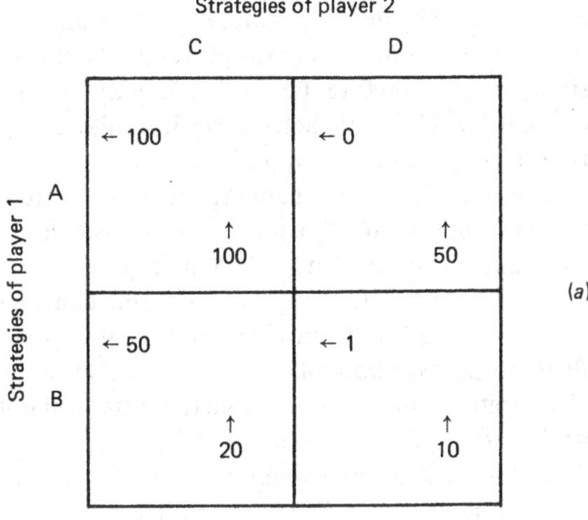

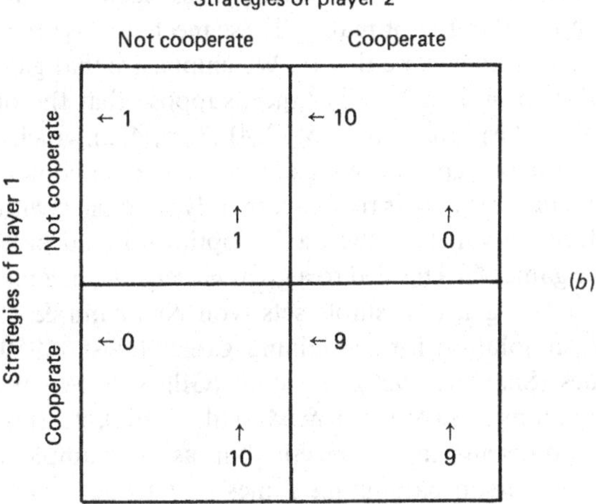

Dilemma which are Pareto-optimal, namely for both players to choose the noncooperative strategy which yields the pair of payoffs (9, 9), or alternatively for one player to play cooperatively and for the other to play noncooperatively, which yields (10, 0), (0, 10).

At first sight, it seems intuitively obvious that both players in the Prisoner's Dilemma should cooperate and thus each receive payoff 9. Although this outcome is Pareto-optimal, there are good reasons for rejecting it as a rational solution. Suppose we were in the position of player 1. Typically, game theory would force us to build expectations about what player 2 will do. However, this is irrelevant for the present case, since whatever player 2 chooses, it is always best for us to choose the noncooperative strategy. The reason is simply that $1>0$ and $10>9$ (see Fig. 1*b*). Exactly the same reasoning applies for player 2. The rational solution, without question, must therefore be for both players to choose the noncooperative strategy. This is, in fact, the Nash equilibrium point for the game and it is clearly not a Pareto-optimal outcome.

In spite of this weakness, modern game theory still centres around the assumption of Pareto optimality. To some extent this can be justified as follows. Suppose that, in the Prisoner's Dilemma game, the two players were allowed to negotiate and to sign an enforceable contract before the game in Fig. 1*b* is actually played. Provided that the penalty for breach of contract is sufficiently high for both players, it is clear that they should both agree to play the Pareto optimum (9, 9). (Note that this need not contradict the more central Nash equilibrium condition because the contract itself must be considered as part of the game.)

The branch of game theory called 'cooperative theory' concerns the search for Pareto-optimal solutions to games (typically there are various other criteria which must also be satisfied, depending on the game). The assumption is made that there are mechanisms, such as contracts, which will always enforce Pareto optimality. These mechanisms are not explicitly modelled in cooperative game theory. We emphasize that games typically have many Pareto optima. For instance, suppose that the only possible pairs of payoffs to the players are (1, 5), (2, 4), (3, 3), (4, 2), which can easily be extrapolated into the continuous case. All these outcomes are Pareto optimal. Cooperative theory is therefore mainly concerned with developing criteria for deciding which of the Pareto optima should be considered a solution of the game. This has led to a great variety of concepts, such as the von Neumann–Morgenstern stable sets (von Neumann & Morgenstern, 1944), the Nash solution for bargaining games (Nash, 1950, 1953), the Shapley values (Shapley, 1953), the core (Gillies, 1959), the Aumann–Maschler bargaining sets (Aumann & Maschler, 1964), etc. It is beyond our present scope to discuss these concepts but, as an example, in the Nash solution for two-person bargaining games (not to be confused with the

Nash equilibrium) the two players would choose the Pareto optimum which maximizes the product $(u-u')(v-v')$; u, v are the payoffs to players 1, 2 and u', v' are the payoffs received if there is no agreement of any sort (e.g. if no trade takes place). Although each of the various solution concepts in cooperative game theory has its own intuitive appeal for specific contexts, the theory on the whole suffers from the lack of a coherent, unifying framework.

Cooperative theory has relevance to cases where commitment to negotiated contracts really can exist. In reality, however, human contracts are often broken (e.g. marriage, publication deadlines, international agreements) and are sometimes highly unenforceable. Even if they are perfectly enforceable, it turns out that cooperative theory can easily lead to unacceptable conclusions. The reason for this is that the contract and bargaining procedure must in principle be regarded as part of the game; when this is done, it turns out that rational bargaining does not necessarily lead to contracts that will be Pareto-optimal (Selten, 1975). This sort of insight is generating a growing movement towards noncooperative modelling of games formerly analysed by the techniques of cooperative game theory. An extensive treatment of this transformation of cooperative games into a branch of noncooperative theory is given in a book currently in preparation by Harsanyi & Selten (see also Harsanyi & Selten, 1980).

In short, the unifying concept in modern game theory is proving to be based on the Nash equilibrium from noncooperative theory. It is perhaps fair to claim that the major step in formulating an idea of rational behaviour in games is due to Nash's (1951) equilibrium concept: a solution should be self-enforcing in the sense that it cannot pay an individual to deviate unilaterally.

Again, we find in noncooperative theory that a game can lead to several different Nash equilibria; roughly, the more strategies in a game, the more Nash equilibria one might expect. The main problem left from Nash's original approach is therefore that of deciding which of the many equilibria should be considered the solution of the game. A comprehensive approach to this problem will be found in the book by Harsanyi & Selten. Their philosophy concerns the consistent application of Bayesian decision theory to interpersonal conflicts; as we shall see, the equivalent problem in evolutionary game theory still remains to be faced.

A brief summary of classical game theory is shown in Table 1. In the next section we argue that evolutionary game theory has strong analogies with the noncooperative part of the schema.

Evolutionary game theory

When von Neumann & Morgenstern (1944) founded game theory with

Table 1. *Summary of the classical theory of games*

	Noncooperative theory	Cooperative theory
Important differences	Seeks to explain cooperative as well as noncooperative behaviour. Strategies are explicitly modelled and account is taken of how payoffs depend on strategies.	Cooperation is to a certain extent assumed and thus not subject to a complete analysis. Often a model only contains sets of achievable payoffs but no strategy sets.
Basic solution concept	Nash equilibrium point: the rational solution of a game should be such that no player has an incentive to deviate from this solution if he expects his opponent to behave according to it.	'Baroque' variety of concepts: von Neumann–Morgenstern stable sets, Nash solution for bargaining games, Shapley value, core, etc.; all based on the idea of Pareto optimality.
Major problem	Many alternative solutions typically exist to a given game.	No coherent theory; no clear reference to individual success.
Approach to solve the problem	Harsanyi & Selten's theory (1981) of equilibrium point selection (N.B. selection by players not by evolution).	Noncooperative modelling of cooperative games, i.e. conversion into a branch of noncooperative theory.

regard to the social sciences, no one would have anticipated any application to biology. The reason for this is perhaps that humans would claim a monopoly on rationality; the notion that animals might act according to the sort of rules in game theory for rational behaviour seems quite untenable. Paradoxically, the literature on experimental games (e.g. Selten, 1979) indicates that humans are rather seldom rational in their behaviour, whereas data for animals look much more promising (Maynard Smith, 1982). This stems from the fact that the games analysed for animals have been natural ones to which they have been adapted by natural selection over thousands of generations. The games analysed for humans have been uniquely set by the experimenters. Although humans are reasonably good generalist problem-solvers, it is unfair of us to compare complex problem-solving by the human brain with the much more specific solutions eventually produced in animals by the powerful process of evolution. However, because of natural selection, it really does appear that the most successful application of game theory will prove to be in the context of evolutionary biology.

In what sense can animals be regarded as rational players in conflicts? We have explained that the concept of rationality has been the main controversy in classical game theory. Similarly, the main task for evolu-

Table 2. *A comparison of evolutionary and noncooperative game theory*

	Evolutionary games	Noncooperative games
The notion of 'success'	'Objective' fitness.	'Subjective' utility.
Who decides?	The process of natural selection.	Fictitious individuals that are completely rational.
Philosophy	Descriptive theory: analyses how animals do behave. Focus on individual selection and the concept of function.	Normative theory: analyses how humans should ideally behave. Focus on individual success and the concept of rationality.
Analogy in central concepts	Evolutionarily stable strategy (ESS).	Nash equilibrium.
Conclusion from analogy	Quasi-rationality of the 'decisions' made by natural selection with regard to animal conflict behaviour and sociality.	Potential for explaining cases of observed rational economic behaviour by quasi-evolutionary processes, including imitation as the analogue of inheritance.
A shared problem	Which of the many alternative ESSs is most likely to be found in nature?	Which of the many alternative Nash equilibria should rational players select?

tionary game theory has been to provide appropriate rules for what will be adaptive behaviour if frequency-dependent selection operates. It is far from trivial that both these developments should lead to a very similar mathematical concept. When John Maynard Smith began to study animal conflict, he saw clearly that the notion of a game would be invaluable for analysing frequency dependence at the phenotypic level, but his idea that ESS is the appropriate rule for adaptation was developed quite independently of any game theoretical solution concept. It turns out that the ESS rule satisfies Nash's equilibrium condition, now the central notion of modern game theory. However, the two are not identical, since though an ESS is always a Nash equilibrium, the reverse need not hold.

The analogies between evolutionary ideas and game theory do not end here. There is a remarkable parallel between the struggle to replace the cooperative theory by noncooperative theory, and the way in which the concept of function in biology has already shifted from one of species advantage to one of individual or gene advantage (e.g. Williams, 1966; Dawkins, 1976). These struggles have in common the idea that maximization of individual benefit has been misrepresented by notions of Pareto optimality on the one hand and group or species advantage on the other. In both cases, the struggle has concerned what is the most useful concept of rationality; both cooperative game theory and naive group selectionism usually *assume* some degree of cooperation instead of attempting to explain it. Since only humans, and not other animals, can negotiate enforceable contracts, cooperative game theory has better reasons than biology for assuming that outcomes will be in some way cooperative.

Having drawn the analogies we must now make two obvious distinctions between classical and evolutionary game theory. Firstly, in classical game theory, a player's payoff is measured on a 'utility scale' determined by his subjective value judgements. In contrast, in evolutionary game theory, there is an objective utility that corresponds to the expected change in fitness due to the interaction. Secondly, there is the difference in mechanism: in classical game theory an outcome depends on a personal decision process, and not on an evolutionary 'decision process'. Thus we should perhaps describe the ESS rule as being 'quasi-rational' rather than rational in the game-theoretical sense.

To reiterate, John Maynard Smith defined a strategy as an ESS if a population adopting it is uninvadable by an initially rare mutant alternative strategy. Imagine a biological game in which I, J, etc. is the set of possible strategies that the animals could play; I, J, etc. is thus the 'material' on which selection operates. Under what conditions will a strategy I be an ESS against all alternative strategies J? An extensive treatment of the conditions is given by Hammerstein (1984). Obviously, we would expect I to be adaptive in a population of I-players, since otherwise some more successful strategy J could start to spread. Thus we could write that

$$W(I, \mathrm{Pop}_I) \geqslant W(J, \mathrm{Pop}_I) \quad \text{for all } J, \qquad (1a)$$

in which $W(I, \mathrm{Pop}_I)$ denotes the fitness payoff to a single I-strategist against a population of I-strategists (called Popeye), and $W(J, \mathrm{Pop}_I)$ denotes the payoff to a J-player against the same population. Before elaborating this condition further, we first consider the specific case of pairwise interactions to which most of evolutionary game theory has so far been applied. Animals contest in random pairs over some fitness-related resource. Again suppose that I, J, etc. are the possible strategies. Let $E(I, J)$ be the payoff to a single I-player against a single J-player. Inequality (1*a*) is now equivalent to

$$E(I, I) \geqslant E(J, I) \quad \text{for all } J, \qquad (1b)$$

because if I is fixed in the population, then opponents can be assumed to play I. Note that the difference between (1*a*) and (1*b*) arises from the fact that W is a payoff for playing against the population, but E is a payoff for playing against a single opponent.

How do conditions (1*a*), (1*b*) relate to the Nash equilibrium concept? First consider (1*a*) for the population game. Strategy I satisfies the Nash equilibrium condition because if the rest of the population of 'players' plays

I, it does not pay to deviate unilaterally from I by shifting to another strategy J. Thus it is a Nash equilibrium for the entire population to play I. Now consider condition (1*b*) for the two-person game. This condition means that if the opponent can be expected to play I, again it can never pay to deviate unilaterally from I. Thus the pair of strategies (I, I) is a Nash equilibrium point.

Before considering an additional condition that an ESS must satisfy, it seems worthwhile mentioning an interesting property of intraspecific biological games. In a natural population, clear phenotypic differences (asymmetries) exist between individuals. A strategy is assumed to be some heritable entity which is, however, allocated independently of all other aspects of phenotype; i.e. each strategy finds itself in a phenotype which is drawn randomly from the phenotype distribution. A phenotype might, for example, relate to an animal's size, condition, age, etc., or might concern 'extended' phenotypic features (see Dawkins, 1982) such as being a prior resident of a territory, etc. Thus a given strategy has random probability of being allocated to individuals differing in 'non-strategic' phenotype. The same strategy can be played against itself in a contest between two animals, since the opponents might have the same genotype. There is thus one set of strategies for all players in a population. This is an assumption which is not typically made in classical game theory, where two opponents usually have different strategy sets. It is emphasized, however, that models of interspecific biological games include different strategy sets for the different species.

A game in which the players share the same strategy set is called *symmetric* in classical game theory (not to be confused with the term 'symmetric contest', used with regard to identical phenotypes in evolutionary game theory). For example, the Prisoner's Dilemma game (Fig. 1*b*) is symmetric, whereas the payoff matrix in Fig. 1*a* is not. In classical game theory, if players 1 and 2 can reference their strategic decisions in relation to some 'label' which distinguishes between them, thus creating a 'phenotypic' difference, then a rational solution can be a Nash equilibrium in which both opponents adopt different strategies depending on their label. Otherwise, only the so-called symmetric Nash equilibrium pairs of strategies can be rational solutions; i.e. both opponents must play the same strategy. ESSs for intraspecific games always correspond to these symmetric Nash equilibria because of the random allocation of strategies to phenotypes, even though strategies can be ESSs in which, for example, the larger opponent behaves differently from his smaller opponent. The point is that such strategies (termed 'conditional' strategies by Dawkins, 1980) prescribe what to do when larger and what to do when smaller. Thus the same conditional strategy can play against itself, although the two phenotypes behave differently. ESSs for interspecific games, however, correspond to asymmetric Nash equilibria, because an individual's strategy is drawn only from the strategy set of its own species; the gene pools are separate.

So far we have discussed only part of what is called Maynard Smith's ESS condition. Why are (1*a*) or (1*b*) incomplete? Suppose again that strategy I satisfies (1*a*) or (1*b*), so that it is adaptive in a population of I-players. But now suppose that an alternative strategy J exists which is also adaptive, since it has an equal payoff in the population of I-players. This will always be the case when I is a mixed strategy (a strategy which selects actions A, B, etc. with probabilities p_A, p_B, etc.), as in the well-known Hawks–Doves game (Maynard Smith & Price, 1973), the war of attrition (Maynard Smith, 1974; Bishop & Cannings, 1978; Hammerstein & Parker, 1982), and many sex ratio games (e.g. Charnov, 1982). An alternative 'best reply' strategy J might spread by drift to reach a significant frequency in the population. Since we are dealing with frequency-dependent selection, this can clearly change the mean fitness of I from what it would be if only a single mutant J were present. For I to be stable, we therefore need selection always to act against J when J begins to spread by drift.

Maynard Smith's second ESS condition is designed specifically to this end. It has been generalized by Hammerstein (1984; see also Maynard Smith, 1982). For I to be an ESS it must satisfy both the Nash equilibrium condition (1*a*) and also the following:

If

$$W(J, \mathrm{Pop}_I) = W(I, \mathrm{Pop}_I),$$

then

$$W(I, \mathrm{Pop}_{[(1-\varepsilon)I+\varepsilon J]}) > W(J, \mathrm{Pop}_{[(1-\varepsilon)I+\varepsilon J]}), \tag{2a}$$

for all sufficiently small ε. The expression $\mathrm{Pop}_{[(1-\varepsilon)I+\varepsilon J]}$ denotes a population playing the mixed strategy $(1-\varepsilon)I+\varepsilon J$, which is to play I with probability $1-\varepsilon$ and J with probability ε (see Hammerstein, 1984, for details). For pairwise contests (2*a*) becomes equivalent to Maynard Smith's original condition:

If

$$E(J, I) = E(I, I),$$

then

$$E(I, J) > E(J, J). \tag{2b}$$

It is sometimes said that evolutionary game theory is simply a restatement of classical noncooperative theory, since an ESS is simply a Nash equilibrium. We feel that this is misleading for two reasons. First, as we have shown, intraspecific biological games have the strong symmetry property that the same set of strategies is available to all players. Second, there exists no parallel to the second ESS conditions (2*a*), (2*b*) in the Nash equilibrium (or elsewhere in classical game theory).

Indeed, the second ESS condition has some rather dramatic consequences. Selten (1980) has used the second condition to show for pairwise contests in which opponents never have the same role, that an ESS must always be a pure (not a mixed) strategy. Two opponents have the same role if the information on which their behaviour is based is the same. For instance, a banana and two identical but hungry chimpanzees are simultaneously thrown into a cage. In the resulting fracas, both chimpanzees must begin the contest with identical information. At least initially they have identical roles, namely: 'I am hungry; I am presented with a banana, a cage, and an opponent like me'. Thus one circumstance where roles are typically equal is when there are objectively no differences between the players, or any differences are too small to perceive. But even if there are objective differences, mistakes can occur in perception so that both opponents estimate themselves to be in the same role. If one chimpanzee were slightly bigger than the other, both opponents would have the same role if each assessed himself to be the larger contestant. Extensive use of the concept of mistakes about roles has been made for the war of attrition (Parker & Rubinstein, 1981; Hammerstein & Parker, 1982), which has a mixed ESS. On the other hand, in asymmetric contests in which role perception is perfect, solutions must be pure ESSs, according to Selten's theorem. Where opponents always have consistent 'views' about existing asymmetries, it can be shown that the evolutionary game decomposes into a set of asymmetric subgames which are then easily analysed (Hammerstein, 1981).

Finally, the second ESS condition (2*a*), (2*b*) has important consequences in games of interspecific interactions. Here solutions must be combinations of ESSs, one for each species. An ESS for one of these species must be a pure strategy if there is no intraspecific frequency dependence of fitness in this population (Hammerstein, 1984). Obvious biological candidates for such a case would be host/parasite or predator/prey interactions. To explain this conclusion, consider as follows. If the result were to be a mixed strategy I, then we know that there must be many alternative strategies J that receive payoffs equal to I (all the pure component strategies of I must do equally well, otherwise some will receive higher payoff than I itself). Now, again assuming that mutations can occur only singly within the monomorphic interacting populations, an alternative best reply J in the I-playing population does not affect the fitnesses of I and J if these change in frequency. It is therefore possible that J invades the population by random drift, and there is now no selective mechanism against J.

Clearly, there are many interspecific biological games which also include intra- as well as interspecific frequency dependence, e.g. root competition for water in communities of desert plants (Riechert & Hammerstein, 1983) and a wide variety of arms races (Parker, 1983). We conclude by noting that the first attempt to investigate interspecific evolutionary games was made by Lawlor & Maynard Smith (1976).

An example: the vigilance game

We now give an example of how evolutionary game theory can be applied to a specific problem in animal behaviour. The example we choose is from a set of models in which two types of behaviour are possible: one which yields only personal benefit and one which yields a benefit to all individuals in a social group (a more detailed treatment of this general type of model will be given elsewhere). In our example, individuals in a group of n animals have two 'tasks', feeding (individual benefit) and vigilance against predators (which can also benefit other group members). We use this example for the following reasons: (i) it is an n-person game which is intermediate between the two-person games and the population games discussed in the previous section; (ii) it shows that game-like conflict arises in situations where cooperation and not fighting is to be analysed; (iii) we wish to emphasize that game theory is not in any way an alternative to the use of kin selection theory, though we may need to consider relatedness in order to calculate an ESS in interaction between relatives.

Our hypothetical vigilance game is similar to that of Pulliam *et al.* (1982) and has the following characteristics. A group of n individuals can expect an attack by a predator but has no knowledge about its timing. For the sake of simplicity we assume that only one attack will occur during the game, and that the predator is successful if no member of the group is vigilant at the time the predator strikes. The game between group members concerns allocation of an individual's time between feeding and vigilance. The predator is not considered to be a player in the game, i.e. we do not analyse his strategic decisions. The game is played for a fixed time, and at any instant there is an equal probability that the attack will occur.

We now specify the strategic choices of an individual in the group. In principle, strategies could be extremely complex, consisting of all sorts of monitoring of behaviour of other group members. However, again for simplicity, suppose that selection can only adjust an individual's probability of being vigilant in a given time unit, and that behaviour (vigilance versus feeding) in two consecutive time units is independent. The strategic parameter is thus the probability v of being vigilant in a given time unit. This strategy v can vary continuously between 0 (no vigilance) and 1 (total vigilance at the exclusion of feeding).

It remains to define payoffs. Fitness W is assumed to be the product of survival probability p and gains through feeding f. Fig. 2(a) illustrates our assumption that gains through feeding decrease monotonically with increasing vigilance. Perhaps the simplest assumption would be that the decrease is linear. However, there are plausible biological reasons to expect diminishing returns to scale with regard to feeding effort; e.g. the value of a food item is greater to an empty stomach, or alternatively, small amounts of vigilance can be accommodated efficiently while changing location during food searching. Therefore we will first use the function $f(v)=1-v^2$ to describe the gains through feeding. Now consider survival probability, p. The assumption is that the predator will capture just one group member provided that no individual is vigilant at the time the strike occurs, otherwise he is unsuccessful. An individual in a group of size n has thus a chance $(n-1)/n$ of surviving even if no group member is vigilant. Let V be the probability that at least one group member is vigilant during the attack. Then an individual's survival probability is $p=(n-1)/n+V/n$. Fitness can therefore be defined as $W=p \cdot s=[(n-1)/n+V/n](1-v^2)$.

The model is a game rather than a simple optimization problem because W depends not only on an individual's strategy v, but also on the strategies played by the other group members (included in V).

Case 1: groups constituted independently of strategy. Suppose that groups are formed randomly within a large population. A rare mutant strategy will

Fig. 2. The vigilance game. (*a*) Two forms of the function $f(v)$. Continuous curve is $f(v)=1-v^2$; broken line is $f(v)=1-v$. (*b*) The ESS v^* for the three cases. (*c*) The total vigilance $n \cdot v^*$ for the three cases.

In (*b*) and (*c*), lower curve (broken): randomly formed groups; middle curve (dash–dot): groups of full sibs; top curve (dotted): species or group optimum.

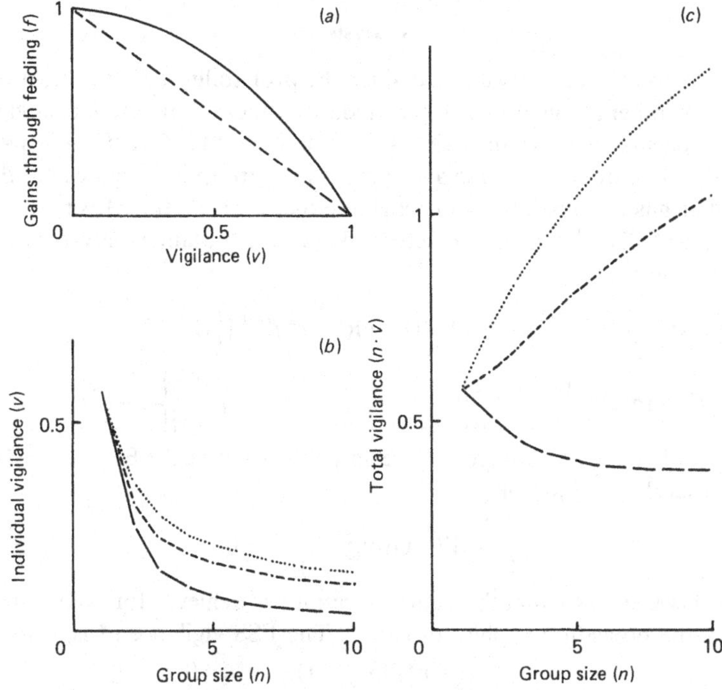

therefore find itself in a group in which other strategies are drawn randomly from the rest of the population; i.e. opponents will almost invariably play the 'wild-type' strategy. In order for a strategy v^* to be an ESS, we require that

$$W(v^*, \text{Group}_{v^*}) \geqslant W(v, \text{Group}_{v^*}), \tag{3}$$

from the first ESS condition (1*a*). We therefore need to calculate the fitness $W(v, \text{Group}_{v^*})$ of a single v-player in a group of v^*-players.

Case 2: groups of full sibs. In a population of sexually reproducing diploids, a dominant mutation arises which alters strategy v. Then the heterozygous mutant parent will mate with a homozygous wild-type parent. A group of their progeny will not play strategies that are a random sample of the population, since every offspring has a probability $\frac{1}{2}$ of playing the mutant strategy. We must now form a condition analogous to (3), but taking into account this relatedness. Thus

$$W(v^*, \text{Group}_{v^*v^*}) \geqslant W(v, \text{Group}_{vv^*}), \tag{4}$$

i.e. the expected fitness of a v^*-player in a randomly drawn group containing v^* must exceed the expected fitness of a v-player in a randomly drawn group containing v. Whereas in (3) other players in the group are taken randomly from the population, in (4) the other players in a group containing v are a highly non-random sample of the population.

Analysis

In order to calculate W we first deduce the probability V that at least one group member is vigilant when the predator attacks. For case 1, a single v-player occurs in a group with $n-1$ v^*-players; therefore $V=1-(1-v)(1-v^*)^{n-1}$. For case 2, a v-player occurs in a group in which each other member has a probability $\frac{1}{2}$ of playing either v or v^*; thus $V=1-(1-v)[1-(v+v^*)/2]^{n-1}$. So the expected fitness of a mutant v-player in a v^* population is

$$W(v, \text{Group}_{v^*}) = \left\{\frac{n-1}{n} + \frac{1}{n}[1-(1-v)(1-v^*)^{n-1}]\right\}(1-v^2) \tag{5}$$

$$W(v, \text{Group}_{vv^*}) = \left\{\frac{n-1}{n} + \frac{1}{n}\left[1-(1-v)\left(1-\frac{v+v^*}{2}\right)^{n-1}\right]\right\}(1-v^2), \tag{6}$$

for cases 1 and 2 respectively. We are now able to derive the ESS value v^* for case 1 as the solution of

$$\left[\frac{\partial}{\partial v} W(v, \text{Group}_{v^*})\right]_{v=v^*} = 0, \tag{7}$$

since it is easy to see that the optimum cannot be achieved for $v=0$ or $v=1$. The same procedure applies for case 2. The ESS vigilance v^* is given by

$$(1-v^*)^n(3v^*+1) - 2v^*n = 0 \tag{8}$$

$$(1-v^*)^n\left[v^*\left(2+\frac{n+1}{2}\right)+\frac{n+1}{2}\right]-2v^*n=0, \qquad (9)$$

for cases 1 and 2 respectively. Solutions for v^* can be derived numerically by standard computer techniques.

Note that the second ESS condition is satisfied because there cannot exist alternative best reply strategies.

Case 3: species or group optimum. Before discussing results, it is interesting to deduce what would be the species or group optimum for the game, i.e. which v^* would maximize the benefit function $b(v)= W(v, \text{Group}_v)$ for the group as a whole. This would be a Pareto optimum, whereas solutions to (4) and (5) are Nash equilibria (see also Pulliam *et al.*, 1982). We present this species optimum for comparative purposes only; it is difficult to see how such a solution could be favoured by evolution unless group selection operates or groups consist of full sibs that are produced asexually. The optimum is found to be the solution of

$$(1-v^*)^n[v^*(2+n)+n]-2v^*n=0. \qquad (10)$$

Conclusions from the model

Results are summarized in Fig. 2(*b*, *c*). Note that vigilance should in all cases decrease with group size. This is due not only to Hamilton's (1971) 'selfish herd' effect (a given individual's probability of capture is diluted by the presence of other individuals). It also stems from the fact that a given individual benefits from the vigilance of other group members. The more group members there are, the greater this effect; hence individual vigilance can be reduced in a larger group. As we would expect, v^* is lowest when groups are formed randomly, and highest at the species optimum (Fig. 2*b*).

However, the distinction between the three cases is more apparent if we compare the total vigilance effort $n \cdot v^*$ of the group (Fig. 2*c*). Rather surprisingly, the total vigilance for case 1 (groups constituted randomly) declines with group size, whereas it increases with group size for cases 2 and 3. Field data generally appear to support relationships of the latter type (e.g. Pulliam *et al.*, 1982), which is a little puzzling unless group members are related, or more complex strategies are operative (see Pulliam *et al.*, 1982), or individuals cannot accurately assess group size.

It is interesting to calculate how an individual's fitness will change with group size if our ESSs apply. For all cases, fitness increases with group size, even though in case 1 the total scanning declines. Thus the drop in total scanning for randomly formed groups could not be interpreted as a mechanism that could counteract Hamilton's selfish herd effect. At the ESS, it always pays to form groups.

Finally, we stress that our results depend critically on the form of $f(v)$. If, for example, we assume that gains through feeding decrease linearly with vigilance (broken line, Fig. 2*a*), then we find for case 1 that the ESS is to play $v^* = 0$ for all groups consisting of at least two members – it pays to watch out for predators only when one is feeding solitarily!

Future problems

The lack of any true genetic framework within evolutionary game theory could present difficulties. On the other hand, the lack of extensive descriptions of the phenotypic interactions is an impediment to classical population genetics theory. It is to be hoped that a marriage of the two disciplines might prove harmonious rather than a mathematical nightmare that stifles any central biological insights; we are not altogether optimistic.

How far has evolutionary game theory attacked the problem of modelling phenotypic interactions in the extensive way that classical game theorists analyse conflicts? In a simple game, the opponents make just one decision each (e.g. to fight or not to fight). By 'extensive' we mean that a whole sequence of decisions is made, depending on what has happened previously in the game. So far, most ESS analyses have not been extensive in form. Perhaps the approach that has come closest to a game in extensive form is that of Axelrod & Hamilton (1981). They gave a biological interpretation of the repeated Prisoner's Dilemma game (following an insight originally due to Trivers, 1971). This game is, in fact, probably the most studied of all games in the game theory literature. Its payoff matrix is given in Fig. 1(*b*). Axelrod & Hamilton (1981) argue that the so-called 'Tit for Tat' strategy is an ESS. This is to cooperate in the first round, and to mimic the opponent's previous decision in subsequent rounds. Unfortunately, their analysis did not take account of the second ESS condition (2*b*). Any mutant strategy which does not initiate the noncooperative behaviour is an alternative best reply to Tit for Tat and can drift into the population without being countered by selection. Tit for Tat therefore violates condition (2*b*). Nevertheless, strategies similar to Tit for Tat can be expected to be stable in models in which players occasionally make some form of 'mistake' about their strategy. This sort of effect is known to stabilize solutions in classical game theory, and Selten (1983) has now developed a systematic framework for its use in evolutionary games in extensive form.

Games in extensive form typically lead to multiple stable solutions. Thus in the step from the 'one-round' Prisoner's Dilemma to the repeated Prisoner's Dilemma, we change from the single Nash equilibrium to a whole array of Nash equilibria ranging from total noncooperation to strategies like Tit for Tat. In the same way, as we make biological games more extensive, we will encounter many different ESSs. For instance, the well-known Hawks–Doves game (Maynard Smith & Price, 1973) has a single ESS, but if similar games are played repeatedly between the same players, many qualitatively different ESSs are possible (Selten & Hammerstein, 1984). In the same way that classical game theorists have the problem of deciding between their various Nash equilibria, a major problem for evolutionary game theory concerns the development of a theory for narrowing down the set of possible ESSs. The two most obvious requirements for such a theory would be (i) to take account of evolutionary history, since the ESS attained generally depends on the 'starting condition'; (ii) to consider what strategies may be most easily achieved by simple biological mechanisms.

Finally, to what extent can evolutionary game theory apply to learning? Recently, Harley (1981) has investigated a model in which learning rules are subject to natural selection. He shows that the learning rules which will evolve are those that, within a generation, take a population to the ESS frequencies (see Maynard Smith, 1982, for a review of Harley's work). His stimulating conclusions probably apply well for simple games though it is as yet unclear how to make progress with learning rules for very complex games. We note the singular lack of ability of humans to adopt rational solutions in many experimental games set by classical game theorists and social psychologists!

Now JMS made us well able
To calculate what may be stable,
He strangled all genes
In his parlour-game scenes....
Could ESS *be* just a fable?

We are also much indebted to Professor R. Selten for further elaborating the parlour games.

References

Andersson, M. (1982). Sexual selection, natural selection and quality advertisement. *Biological Journal of the Linnean Society*, **17**, 375–93.

Aumann, R. J. & Maschler, M. (1964). The bargaining set for cooperative games. In *Annals of Mathematical Studies*, no. 52, ed. M. Dresher, L. S. Shapley & A. W. Tucker, pp. 443–76. Princeton: Princeton University Press.

Axelrod, R. & Hamilton, W. D. (1981). The evolution of cooperation. *Science*, **211**, 1390–6.

Bishop, D. T. & Cannings, C. (1978). A generalised war of attrition. *Journal of Theoretical Biology*, **70**, 85–124.

Charnov, E. L. (1982). *The Theory of Sex Allocation.* Princeton: Princeton University Press.

Clarke, B. & O'Donald, P. (1964). Frequency-dependent selection. *Heredity*, **19**, 201–6.

Dawkins, R. (1976). *The Selfish Gene.* Oxford University Press.

Dawkins, R. (1980). Good strategy or evolutionarily stable strategy? In *Sociobiology: Beyond Nature/Nurture*, ed. G. W. Barlow & J. Silverberg, pp. 331–67. Boulder: Westview Press.

Dawkins, R. (1982). *The Extended Phenotype.* Reading & San Francisco: Freeman.

Dunbar, R. I. M. (1982). Intraspecific variations in mating strategy. In *Perspectives in Ethology*, vol. 5, ed. P. P. G. Bateson & P. H. Klopfer, pp. 385–431. New York: Plenum Press.

Fisher, R. A. (1930). *The Genetical Theory of Natural Selection.* Oxford: Clarendon Press.

Fretwell, S. D. (1972). *Populations in a Seasonal Environment.* Princeton: Princeton University Press.

Gillies, D. (1959). Solutions to general non-zero-games. In *Contributions to the Theory of Games*, vol. IV (Annals of Mathematical Studies, no. 40), ed. A. W. Tucker & R. D. Luce, pp. 47–85. Princeton: Princeton University Press.

Grafen, A. (1979). The hawk–dove game played between relatives. *Animal Behaviour*, **27**, 905–7.

Hamilton, W. D. (1964). The genetical evolution of social behaviour. *Journal of Theoretical Biology*, **7**, 1–52.

Hamilton, W. D. (1967). Extraordinary sex ratios. *Science*, **156**, 477–88.

Hamilton, W. D. (1971). Geometry for the selfish herd. *Journal of Theoretical Biology*, **31**, 295–311.

Hammerstein, P. (1981). The role of asymmetries in animal conflicts. *Animal Behaviour*, **29**, 193–205.

Hammerstein, P. (1984). Evolutionary games with many players. *Working Paper*, no. 142, Institute of Mathematical Economics, University of Bielefeld.

Hammerstein, P. & Parker, G. A. (1982). The asymmetric war of attrition. *Journal of Theoretical Biology*, **96**, 647–82.

Harley, C. B. (1982). Learning the evolutionarily stable strategy. *Journal of Theoretical Biology*, **89**, 611–33.

Harsanyi, J. C. & Selten, R. (1980). A noncooperative solution theory with cooperative applications. *Working Paper*, no. 91, Institute of Mathematical Economics, University of Bielefeld.

Hines, W. G. S. & Maynard Smith, J. (1979). Games between relatives. *Journal of Theoretical Biology*, **79**, 19–30.

Lawlor, L. R. & Maynard Smith, J. (1976). The coevolution and stability of competing species. *American Naturalist*, **110**, 79–99.

Lewontin, R. C. (1961). Evolution and the theory of games. *Journal of Theoretical Biology*, **1**, 382–403.

Lorenz, K. (1965). *Evolution and Modification of Behavior.* Chicago: Chicago University Press.

Maynard Smith, J. (1972). *On Evolution.* Edinburgh University Press.

Maynard Smith, J. (1974). The theory of games and the evolution of animal conflicts. *Journal of Theoretical Biology*, **47**, 209–221.

Maynard Smith, J. (1982). *Evolution and the Theory of Games.* Cambridge University Press.

Maynard Smith, J. & Price, G. R. (1973). The logic of animal conflict. *Nature*, **246**, 15–18.

Nash, J. F. (1950). The bargaining problem. *Econometrica*, **18**, 155–62.

Nash, J. F. (1951). Non-cooperative games. *Annals of Mathematics*, **54**, 286–95.
Nash, J. F. (1953). Two-person cooperative games. *Econometrica*, **18**, 128–40.
Parker, G. A. (1978). Searching for mates. In *Behavioural Ecology: An Evolutionary Approach*, vol. 1, ed. J. R. Krebs & N. B. Davies, pp. 214–44. Oxford: Blackwells.
Parker, G. A. (1982). Phenotype-limited evolutionarily stable strategies. In *Current Problems in Sociobiology*, ed. King's College Sociobiology Group, pp. 173–201. Cambridge University Press.
Parker, G. A. (1983). Arms races in evolution – an ESS to the opponent-independent costs game. *Journal of Theoretical Biology*, **101**, 619–48.
Parker, G. A. (1984). Evolutionarily stable strategies. In *Behavioural Ecology: An Evolutionary Approach*, 2nd edn., ed. J. R. Krebs & N. B. Davies. pp. 30–61. Oxford: Blackwells.
Parker, G. A. & Macnair, M. R. (1979). Models of parent–offspring conflict. IV. Suppression: evolutionary retaliation by the parent. *Animal Behaviour*, **27**, 1210–35.
Parker, G. A. & Rubinstein, D. I. (1981). Role assessment, reserve strategy, and acquisition of information in asymmetric animal conflicts. *Animal Behaviour*, **29**, 221–40.
Pulliam, H. R. & Caraco, T. (1984). Is there an optimal group size? In *Behavioural Ecology: An Evolutionary Approach*, 2nd edn., ed. J. R. Krebs & N. B. Davies. pp. 122–47. Oxford: Blackwell.
Pulliam, H. R., Pyke, G. H. & Caraco, T. (1982). The scanning behaviour of juncos: a game-theoretical approach. *Journal of Theoretical Biology*, **95**, 89–103.
Riechert, S. E. & Hammerstein, P. (1983). Game theory in the ecological context. *Annual Review of Ecology and Systematics*, **14**, 377–409.
Rubinstein, D. I. (1980). On the evolution of alternative mating strategies. In *Limits to Action: The Allocation of Individual Behavior*, ed. J. R. Staddon, pp. 65–100. New York: Academic Press.
Selten, R. (1975). Bargaining under incomplete information – a numerical example. In *Dynamische Wirtschaftsanalyse*, ed. O. Becker & R. Richter, pp. 203–32. Tübingen.
Selten, R. (1979). Experimentelle Wirtschaftsforschung. In *Rheinisch-Westfälische Akademie der Wissenschaften, Vorträge no. 287*. West Deutscher Verlag.
Selten, R. (1980). A note on evolutionarily stable strategies in asymmetric animal conflicts. *Journal of Theoretical Biology*, **84**, 93–101.
Selten, R. (1983). Evolutionary stability in extensive 2-person games. *Working Papers*, nos. 121 & 122, Institute of Mathematical Economics, University of Bielefeld.
Selten, R. & Hammerstein, P. (1984). The evolution of social dominance. *Working Paper*, no. 140, Institute of Mathematical Economics, University of Bielefeld.
Shapley, L. S. (1953). A value for *n*-person games. In *Contributions to the Theory of Games*, vol. II (Annals of Mathematical Studies, no. 28), ed. H. W. Kuhn & A. W. Tucker, pp. 307–17. Princeton: Princeton University Press.
Shaw, R. F. & Mohler, J. D. (1953). The selective advantage of the sex ratio. *American Naturalist*, **87**, 337–42.
Skinner, B. F. (1953). *Science and Human Behavior*. New York: Macmillan.
Tinbergen, N. (1951). *The Study of Instinct*. Oxford University Press.
Trivers, R. L. (1971). The evolution of reciprocal altruism. *Quarterly Review of*

Biology, **46**, 35–57.
von Frisch, K. (1965). *Die Tanzprache und Orientierung der Bienen.* Berlin: Springer.
von Neumann, J. & Morgenstern, O. (1944). *Theory of Games and Economic Behavior.* Princeton: Princeton University Press.
Williams, G. C. (1966). *Adaptation and Natural Selection.* Princeton: Princeton University Press.

7

Behavioural structure and the evolution of cooperation

R. E. MICHOD AND M. J. SANDERSON

Introduction

Most kin selection and group selection models of social behaviour assume that differences in an individual's behaviour result only from differences in an individual's genotype. One of the main conclusions obtained from such models is that population structure among interacting genotypes is necessary for social behaviours, which accrue some cost to the individuals performing them, to spread. However, this necessity of genetic structure stems from the simplifying assumptions concerning the relationship between genotype and behaviour made in these models. In a situation in which individuals have behavioural flexibility and choose from a set of behaviours according to some rule (which may itself be genetically inherited), it is no longer necessary that there be genetic structure for costly behaviours to evolve. Indeed, genotypes can interact completely at random, yet altruistic or cooperative strategies may still be favoured (Brown, Sanderson & Michod, 1982). In order for this to occur, there must be some structure in the associations between the *behaviours* themselves (see appendix). In general, the requirement of genetic structure is replaced by a requirement of *behavioural structure*, so that the behaviours are encountered in frequencies different from their overall frequency in the population. In this chapter, we use a measure of behavioural structure (defined below) to relate the main theories for the evolution of social behaviour.

Discussion of terms

For lack of a better term, we use the term 'behaviour' to refer to the actions and activities which mediate a fitness interaction between two individuals. Consequently, the role of behavioural structure discussed below applies to fitness interactions in general, no matter how they are mediated. In particular, the ideas developed here are not restricted to animals with overt behaviours, but may also apply, for example, to populations of interacting plants.

The term 'altruism' refers to a behaviour which increases the genotypic fitness of other individuals, who are termed the recipients, while decreasing the genotypic fitness of the individual exhibiting the behaviour, who is termed the donor. Consequently, the 'cooperate' behaviour of the 'Prisoner's Dilemma' game (Axelrod & Hamilton, 1981; see Table 1 below) is an altruistic behaviour. In this chapter we study the evolution of a population composed of individuals which display two behaviours, termed 'cooperate' and 'defect', and denoted c and d respectively. Let C be the additive effect of a cooperative behaviour on the individual fitness of the donor and B the additive effect on the individual fitness of the recipient. Defection is assumed to have no effect on the fitness of either interactant.

The conditional distribution of interactions among *genotypes* is the basis of most models of kin selection, and allows different kin selection models to be related to each other (Abugov & Michod, 1981; Michod, 1982). We propose here that an analogous measure defined at the level of the behavioural interactions themselves can be used to relate the various evolutionary mechanisms for the evolution of cooperation, such as kin selection, assortment, reciprocation, and group selection. Let $u_{i|j}$ be the probability that an individual exhibiting behaviour j interacts with an individual exhibiting behaviour i. Let u_i and u_j be the frequencies of behaviour i and j in the population. There is structure in the distribution of behaviours to the extent that $u_{i|j}$ differs from u_i. Consequently, $u_{i|j}$ measures the behavioural structure in the population.

General condition

In this chapter we use the measure of behavioural structure defined above, $u_{c|c}$, to relate the main theories for the evolution of cooperation. We do this by expressing in the following general form the various conditions for increase of a cooperative trait which have been derived by these theories:

$$C/B < ku_{c|c}. \qquad (1)$$

To express the various conditions for increase of a cooperative trait in the form of condition (1), k is defined operationally by first solving for $u_{c|c}$ in the various models and then letting k be whatever is necessary for condition (1) to hold. In all cases considered, k thus operationally defined can be interpreted as the fraction of $u_{c|c}$ which is due to the proximate mechanisms under study, such as family structure, assortment, reciprocation or group structure. Consequently, we refer to this component of behavioural structure as being 'directed' by the various mechanisms under study. In each case, k depends upon a set of constants specific to the mechanism under study and on the mean frequency of cooperators in the population. Although the particular form of k varies from theory to theory, k always equals the magnitude of the directed component of behavioural structure. By representing the various theories in this common framework, the similarities and differences between them can be clearly represented and understood.

Kin selection

It is well known (see Michod, 1982, for review) that the condition for increase of an altruistic trait in a family-structured population is given by Hamilton's rule (1964),

$$C/B < 1/2. \tag{2}$$

This condition applies globally if selection is weak, there is no over- or underdominance and the fitness effects are additive (Michod & Abugov, 1980). If selection is strong, condition (2) gives the condition for increase of an altruistic trait when rare or when common for both the additive and multiplicative (not considered here) models of fitness (Cavalli-Sforza & Feldman, 1978).

We now relate condition (2) to behavioural structure and condition (1). To help fix ideas, consider a simple model of a haploid, sexual population. Assume a single locus with two alleles A and a which cause a difference in behaviour so that A individuals cooperate and a individuals defect on every interaction no matter how their partner behaves. There are three families possible in a sexual, haploid population designated by the genotypes of the parents: $A \times A$, $A \times a$, and $a \times a$. The conditional distribution of interactions between the cooperating behaviours can be calculated as follows. Given that an individual is c, then one of its parents must have been c. The other parent would have been c with probability u_c and d with probability $1 - u_c$, where u_c is the frequency of A in the population. The probability that an offspring receives its allele from any particular parent is, of course, 1/2. Consequently, the probability that a sib of a c individual is also c is

$$u_{c|c} = 1/2 + u_c/2. \tag{3}$$

Hamilton's condition (2) can now be rewritten in the form of condition (1) with

$$k = \frac{1}{1 + u_c}. \tag{4}$$

In Eqn (4), k is the fraction of $u_{c|c}$ which is specifically due to identity by descent and the family structure. Note that the conditional distribution of interactions between cooperators (Eqn 3) contains two components. The first component on the right-hand side of Eqn (3), 1/2, stems directly from identity by descent. This component contributes $1/(1+u_c)$ to the total behavioural structure. The second component in Eqn (3), $u_c/2$, stems from the population frequency of the cooperators and contributes $u_c/(1+u_c)$ to the total behavioural structure. In condition (1) using Eqns (3) and (4), the total behavioural structure is multiplied by the fraction of it which is directed to give the well-known 1/2 in Hamilton's rule (Eqn 2). Our reason for representing Hamilton's rule in the more complex form of condition (1) is to relate kin selection to the other main evolutionary paths to cooperation discussed below.

Assortment of encounters

Eshel & Cavalli-Sforza (1982) have studied a model for the evolution of cooperation which assumes that individuals are either always c or always d and that individuals interact with other like types with probability m and at random with probability $1-m$. They discuss briefly several proximate mechanisms by which individuals may assort with other like types. When interpreted in terms of additive cost and benefit, their condition (3.2) for increase of cooperation becomes

$$C/B<m. \tag{5}$$

When interpreted in terms of the conditional distribution of interactions between cooperators, condition (5) becomes condition (1) with

$$k=\frac{m}{m+(1-m)u_c}. \tag{6}$$

In this case, a c individual expects $m+(1-m)u_c$ encounters with other c individuals. However, of these interactions, only m are due to assortment. Thus $m/[m+(1-m)u_c]$ in Eqn (6) represents the fraction of a c's encounters which are with individuals who cooperate because of assortment.

Reciprocation

The theory of reciprocation (Trivers, 1971), as extended by Axelrod & Hamilton (1981), assumes that individuals can modify their behaviour through information they acquire about their environment. A simple learning rule which incorporates the basic idea of reciprocation is 'Tit for Tat' (or TFT), according to which if an individual recognizes its present partner from a previous interaction, then the individual behaves now as did its partner during that previous encounter. However, if the individual does not recognize its partner, then the individual cooperates. Axelrod & Hamilton (1981) and Brown, Sanderson & Michod (1982) studied the evolution of the TFT reciprocating strategy in competition with a strategy of total defection (called ALL D). Define r as the frequency of TFT individuals in the population, with $1-r$ the frequency of ALL D individuals. Let v be the probability that an individual is recognized from a previous encounter. For the additive model considered here, Brown, Sanderson & Michod (1982, Eqn (30), in which $\beta=\alpha(1-v)$) derived the following condition for increase of the reciprocating strategy

$$C/B < vu_{c|c}, \tag{7}$$

which is in the form of condition (1). Consequently, in this case

$$k=v, \tag{8}$$

and so the fraction of behavioural structure which is directed is simply the probability of recognizing an individual. As shown in Brown *et al.* (1982), condition (7) applies to several multi-partner models of interaction and not just the single-partner model considered by Axelrod & Hamilton (1981). It has also been shown (Brown *et al.*, 1982) that condition (7) applies to a single locus, two allele, diploid population genetic model in which the one homozygote has phenotype TFT, the other homozygote has phenotype ALL D, and the heterozygote behaves according to the TFT rule with probability h and ALL D with probability $1-h$.

For a population composed of TFT and ALL D individuals, $u_{c|c}=r^2/u_c$. This may be derived using the definition of conditional probability. By definition of TFT and ALL D, a c–c interaction occurs if and only if a TFT–TFT interaction occurs. Thus, the joint probability of c–c interactions is r^2 in a population in which individuals or genotypes interact at random. There are again two components to this behavioural structure. Because of the nature of the TFT learning rule, all encounters between two TFT individuals result in mutual cooperation. However, only a fraction v of these are directed in the sense that they are due to *recognition* and, hence, reciprocation *per se*. The remaining fraction, $1-v$, of $u_{c|c}$ arises simply because all TFT individuals cooperate indiscriminately on their first encounter with a stranger.

Group selection

We now apply these ideas of behavioural structure to D. S. Wilson's (1980) 'structured deme' model of group selection, which studies the evolution of two types such as c and d as defined above. Wilson (1980, Chapter 2) has also discussed the relations between behavioural structure and kin selection. Individuals are assumed to interact in the context of local subpopulations of constant size, termed 'trait groups', and to reproduce by a process, such as asexual haploid reproduction, in which the Darwinian fitness of a type is directly related to its frequency in the next generation. Let σ^2 be the variance in the frequency of c types between trait groups and u_c the frequency of c types in the total population. Let $u_{c,i}$ denote the frequency of c

in the ith local population. The fitnesses of the two types in a local population are then

$$W_{c,i}=1-C+u_{c,i}B$$
$$W_{d,i}=1+u_{c,i}B. \tag{9}$$

For Wilson's model, $u_{c|c}$ and $u_{c|d}$ (identical to Wilson's 'experienced frequencies') are given by the following equations (in which $E[x]$ denotes the expected value of x):

$$u_{c|c}=\frac{E[u_{c,i}^2]}{u_c}=u_c+\frac{\sigma^2}{u_c} \tag{10a}$$

$$u_{c|d}=\frac{E[u_{c,i}u_{d,i}]}{u_d}=u_c-\frac{\sigma^2}{u_d} \tag{10b}$$

Following Wright (1951), define F_{ST} as the correlation between types picked at random in the local subpopulations relative to types picked at random from the total population,

$$F_{ST}=\frac{\sigma^2}{u_c u_d}. \tag{11}$$

The correlation F_{ST} (11) is the ratio of the actual variance in frequency of c types to the maximal possible variance if all local subpopulations were fixed for one type or the other. Using Eqns (9)–(11), it can be shown that the expected fitness of the c type will be greater than the expected fitness of the d type if

$$\frac{C}{B}<F_{ST}. \tag{12}$$

To show the relationship of condition (12) to condition (1), write Eqn (10) as (using Eqn(11)):

$$u_{c|c}=F_{ST}+(1-F_{ST})u_c \tag{13a}$$
$$u_{c|d}=(1-F_{ST})u_c. \tag{13b}$$

There are two components of these conditional probabilities which represent the behavioural structure. The second component on the right-hand side of Eqn (13*a*) is experienced by both types and so cannot give rise to any difference in fitness between the types. However, the first component on the right-hand side of Eqn (13*a*), F_{ST}, is directed to the c type alone. Consequently,

$$k=\frac{F_{ST}}{F_{ST}+(1-F_{ST})u_c}. \tag{14}$$

Using the definitions given in Eqns (13*a*) and (14), it is easy to see that $ku_{c|c}=F_{ST}$, which shows that condition (12) is again a special case of condition (1).

Discussion

Through simple algebraic rearrangements, we have expressed in the common form of condition (1) the conditions for evolution of cooperation derived in the theories of kin selection, assortment, reciprocation, and Wilson's structured deme model of group selection. Condition (1) predicts that cooperative behaviours which accrue some cost in individual fitness will increase in frequency if their cost/benefit ratio is less than the directed component of behavioural structure, $ku_{c|c}$. Behavioural structure measures the degree to which a cooperative *behaviour* is associated with other cooperative behaviours. This concept shifts attention from the level of the individual or genotype to the level of the behaviours themselves. The fraction of the total behavioural structure which is directed is k, which depends on the specific mechanisms by which behavioural structure is produced in the four theories studied (Eqns 4, 6, 8 and 13). The non-directed component of the structure does not enter into the condition for selection, because this component is the same for all types in the population and so does not produce any differential fitness effect.

The fraction of the behavioural structure which is directed, k, has a different character in the four theories. In intersibling kin selection, k (Eqn 3) has a minimum value of 1/2, while under the TFT rule of reciprocation, assortment, or group selection, it can become arbitrarily small, as the probability of recognizing an individual, v, the probability of assortative meeting, m, or the variance between local subpopulations, σ^2, get small.

In models which assume that all differences in behaviour are caused by differences in genotype, behavioural structure can only be generated by genetic population structure. This is the case with most models of kin and group selection as well as with Eshel & Cavalli-Sforza's (1982) model of assortment discussed above. However, in the models of reciprocation discussed above, behavioural structure is generated by the learning process. In this case, genotypes may interact completely at random, yet costly social behaviours may still evolve. In such models, differences in genotypic fitness are generated by differences in the learning rules by which genotypes modify their behaviour based on previous experience.

As mentioned in the introduction, the conclusion that genetic structure is necessary for costly, social traits to evolve is a direct consequence of the assumption of a one-to-one correspondence between behaviour and genotype usually made in sociobiological models. Consequently, genetic structure need not be a feature of sociality in the broad sense. In fact, in organisms with some degree of behavioural plasticity, it is more likely that behaviours arise due to an interaction between heritable traits and environmental events (specifically including interactions with conspecifics and learning). This chapter has attempted to show that in such situations

Table 1. *Payoff matrix for the c and d behaviours with additive costs and benefits*

Payoff to behaviour below when encountering behaviour to right	c	d
c	$1+B-C^a$	$1-C$
d	$1+B$	1

a The parameters B and C are the additive effects on the fitnesses of recipient and donor, respectively. See text for further explanation.

the appropriate focus for research is the structure among behaviours in a population and not the genetic structure. Indeed, behavioural structure is necessary for the evolution of costly, social behaviours in all the models studied here. This in no way implies that evolution of these behaviours is somehow removed from genetic evolution, but it does mean that the observable manifestations of such genetic evolution may emerge on the behavioural level independent of considerations such as genetic population structure.

We thank Bob Abugov, Renee D. Rusler, Paul Sherman and Marcy Uyenoyama for discussion and comment on earlier versions of this manuscript. This work was supported, in part, by NSF grant DEB-8118248 to R.E.M.

Appendix

In this appendix we consider the effects of random interactions on individual fitness. We do not address the process by which the traits which mediate the interactions are transmitted from one generation to another. The transmission process is, of course, an integral part of the evolution of any trait, behaviour included. However, our purpose here is more limited in scope than to develop a complete evolutionary description of any particular behaviour. We only wish to illustrate the importance of behavioural structure to individual fitness, realizing that the transmission process will further affect the evolution of the trait in ways not considered here (see, for example, Michod, 1984).

As above, let $u_{i|j}$ be the probability that an individual exhibiting behaviour j interacts with another individual exhibiting i. Let u_i be the frequency of behaviour i in the total population. Consider the case assumed above of two behaviours c and d with additive costs and benefits to fitness. The interactions between individuals result in one of four possible fitness payoffs given in Table 1, which satisfy the conditions of the Prisoner's

Dilemma game (see, for example, Axelrod & Hamilton, 1981; or Brown *et al.*, 1982).

The expected payoffs to individuals exhibiting behaviours c and d are

$$W_c = u_{c|c}(1+B-C)+(1-u_{d|c})(1-C)$$
$$W_d = u_{c|d}(1+B)+(1-u_{d|d})\times 1. \quad \text{(A1)}$$

If encounters among behaviours occur at random, then

$$u_{c|c} = u_c$$
$$u_{d|c} = 1-u_c$$
$$u_{c|d} = u_c$$
$$u_{d|d} = 1-u_c. \quad \text{(A2)}$$

Using Eqns (A2),

$$W_c = u_c(1+B-C)+(1-u_c)(1-C)$$
$$W_d = u_c(1+B)+(1-u_c). \quad \text{(A3)}$$

By inspection, $W_d > W_c$ in all cases. During any interaction an individual adopting c can expect to receive a lower payoff than one adopting d. In many cases of genetic transmission, this should be sufficient to prevent altruism from spreading in a population. However, in some cases of cultural transmission (Cavalli-Sforza & Feldman, 1981), altruism could still be favoured. In such cases, the success of altruism would be due to a bias in the transmission process due to cultural reasons.

References

Abugov, R. & Michod, R. (1981). On the relation of family structured models and inclusive fitness models for kin selection. *Journal of Theoretical Biology*, **88**, 733–42.

Axelrod, R. & Hamilton, W. D. (1981). The evolution of cooperation. *Science*, **211**, 1390–6.

Brown, J., Sanderson, M. & Michod, R. (1982). Evolution of social behavior by reciprocation. *Journal of Theoretical Biology*, **99**, 319–39.

Cavalli-Sforza, L. L. & Feldman, M. (1978). Darwinian selection and 'altruism'. *Journal of Theoretical Population Biology*, **14**, 263–81.

Cavalli-Sforza, L. L. & Feldman, M. (1981). *Cultural Transmission and Evolution.* Princeton: Princeton University Press.

Eshel, I. & Cavalli-Sforza, L. L. (1982). Assortment of encounters and evolution of cooperativeness. *Proceedings of the National Academy of Sciences, USA*, **79**, 1331–5.

Hamilton, W. D. (1964). The genetical evolution of social behavior. *Journal of Theoretical Biology*, **7**, 1–52.

Michod, R. E. (1982). The theory of kin selection. *Annual Review of Ecology and Systematics*, **13**, 23–55.

Michod, R. E. (1984). Genetic constraints on adaptation with special reference to social behavior. In *The New Ecology: Novel Approaches to Interactive Systems*,

ed. P. W. Price, C. N. Slobodchikoff & W. S. Gaud, pp. 253–78. New York: Wiley.
Michod, R. E. & Abugov, R. (1980). Adaptive topography in family-structured models of kin selection. *Science*, **210**, 667–9.
Trivers, R. (1971). The evolution of reciprocal altruism. *Quarterly Review of Biology*, **46**, 35–57.
Wilson, D. S. (1980). *The Natural Selection of Populations and Communities.* Menlo Park, California: Benjamin/Cummings.
Wright, S. (1951). The genetical structure of populations. *Annals of Eugenics*, **15**, 323–54.

Adaptations, constraints and patterns of evolution

8

Evolutionary ecology and John Maynard Smith

R. M. MAY

Any attempt to partition the broad canon of work with which John Maynard Smith is associated into distinguishable categories is necessarily subject to some arbitrariness. If Maynard Smith's own writings were grouped into the three subdivisions adopted in this book, each section would have its own distinct character. 'Population genetics and evolution theory' would embrace a variety of interesting and influential contributions that are mainly within the existing framework of conventional population genetics, as pioneered by Maynard Smith's mentor, J. B. S. Haldane, along with R. A. Fisher and Sewall Wright. This category would also embrace the books (especially *The Theory of Evolution*, published in 1958) and the radio and TV appearances in which Maynard Smith – again following Haldane – has so ably conveyed advances in evolutionary thinking to a wider audience. 'The evolutionary ecology of sex' would comprise those publications in which Maynard Smith (along with Williams and others) has drawn attention to the factor-of-two cost that sexually reproducing organisms often pay in relation to corresponding asexual ones, and to the fact that the advantage gained by the greater genetic variability generated by sexual recombination is often not enough to offset this factor-of-two disadvantage in simple and conventional models. My own guess – very likely wrong – is that this body of work, although of great fascination at the moment, may not be of lasting importance. Conventional '*A*/*a*' population genetics is a very crude caricature of the still-unfolding genetic complexity and diversity exhibited by, for example, the HLA system in humans or the H-2 system in mice. These systems are vital for host defence against a range of pathogens, and sexual recombination plays a central part in their generation and maintenance; it seems to me likely that more realistic computations of the advantages in genetic variability that are conferred by sexual recombination may easily outweigh the twofold demographic disadvantage.

This section, 'Adaptations, constraints and patterns of evolution' covers the far-ranging assortment of Maynard Smith's work that is not tidily pigeonholed under the other two categories. It is among this work – which includes biomechanics, population dynamics, community ecology, and the evolution of behaviour – that I think many of Maynard Smith's most lasting and unique contributions lie. Most notably, the term 'ESS' (evolutionarily stable strategy) is of his coining; it will clearly last as long as people study behavioural ecology. Game theory is a relatively young subject: the first empirical application of which I am aware is Kemeny & Snell's (1957) calculation of the game-theoretically best strategy for playing baccarat (subsequently generalized by Karlin in 1959 to classes of card games not yet adopted by casinos; see May, 1976). Although MacArthur's (1965) discussion of sex ratio appeals to game-theoretic ideas and Hamilton's (1967) paper on extraordinary sex ratios makes explicit use of game theory and the notion of an uninvasible strategy, it was Maynard Smith who systematically studied the way the theory of games can illuminate many aspects of the evolution of plant and animal behaviour. He indicated, moreover, how these ideas may be tested in the field and laboratory. Chapters 6, 7, 9, 10, 11, 13 and 14 describe only a representative few of the growing number of research projects inspired by Maynard Smith's work in this particular area. Harking back to Kemeny & Snell and baccarat, I think Maynard Smith's creative use of the theory of games is to some extent associated with his enthusiasm for games such as bridge and chess; with difficulty, I restrain myself from including a critical analysis of some bridge hands I have seen him play.

The fact that five of the six chapters in this section deal with ESS, in the context either of behaviour or of life history, should not distract us from recognizing that Maynard Smith has done many other important things that would be included in Part 2 were his own output to be partitioned among the three sections of this book. The following examples are chosen from among this work, in each case partly with the view toward making a more general point.

At the level of individual organisms, Maynard Smith's (1968) elegant book on *Mathematical Ideas in Biology* presents a variety of instances where insights can be gained by thinking of organisms as simple machines, obeying and constrained by the laws of classical physics. Typical of these engaging and often surprising insights is the conclusion that, to a rough first approximation, all animals – fleas to kangaroos – will be capable of jumping to the same height, regardless of size. The basic argument is set out by D'Arcy Thompson (1966), who refers to it as 'Borelli's Law', after the individual who wrote on this subject in the 1680s. The energy put into such springing motion can be shown to scale as L^3 (where L characterizes the linear dimensions of the animal, the force exerted on the ground is proportional to power output which scales as L^2, and the total work done depends on this force times the distance, of order L, through which it moves as springing legs extend), and the energy required to raise the animal's centre of gravity, by a height H, scales as L^3H (from mgH), whence H is independent of L. More exact calculations will obviously involve all kinds of details that depend, among other things, on the way a particular animal makes its living. But this simple dimensional argument illuminates the essentials, and gives a first approximation that can be tested against the facts. Indeed, the cover of *Mathematical Ideas in Biology* is adorned with the schematic mouse (cubical body of side length L, springy sticks for hind legs, and so on) whose high-jumping feats are then estimated by elementary mechanics. The cover of the Russian translation of this book has, however, be redrawn to show a Socialist Realist mouse, complete with fur, paws and whiskers (Fig. 1). It is worth emphasizing that this doctrinaire redesign misses the point of the mouse-jumping example in particular (and of much of mathematical biology in general) in an important, but common, way; the essential point is that the mouse's fur, whiskers and other morphological details are irrelevant to a first estimate of how high it can leap. The Commisars are, unfortunately, not alone in thinking it a form of sin to schematize or simplify an animal (or a population, or a community), even though such sacrifice of detail may be a prerequisite to an illuminating mathematical analysis. Such biomechanical studies have been much more fully developed by McMahon (1984), Alexander (1982) and others; I will return to these.

Maynard Smith has also given us insights into the way the behaviour of individuals can be related to the structure and dynamics of populations. The final chapter of his *Models in Ecology* (1974), for example, deals with simple models for territorial behaviour. It is shown that the geometry of territories, in a situation where all territories are established roughly simultaneously (as when migratory birds return together to a patch of woodland), is likely to be different from that in a situation where territory holders arrive seriatim, and that (in either case) the territorial behaviour of individuals has explicit consequences for the overall dynamics of the population.

Among Maynard Smith's contributions to population biology and ecology, I think the most important are his various studies (mainly numerical) which show that simple models for predator–prey associations – or for single populations in which regulatory effects operate with time delays – can exhibit oscillatory dynamics (these are sometimes correctly identified as stable limit cycles, and sometimes described in the earlier work as 'diverging oscillations'). These models are presented in *Mathematical Ideas in Biology* and *Models in Ecology*, and they were in some ways ahead of their time. Although overtaken by a variety of later analytical and numerical studies showing how simple nonlinearities can lead not only to stable cycles but even to apparently random or 'chaotic' dynamics (for a brief review, see May, 1981), Maynard Smith's work was influential in

Дж. Смит

математические идеи в биологии

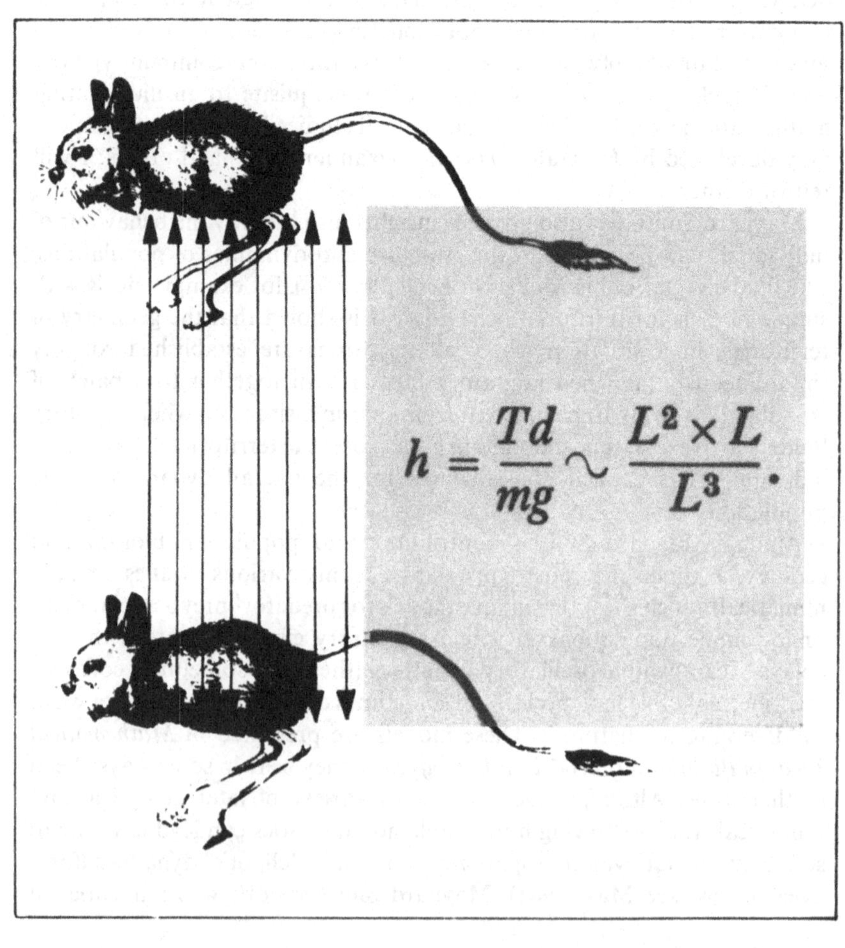

helping to teach us that – to the contrary of the intuition formed by elementary physics and mathematics courses – simple mathematical models can have very complicated dynamics. Given the nonlinearities or density dependences that are present in the simplest models for the dynamics of populations, it should not surprise us to find some populations steady, some showing persistent and regular oscillations, and others fluctuating apparently randomly, even if the world were homogeneous and deterministic.

Remarks on the other chapters in this section

Three of the next six chapters explicitly refer to the concept of ESS. Of these three, two deal with ESS in plant germination and dispersal. León presents a purely theoretical analysis of the factors likely to influence whether the seeds of an annual plant will all germinate in a given year or whether some will remain dormant; he also discusses the circumstances under which those seeds that do germinate in a given year will do so synchronously or spread out over some interval of days or weeks. The factors considered by Leon include different kinds of environmental uncertainty, different assumptions about the plants' ability to detect a 'good' moment for germination, and different modes of dispersal.

Silvertown's chapter complements Leon's, exploring a similar theme but being much more oriented to collating field and laboratory data to test the ideas. Specifically, Silvertown reviews evidence on two kinds of 'mixed strategy' in plants: varied responses in germination and varied dispersal types in seeds. He shows, for example, that when seeds are germinated in the laboratory, not all are triggered by a single treatment and not all are triggered on the same day. Silvertown also surveys data on polymorphisms in seed structure, showing that some polymorphisms appear fixed while others appear to change with plant size and age. The discussion of possible evidence for evolutionary changes in mixed strategies is particularly engaging.

Packer & Pusey assemble and assess evidence from lions, baboons, and chimpanzees to test an idea propounded by Maynard Smith (1982) – that, in stable social groups (where animals have met before), contests are more likely to be settled by observable asymmetries of one kind or another than by actual combat. They find that 'respect of ownership' (Maynard Smith's 'Bourgeois Strategy') varies greatly among species but that, in general, it is most important where fighting is most costly.

The chapters by Gittleman and by Rose do not make explicit appeal to ESS, but they are in the 'Maynard Smithian' spirit in discussing the evolution of behaviour or of life history in terms of the trade-offs between identified costs and benefits. Gittleman observes that communal care of young mammals has not received anything like the attention given to cooperative breeding among birds, and he catalogues different classes of communal care found among mammalian species in the wild (as distinct from in captivity). For these various kinds of social systems, Gittleman reviews data bearing on the ecological and behavioural advantages, and on the associated costs of communal care (conspicuousness to predators, competition, greater prevalence of disease and ectoparasitism).

Rose gives a survey of ideas about the evolution of senescence, emphasizing those aspects in which the costs and benefits of reproductive effort in the present are weighed against the future. He outlines ways in which these ideas have been, or could be, tested against such facts as are available, and indicates some interesting possibilities for gathering data that could provide more clear-cut empirical tests.

In their various ways, each of these five chapters exemplifies the kind of work Maynard Smith has encouraged. León and Rose show how general ideas can be made precise, either by mathematical models (León) or by rigorous verbal argument (Rose). Silvertown, Gittleman, and Packer & Pusey go beyond this to show how systematically compiled data from appropriately chosen groups of species can allow us to test such ideas about the way social or reproductive or other behaviour has evolved.

The chapter by Harvey & Ralls is the wild card in this pack. They use data on the physical characteristics of three species of North American weasels to test ideas about the factors underlying interspecific variation in size and sexual dimorphism among these three congeneric species. To the contrary of Schoener's (1984) similarly rigorous study on hawks, Harvey & Ralls' careful statistical analysis finds no significant degree of geographical covariation in body size among their weasels, which suggests that factors other than niche overlap or limiting similarity are determining size patterns in this group. Although Maynard Smith has not worked in this general area of niche separation and geographical ecology, the chapter by Harvey & Ralls brings a valuable and timely message to this Festschrift. Stimulated by the insights and questions advanced by G. Evelyn Hutchinson, Robert H. MacArthur and others, ecologists over the past two decades have sought to identify systematic patterns in the structure of guilds of competing species, or even in the organization of entire communities. More recently, Daniel Simberloff, Donald Strong and others have pointed out that some of these observed patterns may have been embraced too enthusiastically, and that some apparently orderly relations among the size or geographical distribution of species may in fact be indistinguishable from those expected on statistical grounds from groups of species assembled at random. Although this counter-revolutionary movement to explain apparent community structure by 'neutral' or 'null' hypothesis is (in my view) guilty of its own considerable excesses of enthusiasm, I believe that, when the pendulum has stopped swinging, community ecology will be a better-founded discipline by virtue of the rigorous testing of perceived pattern against properly constructed null models (for fuller discussion of these issues, see Strong *et al.*, 1984; Harvey *et al.*, 1983).

What has all this got to do with behavioural ecology, much less with Part 2 of the present book? I think that the recent ferment of ideas about the way notions of ESS, or inclusive fitness, or optimal foraging may explain aspects of the social organization of animal and plant species is in the midst of a critical phase in which these ideas are being tested against data, both for particular species and for comparisons among groups of species. This enterprise may avoid the pendulum swings and the associated occasional bitterness that have afflicted community ecology, if the search for patterns and for testing of hypotheses is conducted with rigorous attention to statistical design and awareness of alternative (possibly 'neutral') explanations, exemplified in Harvey & Ralls' chapter.

What next?

In conclusion, I indulge in some speculations about the directions that research in the general area of 'Adaptations, constraints and patterns of evolution' may take over the next few years.

One trend that has already been noted, and that is evident in the following chapters, is toward systematic assembly of data to test theoretical ideas about the way evolutionary forces acting on individuals have shaped mating behaviour, group size, and other aspects of behavioural ecology within and among species. This is an important and highly desirable trend but, as such activity burgeons, I hope the practitioners will continue to keep in mind – as they obviously do at present – Darwin's injunction (Gruber & Barrett, 1974, p. 123): 'all observations must be for or against some view if it is to be of any service!'. It is perhaps salutory to remember that (as I see it) the wilderness of meticulous classification and ordination of plant communities, in which plant ecology wandered for so long, began in the pursuit of answers to questions but then became an activity simply for its own sake until relatively recently.

Other areas where I think interesting developments are likely to occur relate to aspects of Maynard Smith's work that are less represented in this book. Thinking of living things as machines that are subject to the same mechanical laws, dimensional considerations and scaling relations as buildings and bridges (as outlined in *Mathematical Ideas in Biology*) is increasingly leading to insights into the way processes of adaptation and development are constrained. This interplay between technical aspects of biomechanical design and grand questions about evolution was, for example, a central theme at the recent Dahlem Conference on 'Evolution and Development' (Bonner, 1982).

The chapters in Part 2 are typical in that they try to understand the behaviour of groups in terms of the natural selection acting on individuals, but make no attempt to carry this analysis to the next hierarchical level in order to understand the dynamics of a population in terms of its behavioural ecology. Conversely, population biologists – whether their goal is basic understanding or practical wildlife management – rarely attempt to relate the demographic parameters in their models to the underlying behaviour of the individual members of the population. As mentioned above (in reference to some of the work in *Models in Ecology*), I see a need for more work that strives toward an explicit understanding of population dynamics in terms of the underlying mating behaviour, foraging behaviour, and general social organization of the species at issue. In functional terms, this means the demographic parameters – birth, death and migration rates, and the like – are not treated as phenomenological quantities, nor are they estimated from population data, but instead are explicitly related to and estimated from the behavioural ecology of the plants or animals involved (for a more full discussion and some examples of existing work of this kind see, for example, Hassell & May, 1984).

In a similar vein, I note that there is a large and growing body of work (well represented in this book) which uses ESS and other evolutionary arguments to explain observed sex ratios. But there are very few studies exploring the way sex ratio can affect population dynamics, especially when the sex ratio depends on population density or on environmental conditions; with few exceptions, population models assume (sometimes explicitly and more often implicitly) a 1:1 sex ratio. Here is another area where the kinds of theoretical and empirical research on sex ratio that are illustrated by Chapter 18 in this volume could be fruitfully carried forward to assess their implications for population dynamics (see, for example, Hassell, Waage & May, 1983).

Maynard Smith and many others have also sought to elucidate the factors influencing the structure of entire communities of interacting populations of plants and animals. In particular, considerable indecisiveness still surrounds the question of why most communities have only three or four, and very rarely more than five, trophic levels (for further discussion of what follows, see Pimm, 1982). Earlier explanations in terms of the efficiency of energy flow from one level to the next are apparently inconsistent with the observations that food chains are not typically longer in more productive environments, and that the structure of food webs with predominantly warm-blooded animals is not significantly different from those with cold-blooded animals, despite the significantly greater energy-transfer efficiency of the latter. An alternative suggestion is that food-chain lengths are determined by dynamical considerations, with excessively long chains leading to intolerably large population fluctuations. This explanation also has problems: for such a mechanism to explain the rough constancy in the number of trophic levels, it is necessary that the time scale characterizing the life history of those organisms which ultimately determine the response time of the ecosystem be roughly the same in all communities. The answer to some of these questions may lie in the mechanical and behavioural factors governing the natural history of the top predators in typical food chains. If so, this area could provide a remarkable instance where community-level phenomena are explicitly related to biomechanics and behavioural ecology (May, 1983).

Finally, there remains an array of questions of fundamental importance that have received essentially no attention, yet where the kind of approach exemplified throughout Maynard Smith's work seems called for. One such topic concerns the distribution of number of species as a function of physical size, both within particular taxonomic groupings and more generally. This topic ultimately relates to the basic question of why the number of species is what it is; I think what we need here are some new ideas, and analysis of dimensional considerations, along the lines in *Mathematical Ideas in Biology.* Another topic concerns the way the geographical distributions of species are divided into size classes: of the totality of terrestrial species, for example, how many have geographical distributions in the range 10^7–10^8 km^2, how many in the range 10^6–10^7 km^2, and so on? Other such broad questions may occur to the reader; they share the property that, although the answers will in the final analysis involve biology, the initial approach is more in the style of physics or engineering.

While I may well be wrong in this catalogue of tomorrow's fashions, I am surely right that the style consistently illustrated by Maynard Smith's work is a useful one: identify a significant problem; translate ideas into a model that includes all the elements thought to be essential, but which is not cluttered with extraneous detail; and give a clear exposition of the conclusions, emphasizing how they relate to the initial assumptions and how they may be tested. Above all, it is Maynard Smith's blend of biological commonsense with an engineer's clarity and rigor that makes his contributions so influential, and such a pleasure to read or listen to.

References

Alexander, R. McN. (1982). *Locomotion of Animals.* New York: Chapman & Hall.

Bonner, J. T. (ed.) (1982). *Evolution and Development.* Berlin: Springer Verlag.

Gruber, H. E. & Barrett, P. H. (1974). *Darwin on Man: A Psychological Study of Scientific Creativity, Together with Darwin's Early and Unpublished Notebooks.* New York: Dutton.

Hamilton, W. D. (1967). Extraordinary sex ratios. *Science*, **156**, 477–88.

Harvey, P. H., Colwell, R. K., Silvertown, J. W. & May, R. M. (1983). Null models in ecology. *Annual Reviews of Ecology and Systematics*, **14**, 189–211.

Hassell, M. P. & May, R. M. (1984). From individual behaviour to population dynamics. In *Behavioural Ecology*, ed. R. Sibley & R. Smith, pp. 3–32. Oxford: Blackwell.

Hassell, M. P., Waage, J. K. & May, R. M. (1983). Variable parasitoid sex ratios and their effect on host–parasitoid dynamics. *Journal of Animal Ecology*, **52**, 889–904.

Karlin, S. (1959). *Mathematical Methods in Games, Programming and Economics.* London: Pergamon.

Kemeny, J. G. & Snell, J. L. (1957). Game theoretic solutions of baccarat. *American Mathematics Monthly*, **64**, 465–9.

MacArthur, R. H. (1965). Ecological consequences of natural selection. In *Theoretical and Mathematical Biology*, ed. T. H. Waterman & H. J. Morowitz, pp. 388–97. New York: Blaisdell.

McMahon, T. A. (1984). *Muscles, Reflexes, and Locomotion.* Princeton: Princeton University Press.

May, R. M. (1976). Mathematics and casinos. *Nature*, **264**, 508–9.

May, R. M. (ed.) (1981). *Theoretical Ecology: Principles and Applications*, 2nd edn. Oxford: Blackwell; and Sunderland, Mass.: Sinauer.

May, R. M. (1983). The structure of food webs. *Nature*, **301**, 566–8.

Maynard Smith, J. (1958). *The Theory of Evolution.* London: Penguin.

Maynard Smith, J. (1968). *Mathematical Ideas in Biology.* Cambridge University Press.

Maynard Smith, J. (1974). *Models in Ecology.* Cambridge University Press.

Maynard Smith, J. (1982). *Evolution and the Theory of Games.* Cambridge University Press.

Pimm, S. L. (1982). *Food Webs.* London: Chapman & Hall.

Schoener, T. W. (1984). Size differences in sympatric bird-eating hawks: a worldwide survey. In *Ecological Communities: Conceptual Issues and the Evidence*, ed. D. R. Strong, D. Simberloff, L. G. Abele & A. B. Thistle, pp. 254–81. Princeton: Princeton University Press.

Strong, D. R., Simberloff, D., Abele, L. G. & Thistle, A. B. (eds). (1984). *Ecological Communities: Conceptual Issues and the Evidence.* Princeton: Princeton University Press.

Thompson, D'A. W. (1966). *On Growth and Form*, abridged edition, ed. J. T. Bonner, pp. 26–8. Cambridge University Press.

9

The evolution of senescence

M. R. ROSE

Over the last century or so, two scientific traditions have developed for the study of senescence: physiological gerontology and evolutionary analysis of age-structured populations. The first research tradition arose fairly naturally from the shift in biology from descriptive naturalist research to manipulative experimental research (cf. Loeb *et al.*, p. 7 in Pearl, 1922). Thus, there was a squadron of physiologists working on senescence from 1870 to 1940: Lankester, Metchnikoff, Minot, Carrel, Child, Bidder, Pearl, and so on (Comfort, 1979, pp. 7–15). As discussed by Medawar (1946) and Comfort (1979), it is extremely doubtful that this research tradition had any real success before 1940.

During this same period, the evolutionary approach also had little success. As discussed by Medawar (1946), Wilson (1974), and Kirkwood & Cremer (1982), early Darwinian analyses put forward by A. R. Wallace and A. Weismann were group selectionist. The individual organism was described as disposable, so far as the propagation of the species was concerned. Thus, once its reproductive role had been discharged, it was discarded lest it interfere with the success of young members of the species. Naturally, this general sort of argument was continued by Wynne-Edwards (1962), who treated senescence as a population-regulation adaptation.

Though these arguments are not as circular as Comfort (e.g. 1979, p. 11) has suggested, they were essentially disposed of by Williams (1957, 1966*a*, pp. 225–6). The chief empirical difficulty for this theory is the low frequency of organisms surviving long enough in wild populations to die of a group-selected death mechanism. The senescent processes observed in laboratory, human, and some domesticated populations occur at ages much greater than are normally achieved under natural conditions. Thus, since the group selection analysis of evolution in age-structured populations was barren, the Darwinian research tradition had little to contribute to the study of senescence before 1940.

Since 1940, both fields have changed radically. The physiological tradition adopted the disciplinary name of gerontology and the style of molecular biology. Physicists and biochemists began to dominate the field, promulgating a great number of general theories. For example, Szilard (1959) proposed that ageing is due to the knocking out of chromosomes in somatic cells, such that all the genes located on a chromosome cease to function together. This theory was inspired by contemporaneous developments in radiobiology, especially ostensible parallels between radiation sickness and the ageing process. Proceeding in a Popperian fashion, Maynard Smith and his students (e.g. Maynard Smith, 1959*a*, 1966; Lamb & Maynard Smith, 1964; Lamb, 1965) demonstrated experimentally that chromosomal ploidy does not have consistent effects on the extent to which lifespan is shortened by exposure to radiation. This work provided ostensible refutation of this somatic mutation theory. As is usually the case, there remained room to modify the theory in an *ad hoc* fashion to escape such refutation. But somatic mutation is no longer considered a preeminent universal mechanism of senescence.

Another notable universal physiological theory of ageing was Orgel's (1963) protein-error catastrophe theory. This theory assumed that ageing is due to the positive feedback of errors in the construction of proteins involved in the translation apparatus, such as amino acyl synthetases, causing a catastrophic accumulation of translation errors. Though Orgel (1963) at first advanced this theory as a general explanation for senescence, he has since retracted such claims to generality (Orgel, 1973). This frank disavowal of generality is laudable, in that proponents of general physiological theories of ageing have typically held on to their views with considerable tenacity.

The overall development of the field of physiological gerontology has left many of its advocates at least somewhat dispirited. Referring to ageing, Burnet (1976, p. 82) wrote: 'No one has yet produced a satisfactory explanation of the whole process, and probably no one ever will.' Comfort (1979, p. 9), perhaps the single person most responsible for the establishment of modern gerontology, proposed that:

> In almost any other important biological field than that of senescence, it is possible to present the main theories historically and to show a steady progression from a large number of speculative ideas to one or two highly probable, main hypotheses. In the case of senescence this cannot be profitably done.

While I would admit the truth of this observation with respect to physiological gerontology, I will argue that (i) the opposite is true of the evolutionary approach to the subject and (ii) perhaps the best way forward for the physiological approach is to exploit the gains made by evolutionary biology.

Development of evolutionary theory for senescence

The development of the evolutionary approach to senescence was at first entirely theoretical, as we have come to expect in evolutionary biology. The first hint of the ideas which would develop came from R. A. Fisher (1930). Fisher (p. 29) alluded to the rough inverse relationship between curves of reproductive value and those of natural death rate, implying a role for natural selection in establishing this pattern, although the role is not specified. As Hamilton (1966) argued, this correspondence is spurious, but it provided a jumping-off point for subsequent hypotheses put forward by Haldane and Medawar.

Haldane (1941, pp. 192–4) was the first to propose an explicit well-formulated hypothesis, concerning the evolution of senescence, which was not based on group selection. Alluding to the decrease in selective significance of decreasing survival probability with increasing age of deleterious gene-effects, Haldane suggested that natural selection has favoured modifier alleles which act to postpone the effects of hereditary diseases. Charlesworth (1980, p. 219) has since pointed out that this mechanism probably has no appreciable effect, because natural selection favouring such modifier alleles affecting rare early-onset genetic diseases is too weak relative to potentially countervailing mutation pressure and any other selective factors arising from pleiotropic effects of the modifier alleles.

It was Medawar (1946, 1952) who provided the essential foundation for all subsequent thinking on the evolution of senescence. Even in the hypothetical complete absence of senescence, if constant fertility is assumed, the reproductive output of each age-class declines with age for the elementary reason that survivorship from birth is a decreasing function of age. Thus, Medawar suggested, the importance of genetic effects declines with age. This conclusion led him to propose a variety of population genetic mechanisms which would foster senescence. Firstly, Medawar reiterated Haldane's point that modifier alleles would tend to postpone deleterious genetic effects, causing their 'recession'. Secondly, Medawar pointed out that genes with sufficiently late deleterious effects would be effectively neutral, so that they would spread through populations in spite of disastrous effects on health late in life. Thirdly, he pointed out that genes enhancing early fitness-components to a small degree may be favoured in spite of large, later, gene effects which act to reduce the survival and/or reproductive success of old individuals. Since Medawar, no one has added any further cogent suggestions for the evolution of senescence. In particular, two subsequent articles, Williams (1957) and Edney & Gill (1968), chiefly expand – albeit with some eloquence and insight – on the third and second mechanisms adduced by Medawar, respectively.

The only additional suggestion of basic significance is due to Hamilton (1966). The idea is that natural selection will establish favourable mutations acting at early ages, but will not act very effectively, if at all, at later ages. Thus, age-specific fitness-components at early ages will be relatively more improved over the evolutionary history of a species. This theory cannot completely account for senescence, since senescent individuals usually have very low absolute reproductive rates, and therefore are hardly likely to represent an unimproved ancestral condition (Charlesworth, 1980, p. 219). However, this theory of Hamilton's is in some ways a natural corollary of Medawar's alternative mechanisms, but viewed as if from the other end of a telescope.

The enduring worth of Hamilton's (1966) paper lies in its development of the mathematical underpinnings necessary to Medawar's verbal and graphical arguments. Assuming that fitness in a population with age-structure is given by the Malthusian parameter, Hamilton (1966) examined the dependence of this parameter on survival probabilities and fecundity. Let: (i) m be the Malthusian parameter, (ii) $\lambda = e^m$, (iii) l_x be the survivorship schedule, (iv) $l_x = \prod_{a=0}^{x-1} p_a$, and (v) F_x be the mean number of same-sex births from a parent in the age-interval $x - \frac{1}{2}$ to $x + \frac{1}{2}$. In this notation, Hamilton (1966) found the following equations:

$$\frac{dm}{d \log p_a} = \frac{\sum_{x=a+1}^{8} \lambda^{-x} l_x F_x}{\sum_{x=1}^{\infty} x \lambda^{-x} l_x F_x} \tag{1}$$

and

$$\frac{dm}{dF_a} = \frac{\lambda^{-a} l_a}{\sum_{x=1}^{\infty} x \lambda^{-x} l_x F_x}, \tag{2}$$

his equations (8) and (25), respectively. As Hamilton pointed out, the denominators are effectively generation-length measures, being the mean age of giving birth associated with the stable age distribution. In the case of Eqn (1), the numerator evidently declines as a increases, because fewer terms are included in the sum. In the case of Eqn (2), if m is non-negative, then the numerator decreases with a, because l_a and λ^{-a} are both decreasing. But if m is negative and of sufficiently large magnitude, dm/dF_a may increase with a for some values. Nonetheless, this applies chiefly to populations declining toward extinction. If the population is to survive, this pattern of selection must change. Thus, it is generally the case that the equations show that the evolutionary importance of modifications to survival probabilities and fecundities declines with the age of modification, given that the Malthusian parameter is in fact the focus of natural selection.

Charlesworth's (e.g. 1970, 1980) contribution was to analyse the conditions required to meet the assumptions which, in turn, underlay Hamilton's (1966) analysis. Charlesworth's work has shown that the usual 'reasonable' conditions of one locus and two alleles or loose linkage between loci and weak selection give rise to selection processes which depend primarily on the Malthusian parameter, providing the population size is sufficiently large and the environment is sufficiently stable to allow achievement of a stable age distribution.

The net upshot of all this work is a sufficient explanation of senescence among the soma of any age-structured population. It implies that senescence is readily moulded by natural selection given suitable genetic variation in life-history attributes. In particular, an externally imposed increase in survivorship will foster the evolution of postponed senescence, with a decrease in survivorship having the converse effect. This corollary is general to all the particular subsidiary population genetic mechanisms of senescence and provides the best focus for tests of the overall evolutionary analysis.

Tests of the general evolutionary theory

Assessing the applicability of a body of formal theory can be done in one of two ways. Firstly, its constituent assumptions may be tested. Secondly, its major corollaries may be tested. If a theory is sufficiently well-developed analytically, the first avenue of empirical examination is preferable, in that theoretical predictions can always be spuriously corroborated, while they are of necessity true if all the assumptions used to derive them deductively are correct. The problem with this approach is that formal analysis usually focusses on cases which have been simplified in order to make the mathematics tractable. Thus, in population genetics, it is a moot point whether or not the typical single diallelic-locus analysis is too simplified. Everyone will admit that it is a gross abstraction from the full complexity of actual genetic systems. But this does not necessarily imply that it is a vitiating simplification in all, or even most, cases.

If a structurally indispensable assumption of a theory is false, then the theory can be shelved. In the present case, the all-important assumption is that genetic variability does in fact affect life-history characters like survivorship and fecundity. A reservation can in fairness be allowed to those who might reasonably wish to exclude lethality or sterility alleles from consideration, on the grounds that they are unlikely to have much relevance to the evolution of life history in wild populations.

Excluding such alleles from consideration, the evidence can be sorted out into three bodies. Firstly, there are the long-standing animal and plant breeding data, which often indicate modest heritabilities for characters such as litter size and egg production (Falconer, 1981, p. 150). Secondly, there is a miscellany of species, chiefly insects, which have been investigated using similar methods, but with more focus on evolutionarily relevant attributes, such as early fecundity and survival to adulthood. These studies also provide evidence for appreciable genetic variability of the required kind (e.g. Istock, Zisfein & Zimmer, 1976; Dingle, Brown & Hegmann, 1977; Derr, 1980). Thirdly, there are studies of *Drosophila* species, frequently covering longevity as well as early life-history attributes, in a long-standing tradition dating back over 60 years (e.g. Pearl, 1922). These studies have shown that (i) inbred lines differ in longevity and often exhibit hybrid vigour for the character (Pearl, 1922; Gowen & Johnson, 1946; Clarke & Maynard Smith, 1955; Giesel, 1979; Giesel & Zettler, 1980) and (ii) selectable quantitative genetic variability for longevity and late fecundity may be present *within* outbred populations (Maynard Smith, 1959*b*; Wattiaux, 1968*a,b*; Rose & Charlesworth, 1981*a,b*).

This pattern of detailed information in *Drosophila* plus a scattering of data for other species is typical in biology. There is no reason to conclude that genetic variability of a similar kind does not arise in other genera, though of course there is no necessity for this inference. In any case, the central assumption of suitable genetic variability has been corroborated. Naturally, this provides no guarantee that such genetic variability will always be present, particularly in systematically inbred lines (nor need it be so for the theory to apply).

Given that the central assumption of the theory is met in appropriate circumstances, empirical evaluation of the general evolutionary theory of senescence turns on the question of whether or not senescence changes in the predicted fashion as externally imposed survivorship and reproductive regimes change. In experiments in which late reproductive opportunities are denied and early reproduction is favoured, it has been found that longevity is indeed reduced (Sokal, 1970; Mertz, 1975), although not all the adduced patterns were statistically significant. In experiments where *early* reproductive opportunities are denied, it is found that senescence is postponed (Wattiaux, 1968*a,b*; Rose & Charlesworth, 1980, 1981*b*; Taylor & Condra, 1980). Thus, the central corollary of the theory has been corroborated in a number of independent laboratory studies.

Even more noteworthy is the evidence of similar processes occurring under more natural conditions. Comparisons of populations of *Poa annua*, the so-called annual meadow grass, show that plants from transient habitats, where survivorship to later ages is reduced, exhibit accelerated senescence compared with plants from permanent pastures when grown under the same conditions (Law, Bradshaw & Putwain, 1977). This is as good a field corroboration as we can expect for a theory of this kind.

There are those who have sought to cast doubt on these results, emphasizing difficulties in the interpretation of the data from experiments of this kind (Lints, 1978; Lints & Hoste, 1974, 1977). But even the experiments that such authors bring forward have results in keeping with the general evolutionary theory of senescence, such as a relative postponement in peak fertility among lines reproduced at relatively later ages (Lints & Hoste, 1977).

Overall, then, the general evolutionary theory of senescence first explicitly propounded by Medawar (1946, 1952) seems as well-developed mathematically and as well-corroborated empirically as could be reasonably expected, given the extremely small number of people that have pursued the neo-Darwinian approach to the study of senescence.

Tests of particular population genetic mechanisms

As discussed above, of the three population genetic mechanisms for senescence put forward by Haldane (1941) and Medawar (1946, 1952), two are entirely reasonable *a priori*: mutation accumulation and antagonistic pleiotropy. These two mechanisms were specially emphasized by Edney & Gill (1968) and Williams (1957), respectively. Both are fully compatible with both the general evolutionary theory of senescence and the evidence which corroborates the general theory. Though these theories are not mutually incompatible, their applicability to particular populations remains an empirical question. Either or both may apply, depending on the pattern of gene action.

The mutation accumulation theory is based on the consequences of the decline with age in the force of natural selection for the evolutionary fate of genes with effects of one direction confined to one age-class. For example, this theory assumes that there are alleles with solely deleterious effects which are confined to late ages. If such alleles arise at all, they would be largely or entirely free of natural selection acting to reduce their frequency. Accordingly, all other things being equal, there should be an increase with adult age in the genetic variability of age-specific life-history characters. This has been tested only for 24-hour fecundity in *Drosophila melanogaster* (Rose & Charlesworth, 1980, 1981*a*). Even after an upward correction favouring the hypothesis, to compensate for the possibility of proportionate gene action, there was no evidence for an increase with age in the genetic variance for daily fecundity in the *Drosophila* population. This is an ostensible falsification of the mutation accumulation theory, notably on the first attempt to test it.

The antagonistic pleiotropy hypothesis leads naturally to the reproductive effort theory (Williams, 1966*a*,*b*), and then on to optimal life-history theory, once something of a growth industry for applied mathematics in population biology (Stearns, 1976; Charlesworth, 1980, pp. 231–51). Thus, there is somewhat more evidence probing the hypothesis that there are genes which enhance early life-history characters at the expense of later life-characters. It should be noted that phenotypic correlations within outbred populations are not relevant. They could be due to physiological interdependence which is strictly environmental. Suitable corroborative evidence has come from three directions: (i) negative correlations between the life-history character means of clones (Snell & King, 1977) and genetically distinct, somewhat inbred, populations grown under standardized conditions (Gowen & Johnson, 1946); (ii) negative additive genetic correlations, or their equivalent, between early and late life-history characters within outbred populations (Law, 1979; Doyle & Hunte, 1981; Rose & Charlesworth, 1981*a*); and (iii) negative correlations in selection response between early and late life-history characters (Wattiaux, 1968*a*; Law *et al.*, 1977; Rose and Charlesworth, 1981*b*).

Giesel (e.g. 1979) has argued against the antagonistic pleiotropy hypotheses on the grounds that his heavily inbred *Drosophila melanogaster* lines exhibited positive genetic correlations in early and late life-history characters. However, as is now well-established in the *Drosophila* literature (cf. Simmons, Preston & Engels, 1980), such positive correlation is expected for low-fitness allelic effects, as arise with new mutations and inbreeding, with negative genetic correlations among segregating high-fitness alleles. Evidently, some alleles will have predominantly positive effects on fitness-components at all ages, and thus approach fixation. It has been shown that artificially inbreeding *Drosophila* populations can produce predominantly positive genetic correlations among life-history characters, even when the ancestral outbred population does not exhibit such predominantly positive correlations (Rose, 1984). Thus Giesel's arguments are not supported by his experiments, because inbreeding produces artifactual positive genetic correlations.

The detected genetic variability for life-history characters is explicable in terms of antagonistic pleiotropy. Protected polymorphism is readily achieved with antagonistic pleiotropy in a variety of population genetic models (Rose, 1982; Rose, 1984). Moreover, the maintained allelic variability gives rise to negative genetic correlations between life-history characters like those found in the studies cited above. To summarise, a small but significant body of evidence corroborates the antagonistic pleiotropy mechanism for the evolution of senescence. As yet, no corroborative evidence has been found for the mutation accumulation mechanism, and there is one ostensible falsification. On the whole, then, the only population genetic mechanism for senescence which might be completely general is the antagonistic pleiotropy evolutionary mechanism.

Prospects for an evolutionarily-based gerontology

As discussed at the outset, physiological gerontology had no success in its search for a single, universal, physiological mechanism of senescence. As Williams (1957) and Maynard Smith (e.g. 1966) have pointed out, this failure is to be expected if one adopts an evolutionary view of senescence. If one physiological mechanism of senescence proceeded much more rapidly than any other, genetic changes which had no deleterious effect other than accelerating the slower physiological mechanisms of senescence would be relatively unopposed by natural selection, if not fostered because of early beneficial effects. Thus, evolution will tend to produce a fair degree of synchrony among diverse senescent physiological processes.

Among evolutionary biologists, Maynard Smith has spent the most time working on gerontological problems. Thus, Maynard Smith frequently had to play the role of spoiler for gerontologists who had become enthusiastic devotees of a particular, universal, physiological theory of senescence. If the past state of affairs were to continue, there would always be a need for a small legion of biologists willing to perform experiments to refute theories concerning putative, universal physiological mechanisms of senescence once their proponents have achieved sufficient nuisance value.

But there are reasons to anticipate that the relationship between the evolutionary and physiological gerontology research traditions will soon change. Firstly, some gerontologists (e.g. Cutler, 1978) are starting to take the *general* evolutionary aspects of senescence seriously, offering the prospect that many of them will eventually accept an overall evolutionary interpretation. Secondly, a still smaller number of gerontologists, the most notable of whom is Kirkwood (e.g. 1977, 1981), are applying cogent evolutionary arguments to *specific* gerontological problems. Thirdly, in my own laboratory, I have been analysing the physiological basis for the antagonistic pleiotropy which appears to underly senescence in a *Drosophila melanogaster* population. To give one example of the sort of findings involved, females from populations which exhibit postponed senescence, as a result of stock maintenance exclusively employing eggs laid by old females, have ovaries which average about *half* the weight of ovaries from females sampled from the appropriate control populations.

This account began by describing the initial profitless efforts to understand senescence made by two estranged research traditions. Perhaps I may be excused if in conclusion I express the hope that these two research traditions will soon ally to achieve substantial success in the unravelling of this intriguing biological problem.

References

Burnet, F. M. (1976). *Immunology, Aging, and Cancer*. San Francisco: W. H. Freeman.

Charlesworth, B. (1970). Selection in populations with overlapping generations. I. The use of Malthusian parameters in population genetics. *Theoretical Population Biology*, **1**, 352–70.

Charlesworth, B. (1980). *Evolution in Age-Structured Populations.* Cambridge University Press.

Clarke, J. M. & Maynard Smith, J. (1955). The genetics and cytology of *Drosophila subobscura.* XI. Hybrid vigour and longevity. *Journal of Genetics*, **53**, 172–80.

Comfort, A. (1979). *The Biology of Senescence*, 3rd edn. Edinburgh: Churchill Livingstone.

Cutler, R. G. (1978). Evolutionary biology of senescence. In *The Biology of Aging*, ed. J. A. Behnke, C. E. Finch & G. B. Moment, pp. 311–60. New York: Plenum Press.

Derr, J. A. (1980). The nature of variation in life history characters of *Dysdercus bimaculatus* (Heteroptera: Pyrrhocoridae), a colonizing species. *Evolution*, **34**, 548–57.

Dingle, H., Brown, C. K. & Hegmann, J. P. (1977). The nature of genetic variance influencing photoperiodic diapause in a migrant insect, *Oncopeltus fasciatus. American Naturalist*, **111**, 1047–59.

Doyle, R. W. & Hunte, W. (1981). Demography of an estaurine amphipod (*Gammarus lawrencianus*) experimentally selected for high 'r': a model of the genetic effects of environmental change. *Canadian Journal of Fisheries and Aquatic Science*, **38**, 1120–7.

Edney, E. B. & Gill, R. W. (1968). Evolution of senescence and specific longevity. *Nature*, **220**, 281–2.

Falconer, D. S. (1981). *Introduction to Quantitative Genetics*, 2nd edn. London: Longman.

Fisher, R. A. (1930). *The Genetical Theory of Natural Selection.* Oxford: Clarendon Press.

Giesel, J. T. (1979). Genetic co-variation of survivorship and other fitness indices in *Drosophila melanogaster. Experimental Gerontology*, **14**, 323–8.

Giesel, J. T. & Zettler, E. E. (1980). Genetic correlations of life-historical parameters and certain fitness indices in *Drosophila melanogaster*: r_m, r_s, diet breadth. *Oecologia*, **47**, 299–302.

Gowen, J. W. & Johnson, L. E. (1946). On the mechanism of heterosis. I. Metabolic capacity of different races of *Drosophila melanogaster* for egg production. *American Naturalist*, **80**, 149–79.

Haldane, J. B. S. (1941). *New Paths in Genetics.* London: Allen & Unwin.

Hamilton, W. D. (1966). The moulding of senescence by natural selection. *Journal of Theoretical Biology*, **12**, 12–45.

Istock, C. A., Zisfein, J. & Zimmer, H. (1976). Ecology and evolution of the pitcher-plant mosquito. 2. The substructure of fitness. *Evolution*, **30**, 535–47.

Kirkwood, T. B. L. (1977). Evolution of ageing. *Nature*, **270**, 301–4.

Kirkwood, T. B. L. (1981). Repair and its evolution: survival versus reproduction. In *Physiological Ecology: An Evolutionary Approach to Resource Use*, ed. C. R. Townsend & P. Calow, pp. 165–89. Oxford: Blackwell Scientific Publications.

Kirkwood, T. B. L. & Cremer, T. (1982). Cytogerontology since 1881: a reappraisal of August Weismann and a review of modern progress. *Human Genetics*, **60**, 101–21.

Lamb, M. J. (1965). The effects of X-irradiation on the longevity of triploid and diploid female *Drorophila melanogaster. Experimental Gerontology*, **1**, 181–7.

Lamb, M. J. & Maynard Smith, J. (1964). Radiation and ageing in insects. *Experimental Gerontology*, **1**, 11–20.

Law, R. (1979). The cost of reproduction in annual meadow grass. *American Naturalist*, **113**, 3–16.

Law, R., Bradshaw, A. D. & Putwain, P. D. (1977). Life-history variation in *Poa annua*. *Evolution*, **31**, 233–46.

Lints, F. A. (1978). *Genetics and Ageing*. Basel: S. Karger.

Lints, F. A. & Hoste, C. (1974). The Lansing effect revisited. I. Lifespan. *Experimental Gerontology*, **9**, 51–69.

Lints, F. A. & Hoste, C. (1977). The Lansing effect revisited. II. Cumulative and spontaneously reversible parental age effects on fecundity in *Drorophila melanogaster*. *Evolution*, **31**, 387–404.

Maynard Smith, J. (1959*a*). A theory of ageing. *Nature*, **184**, 956–8.

Maynard Smith, J. (1959*b*). Sex-limited inheritance of longevity in *Drosophila subobscura*. *Journal of Genetics*, **56**, 1–9.

Maynard Smith, J. (1966). Theories of aging. In *Topics in the Biology of Aging*, ed. P. L. Krohn, pp. 1–35. New York: Interscience.

Medawar, P. B. (1946). Old age and natural death. *Modern Quarterly*, **1**, 30–56.

Medawar, P. B. (1952). *An Unsolved Problem in Biology*. London: H. K. Lewis.

Mertz, D. B. (1975). Senescent decline in flour beetles selected for early adult fitness. *Physiological Zoölogy*, **48**, 1–23.

Orgel, L. E. (1963). The maintenance of the accuracy of protein synthesis and its relevance to aging. *Proceedings of the National Academy of Sciences, USA*, **49**, 517–21.

Orgel, L. E. (1973). Ageing of clones of mammalian cells. *Nature*, **243**, 441–5.

Pearl, R. (1922). *The Biology of Death*. Philadelphia: J. B. Lippincott.

Rose, M. R. (1982). Antagonistic pleiotropy, dominance, and genetic variation. *Heredity*, **48**, 63–78.

Rose, M. R. (1984). Genetic covariation in *Drosophila* life history: untangling the data. *American Naturalist*, **123**, 565–9.

Rose, M. & Charlesworth, B. (1980). A test of evolutionary theories of senescence. *Nature*, **287**, 141–2.

Rose, M. R. & Charlesworth, B. (1981*a*). Genetics of life-history in *Drosophila melanogaster*. I. Sib analysis of adult females. *Genetics*, **97**, 173–86.

Rose, M. R. & Charlesworth, B. (1981*b*). Genetics of life-history in *Drosophila melanogaster*. II. Exploratory selection experiments. *Genetics*, **97**, 187–96.

Simmons, M. J., Preston, C. R. & Engels, W. R. (1980). Pleiotropic effects on fitness of mutations affecting viability in *Drosophila melanogaster*. *Genetics*, **94**, 467–75.

Snell, T. W. & King, C. E. (1977). Lifespan and fecundity patterns in rotifers: the cost of reproduction. *Evolution*, **31**, 882–90.

Sokal, R. R. (1970). Senescence and genetic load: evidence from *Tribolium*. *Science*, **167**, 1733–4.

Stearns, S. C. (1976). Life-history tactics: a review of the ideas. *Quarterly Review of Biology*, **51**, 3–47.

Szilard, L. (1959). On the nature of the aging process. *Proceedings of the National Academy of Sciences, USA*, **45**, 30–45.

Taylor, C. E. & Condra, C. (1980). *r*- and *K*-selection in *Drosophila pseudoobscura*. *Evolution*, **34**, 1183–93.

Wattiaux, J. M. (1968*a*). Cumulative parental age effects in *Drosophila subobscura*. *Evolution*, **22**, 406–21.

Wattiaux, J. M. (1968*b*). Parental age effects in *Drosophila pseudoobscura*. *Experimental Gerontology*, **3**, 55–61.

Williams, G. C. (1957). Pleiotropy, natural selection, and the evolution of senescence. *Evolution*, **11**, 398–411.

Williams, G. C. (1966*a*). *Adaptation and Natural Selection*. Princeton: Princeton University Press.

Williams, G. C. (1966*b*). Natural selection, the costs of reproduction, and a refinement of Lack's principle. *American Naturalist*, **100**, 687–90.

Wilson, D. L. (1974). The programmed theory of aging. In *Theoretical Aspects of Aging*, ed. M. Rockstein, M. L. Sussman & J. Chesky, pp. 11–21. New York: Academic Press.

Wynne-Edwards, V. C. (1962). *Animal Dispersion in Relation to Social Behaviour*. Edinburgh: Oliver & Boyd.

10

Germination strategies

J. A. LEÓN

John Maynard Smith and Dan Cohen are two of the most creative biologists I have ever met. I can imagine what an interminable dance of ideas would adorn a meeting between them. For want of the opportunity to watch that feast, I substitute here a 'play': an encounter between their themes. I shall review a problem first posed by Cohen, and explore some additional aspects of it. I shall then formulate an analogous problem in a different temporal scale. I will conclude by applying within this framework procedures of evolutionary game theory developed by Maynard Smith. Some relevant factual material is discussed by Silvertown in Chapter 11.

Germination or dormancy in uncertain environments

Suppose there is uncertainty about the yield, Y(the probability of surviving to reproduce multiplied by the number of seeds produced) which a seed of an annual plant can obtain by germinating in any particular year. Should the seed germinate anyway, or would it do better by having the probability G of germination somehow related to the probabilities of the possible yields? The option in the latter case is of course to remain dormant, with probability D ($G+D=1$), facing risks in the soil where the probability of surviving (i.e. viability) is V_t. The recurrence equation connecting numbers of seeds in successive years is

$$N_{t+1}=N_t\{DV_t+GY_t\}. \tag{1}$$

The answer to the question in the previous paragraph, first given by Cohen (1966, 1968), is obtained by identifying the germination probability $\hat{G}$ which maximizes the expectation of the logarithm of λ_t, the interannual growth rate. The form of λ_t for this problem is given between braces in Eqn (1). It is assumed that λ_t is a stochastic process independently and identically distributed over years. Of course, if both V_t and Y_t vary randomly, λ_t is a function of a vector stochastic process.

A genetic justification for the optimization procedure can be envisaged as follows. If the plant considered is apomictic, and the germination fraction G is controlled genetically, $\hat{G}$ corresponds to the genotype with maximal value of $E(\ln \lambda_t)$. In sexual plants, $E(\ln \lambda_t)$ can again be used as an index of genotypic fitness, applying to this case the kind of arguments given in detail by Gillespie (1973) and Karlin & Lieberman (1974) for selection in random environments. Yet, with sexual reproduction, a monomorphic solution, $\hat{G}$, will not necessarily prevail. Despite this, a strategic analysis is still defensible, along the lines discussed by Charlesworth & León (1976), León (1976) and Maynard Smith (1978, 1982).

To obtain the optimal G the derivative of the expectation needed:

$$\frac{\mathrm{d}}{\mathrm{d}G} E(\ln \lambda_t) = E\left(\frac{Y_t - V_t}{V_t + G(Y_t - V_t)}\right). \tag{2}$$

This derivative is a monotonically decreasing function of G. Therefore, it can cross the horizontal G axis only once. If this intercept occurs at or before $G=0$, then $\hat{G}=0$ and the species is doomed to extinction. If this intercept, i.e. the unique zero of the derivative, occurs within the interval $0<G<1$, it corresponds to an 'internal' maximum of $E(\ln \lambda_t)$. If it is located at or beyond $G=1$, then $\hat{G}=1$ is the optimal strategy, with no seeds left dormant.

Conditions for any of these outcomes can be obtained as follows. The derivative is negative or zero at $G=0$ (and so $\hat{G}=0$), if and only if:

$$E\left\{\frac{Y_t}{V_t}\right\}_{G=0} \leqslant 1. \tag{3}$$

The derivative is zero or positive at $G=1$ (and, so $\hat{G}=1$) if and only if:

$$E\left\{\frac{V_t}{Y_t}\right\}_{G=1} \leqslant 1. \tag{4}$$

Reversing both inequalities gives the conditions for an 'internal' optimum ($0<\hat{G}<1$):

$$E\left\{\frac{Y_t}{V_t}\right\}_{G=0} > 1 \quad \text{and} \quad E\left\{\frac{V_t}{Y_t}\right\}_{G=1} > 1. \tag{5}$$

All these expectations are of course calculated using the joint probability distribution $P(Y_t, V_t)$. If Y_t and V_t are assumed to be independent, the inequations can be rearranged as:

$$E(Y_t) \leqslant H(V_t) \qquad (\hat{G}=0) \tag{6}$$

$$E(V_t) \leqslant H(Y_t) \qquad (\hat{G}=1) \tag{7}$$

$$\left.\begin{array}{l} E(Y_t) > H(V_t) \\ E(V_t) > H(Y_t) \end{array}\right\} \quad (0<\hat{G}<1) \tag{8}$$

where $H(Z)$, the harmonic mean of a random variable Z_t, is defined as $H(Z) \equiv 1/[E(1/Z_t)]$.

The harmonic mean can be approximated as follows:

$$H(Z) \approx E(Z)\left(1 - \frac{\sigma_Z^2}{E^2(Z)}\right). \tag{9}$$

Therefore, the inequations (6), (7) and (8) can be rewritten as:

$$E(V) - E(Y) \geqslant \frac{\sigma_V^2}{E(V)} \qquad (\hat{G}=0) \tag{10}$$

$$E(Y) - E(V) \geqslant \frac{\sigma_Y^2}{E(Y)} \qquad (\hat{G}=1) \tag{11}$$

$$\left.\begin{aligned} E(V) - E(Y) &< \frac{\sigma_V^2}{E(V)} \\ E(Y) - E(V) &< \frac{\sigma_Y^2}{E(Y)}. \end{aligned}\right\} \qquad (0 < \hat{G} < 1) \tag{12}$$

Since V is a probability, it is restricted to $0 \leqslant V \leqslant 1$. So $E(V)$ is less than one and inequality (10) can hold only in the extreme situation in which most of the environmental states that can supervene allow only a yield lower than one (or lower than the viability of the dormant seeds that year, according to (3)). Such a harshness precludes life altogether. Consistently high yields (i.e. high $E(Y)$ and low σ_Y^2) favour full germination. Some dormancy will evolve, though, when the yields show substantial uncertainty and a moderate expectation, whereas the fate of seeds in the soil is relatively hopeful.

A particularly interesting case, which can be fully solved, is that in which Y_t is either Y with probability p, or 0 with probability $q = 1 - p$, and V is a constant. The solution here is

$$\hat{G} = \frac{pY - V}{Y - V}. \tag{13}$$

Since pY can never be greater than Y, full germination ($G = 1$) is precluded. Some adaptive dormancy will ensue, insofar as $pY > V$. Of course, full dormancy and extinction supervene when $pY \leqslant V$. Notice that (13) can be rewritten as:

$$\hat{G} = p - q\,\frac{V}{Y - V}, \tag{14}$$

so that $\hat{G}$ becomes approximately equal to p when Y is large or when either q or V are close to zero.

These results of Cohen (1966, 1968) can be expressed in descriptive terms recently coined by León (1983) in another context. The situation in which $E(Y)$ is much higher than $E(V)$ and where σ_Y^2 is low may be called 'promising uncertainty': the gains when favourable conditions occur, and the frequency of these occurrences, are both high enough to overcompensate for the losses inflicted by adverse events (gains and losses being defined here with respect to a reference level, Y_0, the yield in the mean environment). The strategy adopted in this case, therefore, is 'risk incurrence': full germination ($\hat{G}=1$). When the gains are never too high, or are too infrequent, so that $E(Y)$ exceeds $E(V)$ in an amount lower than $\sigma_Y^2/E(Y)$, the situation may be named 'threatening uncertainty'. The pertinent strategy is then 'risk avoidance': some dormancy as a hedge against an insecure life. This is the case in the two-states environment (Y_t only 0 and Y) envisaged previously.

Comprehensive dispersal and the geometric vs. arithmetic mean dilemma

MacArthur (1972) reproached Cohen for using the geometric mean. He suggested that this would be valid in the case of just one special realization of the stochastic process – namely, that the different possible states of the environment occur with frequencies exactly equal to their probabilities. But if all the replicates are to be taken into account, the expectation (that is, the arithmetic mean) is obligatory. This is not correct. It would be right if the issue were *total* population growth, but what matters in natural selection is *differential* growth of genotypic or genic subpopulations. To see this, consider two genotypes (1 and 2) of an apomictic annual plant. After T generations the expected value of the ratio of their abundances is:

$$E\left(\frac{n_1(T)}{n_2(T)}\right)=\frac{n_1(0)}{n_2(0)}\,E\left(\prod_{t=1}^{T}\frac{\lambda_1(t)}{\lambda_2(t)}\right). \tag{15}$$

If the λs are independent and identically distributed:

$$E\left(\frac{n_1(T)}{n_2(T)}\right)=\frac{n_1(0)}{n_2(0)}\left[E\left(\frac{\lambda_1}{\lambda_2}\right)\right]^T. \tag{16}$$

But rewriting (λ_1/λ_2) as $\exp[\ln(\lambda_1/\lambda_2)]$ and using a Taylor expansion (on the assumption that $\lambda_1(t)$ and $\lambda_2(t)$ differ from each other by little) we get:

$$E\left(\frac{n_1(T)}{n_2(T)}\right)=\frac{n_1(0)}{n_2(0)}\,E[1+(\ln\lambda_1-\ln\lambda_2)+\ldots]. \tag{17}$$

This shows that the genotype with higher $E(\ln\lambda)$ is favoured by selection in a random environment.

Nevertheless, if, within any generation, both genotypes have indiscriminate access to all the possible states of the environment in proportion to their frequency of occurrence, the arithmetic mean should be used instead of the expectation of the logarithm. I call this situation comprehensive dispersal. It would apply, for instance, if there is a large variety of patches which can be reached by seeds coming from any of them, or if seeds are collected each year in a common pool and redistributed. The pertinence here of the expectation can be seen by replacing the λs in the foregoing argument with the $E(\lambda)$s. The equation equivalent to (16) becomes:

$$E\left(\frac{n_1(T)}{n_2(T)}\right)=\frac{n_1(0)}{n_2(0)}\left[E\left(\frac{E(\lambda_1)}{E(\lambda_2)}\right)\right]^T, \tag{18}$$

and, since $E(\lambda_1)$ and $E(\lambda_2)$ are numbers, this is equivalent to

$$E\left(\frac{n_1(T)}{n_2(T)}\right)=\frac{n_1(0)}{n_2(0)}\left(\frac{E(\lambda_1)}{E(\lambda_2)}\right)^T. \tag{19}$$

Therefore, under conditions of comprehensive dispersal, the genotype endowed with maximal $E(\lambda)$ will predominate.

Under conditions that are appropriate for the maximization of the arithmetic mean, the optimal solution is always $\hat{G}=1$, full germination. Venable & Lawlor (1980) present an important contribution to the geometric vs. arithmetic mean dilemma. Notice also the similarity with the coarse-grained vs. fine-grained distinction formulated by Levins (1968).

Predictive and innate germination

Cohen (1967) also proposed a model in which seeds are able to adjust their germination to cues which are indicative of future conditions. This is formalized such that, at the time of seed germination, the environment can adopt any of a set of states which can serve as signals, S. These are presumably correlated with the states prevalent at the time of plant growth and eventual reproduction, which will determine the yields, Y. Also the viabilities of dormant seeds are correlated with the signals. It is further assumed that the germination fraction G varies in response to the signals, so that a certain G_S corresponds to each level of the variable S.

The expectation of the logarithm is in this model:

$$E(\ln \lambda)=\sum_S \sum_V \sum_Y P(S, V, Y) \ln [V+G(Y-V)], \tag{20}$$

and, since $P(S, V, Y)=P(S)P(V, Y \mid S)$, we have:

$$E(\ln \lambda)=\sum_S P(S) \sum_V \sum_Y P(V, Y \mid S) \ln [V+G_S(Y-V)], \tag{21}$$

or equivalently:

$$E(\ln \lambda)=\sum_S P(S)E\{\ln [V+G_S(Y-V)] \mid S\}, \tag{22}$$

where the vertical bar serves to indicate conditional probabilities and a conditional expectation, as is usual in probability theory.

The partial derivative with respect to any G_S is:

$$\frac{\partial E(\ln \lambda)}{\partial G_S}=P(S)E\left(\frac{Y-V}{V+G_S(Y-V)}\middle| S\right); \tag{23}$$

that is:

$$\frac{\partial E(\ln \lambda)}{\partial G_S} = P(S) \sum_V \sum_Y P(V, Y \mid S) \left(\frac{V - V}{V + G_S(Y - V)} \right). \tag{24}$$

The optimal response $\hat{G}_S$ to the signal S is determined by the location of the zeros of the corresponding partial derivative (24), in a way strictly parallel to what was extensively discussed after Eqn (2). Clearly, optimizing the response to any particular signal is independent of the response to all the other signals. Therefore, for each optimization (i.e. for each signal) we get conditions analogous to those expressed by the sets of equations going from (3)–(4)–(5) to (10)–(11)–(12). The difference, of course, is that the expectations and harmonic means are now replaced by conditional expectations and harmonic means (Cohen, 1967).

As an illustrative – though drastically simplified – example, consider two signals, interpreted as indicating a good (g) or a bad (b) environment for growth and reproduction. The actual possible outcomes are yields Y or 0. Viability in the soil (V) is constant. The joint probability distribution is specified by the matrix

$$\begin{pmatrix} P(\mathrm{g}, Y) & P(\mathrm{g}, 0) \\ P(\mathrm{b}, Y) & P(\mathrm{b}, 0) \end{pmatrix}. \tag{25}$$

For the signals to have some predictive value it is required that

$$P(\mathrm{g}, Y) > P(\mathrm{g}, 0)$$
$$P(\mathrm{b}, Y) < P(\mathrm{b}, 0). \tag{26}$$

The optimal germination fractions, matched to the occurrence of the two signals, are:

$$\hat{G}_\mathrm{g} = \frac{P(Y \mid \mathrm{g})Y - V}{Y - V} \tag{27}$$

$$\hat{G}_\mathrm{b} = \frac{P(Y \mid \mathrm{b})Y - V}{Y - V}, \tag{28}$$

where the conditional probabilities are:

$$P(Y \mid \mathrm{g}) = \frac{P(\mathrm{g}, Y)}{P(\mathrm{g})} = \frac{P(\mathrm{g}, Y)}{P(\mathrm{g}, Y) + P(\mathrm{g}, 0)} \tag{29}$$

$$P(Y \mid \mathrm{b}) = \frac{P(\mathrm{b}, Y)}{P(\mathrm{b})} = \frac{P(\mathrm{b}, Y)}{P(\mathrm{b}, Y) + P(\mathrm{b}, 0)}. \tag{30}$$

Notice that if both signals are reliable predictors – so that $P(Y \mid \mathrm{g})$ is high and $P(Y \mid \mathrm{b})$ low – and the yield Y is moderate, the policy obtained will be a switching between full germination in years looking good ($\hat{G}_\mathrm{g} = 1$) and no germination when bad prospects show up ($\hat{G}_\mathrm{b} = 0$).

Venable & Lawlor (1980) envisaged a model analogous to this one, but with the optimization being conditional on the good signal, and with the announcement of possible bad times always switching on full dormancy. They call their equivalent to $\hat{D}_g$ 'innate dormancy' after Harper (1977); $\hat{D}_b = 1$ is named 'predictive dormancy' (remember $G + D = 1$) and claimed to be similar to Harper's 'enforced dormancy'. But notice that: (i) The two strategies ($\hat{G}_g$ and $\hat{G}_b$) are 'innate-predictive' in the (two-signals–two outcomes) version of Cohen's model presented here; (ii) this model gives results equivalent to the Venable–Lawlor model when $P(Y \mid b)$ is small enough (for instance, as $P(b, 0)$ increases, for a given $P(0)$ $[= P(g, 0) + P(b, 0)]$, $P(g, 0)$ decreases and therefore $P(Y \mid g)$ goes upward, so increasing $\hat{G}_g$); (iii) the capacity to experience enforced dormancy has to evolve, and as such it is perhaps better explained as a case of our $\hat{G}_b$ – namely: $\hat{G}_b = 0$.

An important issue, mentioned but not developed theoretically by Cohen, is 'the cost of perceiving and processing external signals, and of the elaborate control systems necessary to carry out the response' (Cohen, 1967). One would expect that the amount of external information actually used will increase up to the point at which the extra gain derived from its use is balanced by the extra cost.

Competition and dormancy

'Of course in wholly homogeneous environments dormancy has no role' according to Harper (1977). The current ecological wisdom sees dormancy as escape in time induced by local uncertainty. Similarly, dispersal had been regarded as adaptive escape in space in response to uncertainty, until Hamilton & May (1977) showed that even in a constant saturated environment dispersal may be favoured. In fact, Gill (1978) suggested in a discussion of K-selection that both dormancy and dispersal could be escapes to avoid local density-dependent environmental deterioration, but he did not produce a precise model to back his hints. Here, a simple model will be presented to show that competition without uncertainty may also favour dormancy.

Consider again Eqn (1). The larger the number GN_t of seeds that germinate, the keener competition will be. This turns the yield Y into a decreasing function of GN_t.

$$N_{t+1} = \{V(1 + G) + Gy(GN_t)\}N_t. \tag{31}$$

To obtain the evolutionarily stable germination strategy, $\hat{G}$, we must look for a maximum of the density-dependent fitness $\lambda(N)$ – which is the quantity between braces in (31) – that simultaneously implies $\lambda(N) = 1$. That is, $\hat{G}$ is a strategy such that any other G, at the saturation density corresponding to $\hat{G}$, will have $\lambda(\hat{N}) < 1$. A graphical technique is used in Fig. 1 to locate $\hat{G}$.

Clearly, the density- and frequency-dependence produces a bending of the product GY, which otherwise would be a straight line. This effect brings about an optimal germination fraction $\hat{G}<1$, and therefore a dormancy fraction $\hat{D}>0$, where of course $\hat{D}=1-\hat{G}$.

Some additional insight can be obtained by treating λ as a function of the two variables G and D, and formulating the conditions for a constrained maximum. The Lagrangian function is: $L=VD+y(G)G+\mu(1-D-G)$, where μ is a Lagrangian multiplier. Making equal to zero the three partial derivatives $\partial L/\partial D$, $\partial L/\partial G$ and $\partial L/\partial \mu$ we get, besides the constraint $D+G=1$, the condition

$$\frac{\partial(VD)}{\partial D}=\frac{\partial(YG)}{\partial G}. \tag{32}$$

To see the meaning of this, notice that there are two ways of augmenting fitness, that is, of increasing the number of seeds passed on to the next generation. One is to germinate, grow and produce new seeds; the other is to remain as seed, risking mortality in the soil, until next year. Condition

Fig. 1. The point $\hat{G}$ makes the difference $Gy-V(G-1)$ maximal and equal to one.

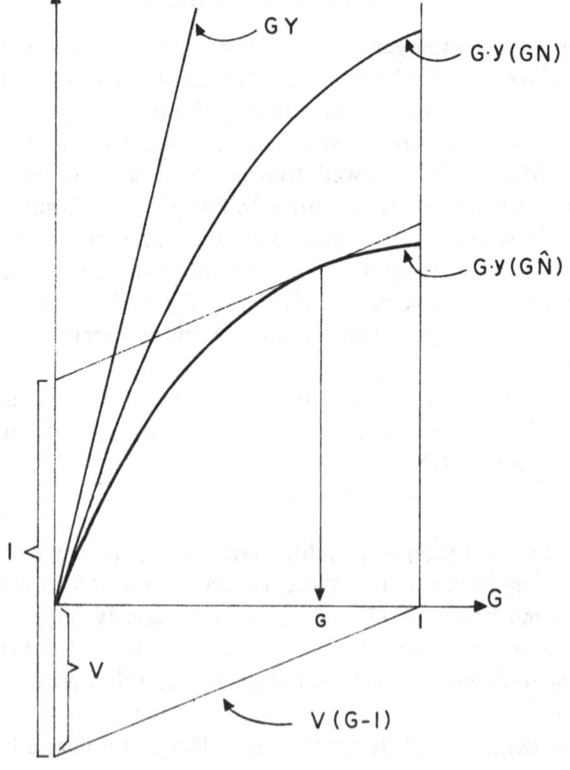

(32) says that it is optimal to equilibrate the marginal gains coming from both routes.

The timing of germination in uncertain seasonal environments

Suppose a seasonal environment in which the date of onset of the favourable season varies randomly over a certain interval. Should plants programme their seeds so as to choose some optimal day for germinating every year all at once (and then, which day?), or should they programme a probability distribution in the seeds, so that a certain fraction germinates each day over an interval? (and then, what fraction?, which interval?). These two possibilities will be called *synchronic* and *diachronic* strategies respectively, following a manuscript by León and Iwasa (unpublished results) on which the present account is based. Notice that the problem shows some resemblance to Cohen's formulation. But what was before postponement until next year is here delay until tomorrow, or the day after, and so on.

Consider an annual plant which spends the unfavourable season as a seed. The favourable season can start any time in an interval $[\alpha, \beta]$. In any particular year it begins at t ($\alpha \leqslant t \leqslant \beta$), and this occurs historically with frequency $f(t)$. This $f(t)$ specifies the stationary distribution of the pertinent stochastic process.

The function $g(x)$ is the probability that a seed will germinate on day x. If this day is before the onset, t, of the good season for that year, the seedling dies. If x occurs after t, the plant can achieve a yield $Y(x)$. This $Y(x)$ is assumed to be a decreasing function, since the earlier germination occurs the longer will be the opportunity for growth and reproduction. In a particular year, therefore, the expected fitness of a seed endowed with a germination policy $g(x)$ will be:

$$\lambda(t) = \int_t^\beta g(x) S(x)\, \mathrm{d}x. \tag{33}$$

But in a randomly changing environment some appropriate expectation must be used for the strategic analysis. As discussed in a previous section, the arithmetic mean is pertinent when there is comprehensive dispersal and the good season begins at different days in different available areas. Without dispersal, on the other hand, the expectation of the logarithm must be used. The two situations will be called here *fine-grained* and *coarse-grained* uncertainty, respectively.

(1) *Fine-grained uncertainty*. The function to be maximized, $E[\lambda(t)]$, can be written explicitly as:

$$\phi_{\mathrm{FG}}[g(\cdot)]=\int_{\alpha}^{\beta} f(t)\left[\int_{t}^{\beta} g(x)S(x)\,\mathrm{d}x\right]\mathrm{d}t, \tag{34}$$

under the constraint:

$$\int_{\alpha}^{\beta} g(t)\,\mathrm{d}t=1. \tag{35}$$

The maximization is easily accomplished. Interchanging the integration limits in (34) we get

$$\phi_{\mathrm{FG}}[g(\cdot)]=\int_{\alpha}^{\beta} g(t)S(t)\left[\int_{t}^{\alpha} f(x)\,\mathrm{d}x\right]\mathrm{d}t. \tag{36}$$

Substituting $g(t)=\delta(t_0-t)$, where δ is the Dirac delta generalized function, and remembering the following property of δ:

$$\int_{-\infty}^{\infty} \delta(t_0-t)F(t)\,\mathrm{d}t=F(t_0), \tag{37}$$

we obtain

$$\phi_{\mathrm{FG}}[\delta(t_0-t)]=S(t_0)\left[\int_{\alpha}^{t_0} f(x)\,\mathrm{d}x\right], \tag{38}$$

so that to maximize ϕ_{FG} we must choose a (t_0) to maximize the expression in the right-hand side of Eqn (38) and concentrate all the probability on that day (t_0).

Thus, the optimal strategy in situations with all-around dispersal is to *germinate synchronously* on a day which maximizes the product of the *reproductive success* gained by germinating that day, multiplied by the *chance that the favourable season begins before it.*

(2) *Coarse-grained uncertainty.* In this case, $E[\ln \lambda(t)]$ has to be maximized, and it is fully written now:

$$\phi_{\mathrm{CG}}[g(\cdot)]=\int_{\alpha}^{\beta} f(t)\left[\ln \int_{t}^{\beta} g(x)S(x)\,\mathrm{d}x\right]\mathrm{d}t. \tag{39}$$

The procedure to maximize this functional is rather involved. It is detailed in an appendix of the unpublished manuscript by León & Iwasa. The gist is to introduce a variation in the function $g(\cdot)$ and to obtain its effects on the Lagrangian functional formed by ϕ_{CG} plus the constraint (35). A sequence of steps including integrations by parts, a MacLaurin expansion, and so on, lead to an expression for the variation of the Lagrangian. Making this everywhere equal to zero is a necessary condition for a maximum. Then following some differentiations which, at last, lead to a solution $\hat{g}(t)$:

$$\hat{g}(t)=-\frac{1}{vS(t)}\frac{\mathrm{d}}{\mathrm{d}t}\left[\frac{f(t)}{\frac{\mathrm{d}}{\mathrm{d}t}\left(\frac{1}{S(t)}\right)}\right], \tag{40}$$

where v is a constant Lagrangian multiplier introduced for combining (35) with (39).

To interpret this solution (40), suppose the stationary distribution $f(t)$ is centred in a mode, M, and perhaps symmetric (the last requirement is not necessary). Then $f(t)$ is increasing ($df/dt>0$) on the interval (α, M) and decreasing ($df/dt<0$) on (M, β). As is proved by León & Iwasa (unpublished results), the solution $\hat{g}(t)$ can be characterized in reference to these intervals, as follows:

(i) When $S(t)$ is decreasing, concave $[S'(t)<0, S''(t)<0]$, linear or even slightly convex, the solution $\hat{g}(t)$ is positive only on (M, β).

(ii) When $S(t)$ is strongly convex $[S'(t)<0, S''(t)\gg 0]$, the solution $\hat{g}(t)$ is positive on a wider interval, including (M, β) and part of (α, M).

Here, again, we have 'risk avoidance' and 'risk incurrence', the two strategies described by León (1983) and mentioned before. The former is used when uncertainty is 'threatening', 'neutral' or slightly 'promising'. The latter is adopted when facing strongly 'promising' uncertainty. When $S(t)$ is concave the benefit of germinating early increases at a decreasing rate. It is therefore not worth risking germination before the most common day, M, of onset of the favourable season. The concavity of the logarithm induces the same effect when $S(t)$ is linear or slightly convex. But when $S(t)$ is strongly convex, this promises enough to promote taking the risk of early germination: it gives a gain that grows at an increasing rate.

Diachronic germination and competition between seedlings

We saw before that competition in a crowded environment can, by curbing yield, make it favourable to attempt dormancy as a partial route towards the next year. Similarly, competition between plantulae may cause a spread in germination time.

Consider a strictly seasonal environment in which the favourable season begins on day 0 and ends at τ. Reproduction begins on day T, being exogenously synchronized. A seed that germinates on day t has a fitness $\phi(t)$:

$$\phi(t)=\int_T^\tau l(t, y)b(t)\,\mathrm{d}y, \tag{41}$$

where $l(t, y)$ is the survival probability from germination to dat t, and $b(t)$ is the birth rate of an individual that germinated on day t.

If we assume constant mortalities of seeds, plantulae and reproductives – μ_0, μ_j, and μ_A respectively – survivorship is:

$$l(t, y)=\mathrm{e}^{-\mu_0 t}-\mu_j^{(T-t)}-\mu_A^{(\tau-y)}. \tag{42}$$

The birth rate is supposed to be proportional to the size W reached at maturation (on day T, when growth is assumed to cease),

$$b(t)=\beta W(t, T), \tag{43}$$

where β is a conversion factor. Here competition is introduced by making the growth rate proportional to resources available *per capita*, and to the size already attained at time X:

$$\frac{\mathrm{d}W(x)}{\mathrm{d}x}=KW(x)\frac{R(x)}{n(x)}. \tag{44}$$

Germination since the start of the season and mortality since then are the two forces whose balance determines the number of plantulae competing at day X, $n(x)$:

$$n(x)=\int_0^x g(s)N\,\mathrm{e}^{-\mu_0 S}\,\mathrm{e}^{-\mu_j(X-S)}\,\mathrm{d}s, \tag{45}$$

where $g(s)$ is the fraction of seeds germinating on day S, and N is the total number of seeds expected to germinate. Integrating (44) and substituting in (43) we obtain the birth rate:

$$b(t)=\beta\exp\left[\int_t^T KR(x)/n(x)\,\mathrm{d}x\right] \tag{46}$$

Eqns (42), (45) and (46) can be used to convert (41) into a detailed expression for fitness.

To look for the optimal strategy $\hat{g}(t)$ we resort now to game theory (Maynard Smith, 1982). The intensity of competition impinging on a plantula at any moment depends on how many competitors have already emerged, i.e. upon the germination strategy adopted by other competitors. This modulating effect of competition allows us to find a germination probability function $\hat{g}(t)$ which endows the *same* fitness to individuals germinating at any of a set of *different* days. Such a requirement is, by the Bishop–Cannings theorem (1978), a necessary condition for $\hat{g}(t)$ to be a 'mixed' evolutionarily stable strategy (ESS), that is, a 'time distributed' strategy which cannot be invaded by mutant strategies (Maynard Smith, 1974).

Thus impose the requirements:

$$\begin{aligned}\phi(t)&=\lambda \text{ (constant)} && \text{if } g(t)>0\\ \phi(t)&<\lambda && \text{if } g(t)=0,\end{aligned} \tag{47}$$

together with the constraints:

$$\phi(\hat{N})=1 \tag{48}$$

and

$$\int_0^\Omega g(t)\,\mathrm{d}t=1, \tag{49}$$

where Ω is the last germination day ($\Omega\leqslant T$). The constraint (48) is analogous to the one used in the section on competition and dormancy.

For applying condition (47) we need the derivative $\mathrm{d}\phi/\mathrm{d}t$, which is:

$$\frac{\mathrm{d}\phi}{\mathrm{d}t}=\int_T^\tau\left(\frac{\mathrm{d}l}{\mathrm{d}t}\cdot b+l\cdot\frac{\mathrm{d}b}{\mathrm{d}t}\right)\mathrm{d}y, \tag{50}$$

and, after some manipulation, using, of course, Eqns (41), (42), (45) and (46), we get:

$$\frac{\mathrm{d}\phi}{\mathrm{d}t}=\phi\left((\mu_j-\mu_o)-\beta K\,\frac{R(t)}{n(t)}\right). \tag{51}$$

Applying condition (47), we make the derivative (51) equal to zero and obtain:

$$n(t)=\frac{\beta K}{\mu_j-\mu_o}\,R(t). \tag{52}$$

Calculating $\mathrm{d}n/\mathrm{d}t$ from (45):

$$\frac{\mathrm{d}n}{\mathrm{d}t}=-\mu_j n(t)+g(t)N\,\mathrm{e}^{-\mu_o t}, \tag{53}$$

which, after multiplying all terms by $\exp(\mu_j t)$, can be rearranged as:

$$g(t)=N^{-1}\,\mathrm{e}^{-(\mu_j-\mu_o)t}\,\frac{\mathrm{d}}{\mathrm{d}t}\left[n(t)\,\mathrm{e}^{\mu_j t}\right]. \tag{54}$$

Using (52) we find the ESS:

$$\hat{g}(t)=\frac{C}{\hat{N}}\,\mathrm{e}^{-(\mu_j-\mu_o)t}\,\frac{\mathrm{d}}{\mathrm{d}t}\left[R(t)\,\mathrm{e}^{\mu_j t}\right], \tag{55}$$

where $C=\beta K/(\mu_j-\mu_o)$, which is positive if $\mu_j>\mu_o$.

This solution can also be expressed as:

$$\hat{g}(t)=\frac{C}{\hat{N}}\,\mathrm{e}^{\mu_o t}\left(\frac{\mathrm{d}R}{\mathrm{d}t}+\mu_j R(t)\right). \tag{56}$$

To interpret this result, just imagine that the resource supply remains constant throughout the favourable season. In such a case, $\hat{g}(t)$ grows exponentially:

$$\hat{g}(t)=\frac{C}{\hat{N}}\,\mu_j R\,\mathrm{e}^{\mu_o t}. \tag{57}$$

Thus the optimal germination fraction increases until day Ω, when the integral of $\hat{g}(t)$ becomes equal to one, as imposed by constraint (49). The higher the mortality of seeds, the sooner Ω arrives. Thus a high μ_o favours a concentration of germination at the onset of the season. A small difference $\mu_j-\mu_o$ and a high KR produce similar effects. The last result emphasizes that time-spreading is a strategy induced by competition, since relaxing the latter when the resource supply, R, is abundant promotes time-concentration.

Combining selective pressures

Two problems were discussed herein: (i) full vs. partial germination in any year and (ii) full germination a certain day vs. time-spread germination. In both cases, the separate effects of uncertainty without dispersal and with encompassive dispersal, and of competition, were considered. What has yet to be tackled is the combined consequences of these factors. Venable & Lawlor (1980) tried germination and moderate dispersal. Dan Cohen told me that he is working with Simon Levin on some models which include germination, dispersal and competition. The rest is silence, so far, but the door is presumably open. Invitations are not required to enter, I suppose.

References

Bishop, D. T. & Cannings, C. (1978). A generalized war of attrition. *Journal of Theoretical Biology*, **70**, 85–124.

Charlesworth, B. & León, J. A. (1976). The relation of reproductive effort to age. *American Naturalist*, **110**, 449–59.

Cohen, D. (1966). Optimizing reproduction in a randomly varying environment. *Journal of Theoretical Biology*, **12**, 119–29.

Cohen, D. (1967). Optimizing reproduction in a randomly varying environment when a correlation may exist between conditions at the time a choice has to be made and the subsequent outcome. *Journal of Theoretical Biology*, **16**, 1–14.

Cohen, D. (1968). A general model of optimal reproduction in a randomly varying environment. *Journal of Ecology*, **56**, 219–28.

Gill, D. E. (1978). On selection at high population density. *Ecology*, **59**, 1289–91.

Gillespie, J. H. (1973). Natural selection with varying selection coefficients – a haploid model. *Genetical Research, Cambridge*, **21**, 115–20.

Hamilton, W. D. & May, R. M. (1977). Dispersal in stable habitats. *Nature*, **269**, 578–81.

Harper, J. L. (1977). *The Population Biology of Plants.* New York: Academic Press.

Karlin, S. & Liberman, U. (1974). Random temporal variation in selection intensities: case of large population size. *Theoretical Population Biology*, **6**, 355–82.

León, J. A. (1976). Life histories as adaptive strategies. *Journal of Theoretical Biology*, **60**, 301–35.

León, J. A. (1983). Compensatory strategies of energy investment in uncertain environments. In *International Conference on Population Biology: Lecture Notes in Biomathematics*, ed. H. I. Freedman, **52**, 85–90. Berlin: Springer-Verlag.

Levins, R. (1968). *Evolution in Changing Environments.* Princeton: Princeton University Press.

MacArthur, R. H. (1972). *Geographical Ecology.* New York: Harper & Row.

Maynard Smith, J. (1974). The theory of games and the evolution of animal conflicts. *Journal of Theoretical Biology*, **47**, 209–21.

Maynard Smith, J. (1978). Optimization theory in evolution. *Annual Review of Ecology and Systematics*, **9**, 31–56.

Maynard Smith, J. (1982). *Evolution and the Theory of Games.* Cambridge University Press.

Venable, D. L. & Lawlor, L. (1980). Delayed germination and dispersal in desert annuals: escape in space and time. *Oecologia*, **46**, 272–82.

11

When plants play the field

J. SILVERTOWN

Although mixed evolutionarily stable strategies (ESSs) are of uncertain importance in animal behaviour (Maynard Smith, 1982) there are many ways in which plants simultaneously or successively display multiple growth strategies (Lloyd, 1984). In this chapter I review some examples of two of the commonest kinds of multiple strategies in plants: multiple germination responses and multiple dispersal types in seeds.

Seed germination

Germinating seeds very often show the type of behavioural variety which suggests that a mixed strategy of response is favoured by natural selection. In laboratory tests it is unusual to find that a single type of treatment produces germination in all the viable seeds of a sample (e.g. Grime *et al.*, 1981) and even amongst those seeds which do respond it is rare for them all to germinate simultaneously. Field studies of germination frequently report a staggered pattern of seedling emergence which in populations of annual plants is often separated into distinct cohorts that emerge at different seasons of the year.

In the vast majority of populations displaying this type of behaviour, we do not know whether the observed variety is the product of variety within the seed clutches of individual parents or is due to differences between the seed clutches of different parents or both. Such data as exist suggest that the former is the likeliest source of variable behaviour in most cases (Silvertown, 1983). A number of theoretical studies have shown how environments which vary in time and space may select for bet-hedging and differential seed dormancy amongst a plant's progeny (Cohen, 1966, 1967; MacArthur, 1972; Jain, 1979; Venable & Lawlor, 1980; Schoen & Lloyd, 1983; León, this volume).

Table 1. *Pay-off matrix for a contest between seedlings emerging in two cohorts 30 days apart*[a]

	Early	Late
Early	0.32	1
Late	0.01	0.45

Data from Abul-Fatih & Bazzaz, 1979.
[a] Pay-offs are values of relative fitness calculated from the life-time sum of $l_x \cdot m_x$.

Seedling emergence and plant fitness

Parallels exist between game theory treatments of animal conflict, in those cases where strategies are played against the field, and the situation for seeds germinating in a population with mixed environmental response. Established plants occupy space and thereby hold access to resources which are denied to other individuals in a very similar manner to the way in which territory holders defend space against intruders. Contests between territory holders or established plants and intruders are both strongly asymmetric. An example of a pay-off matrix for a contest between early- and late-germinating seedlings is given in Table 1. The pay-offs in the matrix are actual values of relative fitness (calculated from the lifetime sums of $l_x \cdot m_x$) for seedlings of an annual plant, *Ambrosia trifida*, emerging in two cohorts 30 days apart. Although these values are derived from an experiment in which the alternative 'strategies' of early and late germination are artificially produced by sowing the two cohorts at different times, the result typifies field situations in which differences in emergence time occur naturally. A difference of as little as one day in emergence time between seedling cohorts can have a significant effect on the fitness of the later-emerging seedlings (see Silvertown, 1982, for examples). *Other things being equal*, an early-germinating strategy is the ESS.

In such circumstances, where contests are strongly asymmetric, game theory predicts that the asymmetry should be used as a cue to settle the conflict (Maynard Smith & Price, 1973). In the present plant case the penalties for a conflict between early- and late-emerging seedlings fall exclusively on the later contestant. Thus, in many species, there is a mechanism which induces dormancy in a seed when it is exposed to the characteristic spectral quality of light filtered by a leaf canopy (Gorski, 1975; Silvertown, 1980). Leaf-canopy-induced dormancy is a seed's way of assessing an asymmetric contest and withdrawing from it. León (this volume) presents a model which predicts that seed dormancy may be favoured in competitive situations.

Germinating seeds have to contend with environmental hazards as well as playing against a field of alternative germination strategies. Both physical hazards and predation are often seasonal and place costs on an early-germination strategy which may alter or even reverse the balance of advantage which usually favours early germinators. In experimental populations in Alaska, early-germinating plants of *Thlapsi arvense* reached greater size than late-germinating ones and had more than twice the fecundity, but fitnesses were nearly equal due to high mortality amongst early germinators (Table 2).

In other cases, the variety of germination responses found among the seed progeny of individual plants (which we suppose is the main source of variety in seed populations) clearly creates large differences in fitness between siblings (Table 2). Since the germination behaviour of seeds is under the control of the maternal tissue in which the embryo is encased, variable behaviour in a clutch can be viewed as a parental strategy. This may conflict with the best strategy of the embryo (Westoby, 1981; Queller, 1983).

However, apparently diverse germination behaviour in a clutch might also be the result of some variance in germination response with low heritability. For example, seeds in enforced dormancy (Harper, 1977) may germinate whenever soil moisture and temperature rise above a threshold. If the speed of germination response is normally distributed, individuals in one tail of the distribution may germinate as soon as conditions permit, say the autumn, but those in the rest of the distribution might miss a brief period of suitable weather and actually not emerge till the spring. It would be wrong to interpret a bimodal distribution of seedling emergence as evidence of a mixed parental strategy if it is simply the consequence of a bimodal distribution of soil temperature (Popay & Roberts, 1970).

Mixed germination strategies

How then are we to distinguish a genuine mixed parental germination strategy and a true ESS from a situation of Hobson's choice in which parental fitness would be raised if all seeds germinated early but in which some simply fail to manage it? For some plants (e.g. those in Table 2) the resolution of this problem is just as difficult as it is for analogous cases in animal behaviour (Maynard Smith, 1982) but there is a category of plants which are unmistakably mixed strategists. These are the species with strong structural dimorphisms among the progeny of a single clutch. Structural seed dimorphisms are almost invariably linked to germination differences (but not vice versa) and/or to differences in their dispersal and longevity in

Table 2. *Relative fitness for short-lived plants (annuals or biennials) germinating in early (1) or late (2) cohorts in the same populations*

Species	Type of study population	Relative fitness (cohort 1: cohort 2)	Cohort 1			Cohort 2			Source
			l_x[a]	m_x[b]	$l_x \cdot m_x$	l_x[a]	m_x[b]	$l_x \cdot m_x$	
Ambrosia trifida	E[c]	1:0.01	0.22	320	70.4	0.13	5	0.65	Abul-Fatih & Bazzaz (1979)
Lactuca serriola	N[d]	0.91:1	0.10	1226	122.6	0.28	482	135	Marks & Prince (1981)
Papaver dubium	E	1:0.02	0.84	1206	1013	0.98	17	16.7	Arthur, Gale & Lawrence (1973)
Thlapsi arvense	E	1:0.99	0.50	11 220	5610	1	5565	5565	Klebesadel (1969)
Leavenworthia stylosa	N	1:0.27	0.38	45	17.1	0.78	6	4.7	Baskin & Baskin (1972)
Lobelia gattingeri	N	0.13:1	0.05	8.5[f]	0.43	0.36	9.1[f]	3.3	Baskin & Baskin (1979*a*)
Vulpia fasiculata	N	1:0.57	0.86	4.09	3.52	0.20	2	0.4	Watkinson (1981)
Pastinaca sativa	N	1:0.50	0.02	*[e]	*	0.01	*	*	Baskin & Baskin (1979*b*)
Senecio vulgaris	N	1:0.05	0.84	*	*	0.04	*	*	Putwain, Scott & Holliday (1982)

[a] l_x = Survivorship from seedling emergence to flowering.
[b] m_x = Fecundity (number seeds or f = number of fruits).
[c] E = Experimentally sown population.
[d] N = Natural population.
[e] *No fecundity data, fitness values are based on relative values of l_x only.

Table 3. *Number of European taxa in the genus* Crepis *with monomorphic or polymorphic achenes, divided by lifespan* ($p<0.001$, $G=12.93$, *G Test*)

	Lifespan	
	Annual or annual/biennial	Perennial or perennial/biennial
Number of achene types in a clutch		
1	5	15
2 or 3	6	0

Data from *Flora Europaea*, vol. IV (Tutin *et al.*, 1976).
[a] The editorial rules used in this compilation were: 1. All species in the same section of the genus and having the same syndrome were scored as a single case. 2. Species with different syndromes in the same section were scored independently. 3. Species with the same syndrome in different sections were scored independently. There are 71 species in 21 sections of *Crepis* in Europe.

the soil (Becker, 1913; Silvertown, 1983). Many examples are found in the Compositae and amongst annual and short-lived plants in particular.

I have tabulated the occurrence of structural seed polymorphism in the European members of the genus *Crepis* (Compositae) in Table 3. It clearly illustrates the occurrence of mixed parental strategies in short-lived plants as predicted by Cohen's (1967) theoretical treatment of bet-hedging. Structural polymorphism within seed clutches also occurs in many species in the genus *Leontodon* and in the tribe *Calenduleae* (both Compositae), but it does not correlate with lifespan in either of these taxa (Silvertown, unpublished data; Norlindh, 1943; Heyn, Dagan & Nachman, 1974). Correlations between polymorphism and lifespan do crop up in other plant families but these also contain exceptions. The constraints which operate on this kind of mixed strategy require investigation.

Seed dispersal

Plastic adjustments of investment in alternative strategies

The existence of clearly identifiable mixed strategies in plants and the fact that parental investment in each of the alternatives can be measured by counting the number of seeds of each type in a clutch, means that we can begin to ask questions about what determines the level of relative investment in each. Schoen & Lloyd (1984) have approached this problem with a model that assumes differential dispersal distances for two seed types. As clutch size increases, the seed type dispersed near the plant rapidly fills available space and encounters density-dependent competition from sibs. Thus the fitness gain by a parent which produces more near-dispersed morphs is a decreasing function of clutch size and the ESS is to bias the ratio in favour of the far-dispersed morph as seed production increases.

Plants producing a polymorphic clutch can be divided into two groups according to how they alter the ratio of seed types as reproductive expenditure (usually related to plant size) rises. In the first group the ratio of morph types remains constant and in the second it changes in favour of one particular morph.

Likely candidates for species in the constant-ratio group are plants such as *Cakile maritima* (Cruciferae) which produces a bi-locular fruit containing one small seed and one large one and the grass *Aegilops kotschyii* which bears a large and a small caryopsis with different germination behaviour in the same spikelet (Wurzburger & Lesham, 1967). It might be difficult for such plants to vary the ratio of morph types allometrically with the size (and reproductive expenditure) of the plant, even if this could raise parental fitness, because of close developmental association of the two seed types. On the other hand, there is at least one species which could easily alter its morph ratio as it grows larger but which has an elaborate mechanism that appears to conserve the ratio. *Aethionema saxatile* (Cruciferae) produces indehiscent one-seeded fruit (nucamenta) and dehiscent many-seeded fruit (silicules) in alternating groups along its branches as they grow (Puech, 1970). No neo-Darwinist could resist speculating that such disciplined behaviour raises parental fitness in some way! The questions of whether it does and, if so, how, are ripe for experiment.

Changes in morph ratio with clutch size have been studied in the annual *Hypochoeris glabra* (Compositae) which produces fruit (achenes) with dimorphic dispersal abilities but similar weight and germination responses. Small plants produce small clutches with a ratio of about 2:1 in favour of the more dispersable morph. In plants with a larger clutch, the number of poorly dispersed morphs rises only slightly and the ratio of achene types is about 3:1 in favour of the more dispersable morph (Baker & O'Dowd, 1982). This shift in morph ratio with clutch size accords well with Schoen & Lloyd's (1984) prediction that sib competition amongst poorly dispersed morphs should favour the far-dispersal strategy as clutch size increases.

The two-achene types produced by *H. glabra* develop from different flowers in the fruiting head (capitulum). The near-dispersed achenes develop around the circumference of the capitulum and far-dispersed ones from the disc. Dimorphic achenes with differential dispersal ability have evolved independently in a number of other Composite genera and in all of these show the same arrangement of dispersal type within the capitulum

Table 4. *Pay-off matrix for alternative parental strategies of seed production by Emex spinosa at low density*[a]

	Subterranean	Aerial
Subterranean	0.90	0.90
Aerial	0.02	1

Data from Weiss, 1980.
[a] Pay-offs are values of relative partial fitness calculated from total seed production (m_x) per plant.

(Zohary, 1950). The contrast in allometric relationships between achene number and clutch size for the two-achene types in *H. glabra*, which is responsible for the shift in the ratio of types, is the result of their different positions in the capitulum. A small increase in the radius of a capitulum allows only a small increase in its circumference and in the number of marginal achenes. However, the area of the capitulum disc and hence the number of disc achenes, increases with the square of the radius. Thus ratio adjustment with clutch size has an architectural basis. The similarity of structure between *H. glabra* and other composites with dimorphic fruit suggest that there are many independent examples of morph ratio adjustment in this family which all corroborate Schoen & Lloyd's (1984) hypothesis.

This model is also excellently corroborated by another, taxonomically very heterogeneous, group of species which vary morph ratio with clutch size. Plants which produce subterranean fruit and also aerial ones from separate flowers (behaviour known as amphicarpy) occur in at least nine different families but show remarkable similarities. Nearly all appear to be annual, have subterranean fruit larger than aerial ones and, where investigated, initiate their clutch by producing near-dispersed (subterranean) fruit and progressively change the ratio in favour of far-dispersed (aerial) fruit as clutch-size increases (Zeide, 1978; Loria & Noy-Meir, 1980; Cheplick & Quinn, 1982, 1983).

Thus far, plant behaviour agrees with theoretical prediction but it still remains to test the change in parental fitness as the morph ratio shifts with clutch size and seed density, and to establish whether an ESS is achieved. The only remotely relevant experiment published is for *Emex spinosa* (Polygonaceae) a Mediterranean amphicarp studied by Weiss (1980) in Australia, where it has become a weed. A replacement series competition experiment between seedlings from subterranean and aerial fruit demonstrated that subterranean fruit production is an ESS (Table 4). This experiment was conducted only at low density and so, as far as it goes, the observed result is consistent with theory. A complete test of Schoen & Lloyd's (1984) model would involve measurements of fitness at a range of densities and proportions of the two seed types.

Evolutionary changes in mixed strategies

Taxonomists have paid a great deal of attention to angiosperm fruit structure and diversity and have traced putative lines of evolutionary relationship between species with monomorphic and with polymorphic fruits or seeds. The studies reveal a recurring association between the evolution of the annual habit and mixed strategies in seeds. The 50 or so species in the genus *Aethionema* divide into a group of perennials, mostly with a single fruit type (silicules), and two groups of annual which are considered more recently evolved. Species in one annual group have two fruit types, dehiscent silicules and indehiscent nucamenta (e.g. *A. saxatile*, above); those in the other annual group have only nucamenta. The same evolutionary trend appears in the family Fumariaceae, where a perennial genus with monomorphic dehiscent fruit (*Corydalis*) is thought to be antecedent to annuals in the genus *Ceratocapnos* – which have both dehiscent and indehiscent fruit – and annuals in the genus *Fumaria* which have only indehiscent fruit. *Fumaria* is the most recently evolved of the three genera (Zohary & Fahn, 1950).

The evolutionary association of mixed seed strategies with the annual habit is predicted by the theory of bet-hedging (see above) but the loss of one strategy type in some annual species seems, at first sight, surprising. The annuals with monomorphic fruit have lost the dehiscent fruit type which aids the long-range dispersal of seeds. This would normally be expected to increase sib competition and hence to lower fitness compared with the annuals with polymorphic fruit from which they appear to have evolved. The evolution of a monomorphic, near-dispersal strategy could be favoured in conditions where sib competition is reduced. Ellner & Shmida (1981) have suggested that this is the case in deserts, where density-independent mortality is severe. A reasonable prediction, not yet tested, is that density-independent mortality is greater, and sib competition is less, in natural populations of both *Aethionema* with monomorphic fruit and *Fumaria* than in natural populations of their dimorphic relatives. Annuals with monomorphic fruit do occur in deserts (Ellner & Shimida, 1981). This does not contradict the expectation that desert annuals should hedge bets by a mixed-germination strategy since polymorphic seed behaviour may be selected independently of dispersal type.

Polymorphic fruit types (and by inference mixed-germination or seed-dispersal strategies) have evolved repeatedly, have been lost, and have evolved again in the Compositae. These changes have been traced in the taxonomy of the tribe Calenduleae by Norlindh (1943). The genus *Dimorphotheca* contains both annual and perennial species which all have two (sometimes more) fruit (achene) types within a head (capitulum). Different types of achenes are produced from female ray florets around the circumference of the capitulum and from hermaphrodite disc florets.

Two lines of development have occurred from this 'primitive' state. In the genus *Castalis*, with three perennial species, the female ray florets have become sterile and only one achene type is produced by the disc florets. In the four remaining genera of the tribe the disc florets are female-sterile and there are varying degrees of polymorphism in the achenes produced by ray florets. *Gibbaria* species are perennial and have only one achene type. There is no correlation between lifespan and the occurrence of achene polymorphism among either *Calendula* spp. or *Osteospermum* spp., which are variable for both these traits. The fifth genus produces drupaceous animal-dispersal fruits.

The Calenduleae illustrate the complex shifts in the occurrence of mixed strategies and in the ratio of types of strategy which have occurred during plant evolution. The causes behind some of these kinds of changes may yet be revealed as we learn more about what happens when plants with different strategies play the field.

References

Abul-Fatih, H. A. & Bazzaz, F. A. (1979). The biology of *Ambrosia trifida* L. II. Germination, emergence, growth and survival. *New Phytologist*, **83**, 817–27.

Arthur, A. E., Gale, J. S. & Lawrence, K. J. (1973). Variation in wild populations of *Papaver dubium*. VII. Germination time. *Heredity*, **30**, 189–97.

Baker, G. A. & O'Dowd, D. J. (1982). Effects of parent plant density on the production of achene type in the annual *Hypochoeris glabra*. *Journal of Ecology*, **70**, 201–15.

Becker, W. (1913). Uber die kemung verschiedenartiger fruchte und samen der selben spezies. *Beihefle zum Botanischen Centralblatt*, **29**, 21–143.

Baskin, J. M. & Baskin, C. C. (1972). Influence of germination date on survival and seed production in a natural population of *Leavenworthia stylosa*. *American Midland Naturalist*, **88**, 318–323.

Baskin, J. M. & Baskin, C. C. (1979*a*). The ecological life cycle of the cedar glade endemic *Lobelia gattingeri*. *Bulletin of the Torrey Botanical Club*, **106**, 176–81.

Baskin, J. M. & Baskin, C. C. (1979*b*). Studies on the autecology and population biology of the weedy monocarpic perennial, *Pastinaca sativa*. *Journal of Ecology*, **67**, 601–10.

Cheplick, G. P. & Quinn, J. A. (1982). *Amphicarpum purshii* and the pessimistic strategy in amphicarpic annuals with subterranean fruit. *Oecologia*, **52**, 327–32.

Cheplick, G. P. & Quinn, J. A. (1983). The shift in aerial/subterranean fruit ratio in *Amphicarpum purshii*: causes and significance. *Oecologia*, **57**, 374–9.

Cohen, D. (1966). Optimizing reproduction in a randomly varying environment. *Journal of Theoretical Biology*, **12**, 119–29.

Cohen, D. (1967). Optimizing reproduction in a randomly varying environment when a correlation may exist between the conditions at the time a choice has to be made and the subsequent outcome. *Journal of Theoretical Biology*, **16**, 1–14.

Ellner, S. & Shmida, A. (1981). Why are adaptations for long-range seed dispersal rare in desert plants? *Oecologia*, **51**, 133–44.

Gorski, T. (1975). Germination of seeds in the shadow of plants. *Physiologia Plantarum*, **34**, 342–6.

Grime, J. P., Mason, G., Curtis, A. V., Rodman, J., Band, S. R., Mowforth, M. A. G., Neal, A. M. & Shaw, S. (1981). A comparative study of germination characteristics in a local flora. *Journal of Ecology*, **69**, 1017–59.

Harper, J. L. (1977). *Population Biology of Plants*. London: Academic Press.

Heyn, C. C., Dagan, O. & Nachman, B. (1974). The annual *Calendula* species: taxonomy and relationships. *Israel Journal of Botany*, **23**, 169–201.

Jain, S. K. (1979). Adaptive strategies: polymorphism, plasticity and homeostasis. In *Topics in Plant Population Biology*, ed. O. T. Solbrig, S. Jain, G. B. Johnson & P. H. Raven, pp. 160–87. London: Macmillan.

Klebesadel, L. J. (1969). Life cycles of field pennycress in the subarctic as influenced by time of seed germination. *Weed Science*, **17**, 563–6.

Lloyd, D. G. (1983). Variation strategies of plants in heterogeneous environments. *Biological Journal of the Linnean Society*, **27**, 357–85.

Loria, M. & Noy-Meir, I. (1980). Dynamics of some annual populations in a desert loess plain. *Israel Journal of Botany*, **28**, 211–26.

MacArthur, R. H. (1972). *Geographical Ecology*. New York: Harper & Row.

Marks, M. & Prince, S. (1981). Influence of germination date on survival and fecundity in wild lettuce *Lactuca serriola*. *Oikos*, **36**, 326–30.

Maynard Smith, J. (1982). *Evolution and the Theory of Games*. Cambridge University Press.

Maynard Smith, J. & Price, G. R. (1973). The logic of animal conflict. *Nature*, **246**, 15–18.

Norlindh, T. (1943). *Studies in the Calenduleae*. Lund: Gleerup.

Popay, A. I. & Roberts, E. H. (1970). Ecology of *Capsella bursa pastoris* (L) Medk. and *Senecio vulgaris* L. in relation to germination behaviour. *Journal of Ecology*, **58**, 123–39.

Puech, S. (1970). Hétérocarpie rythmique dans une population cévénole d'*Aethionema saxatile* (L). R. Br. Reineignments appartés par les cultures experimentales. *Bulletin Societé Botanique de France*, **117**, 505–31.

Putwain, P. D., Scott, K. R. & Holliday, R. J. (1982). The nature of resistance to triazine herbicides: case histories of phenology and population studies. In *Herbicide Resistance in Plants*, ed. H. M. Lebaron & J. Gressel, pp. 99–115. Chichester: John Wiley & Sons.

Queller, D. C. (1983). Kin selection and conflict in seed maturation. *Journal of Theoretical Biology*, **100**, 153–72.

Schoen & Lloyd, D. G. (1984). The selection of cleistogamy and heteromorphic diaspores. *Oecologia* (in press).

Silvertown, J. W. (1980). Leaf-canopy induced seed dormancy in a grassland flora. *New Phytologist*, **85**, 109–18.

Silvertown, J. W. (1982). *Introduction to Plant Population Ecology*. London & New York: Longman.

Silvertown, J. W. (1984). Phenotypic variety in seed germination behaviour: the ontogeny and evolution of somatic polymorphism in seeds. *American Naturalist*, **124**, 1–16.

Tutin, T. G., Heywood, V. H., Burges, N. A., Moore, D. M., Valentine, D. H., Walters, S. M. & Webb, D. A. (1976). *Flora Europaea*, volume IV. Cambridge University Press.

Venable, D. L. & Lawlor, L. (1980). Delayed germination and dispersal in desert annuals: escape in space and time. *Oecologia*, **46**, 272–82.

Westoby, M. (1981). How diversified seed germination behaviour is selected. *American Naturalist*, **118**, 882–5.

Watkinson, A. R. (1981). The population ecology of winter annuals. In *The Biological Aspects of Rare Plant Conservation*, ed. H. Synge, pp. 253–64. Chichester & New York: Wiley.

Weiss, D. W. (1980). Germination reproduction and interference in the amphicarpic annual *Emex spinoza* (L.) Campd. *Oecologia*, **45**, 244–51.

Wurzburger, J. & Leshem, Y. (1967). Gibberellin and hull-controlled inhibition of germination in *Aegilops rotachyi* Boiss. *Israel Journal of Botany*, **16**, 181–6.

Zeide, B. (1978). Reproductive behaviour of plants in time. *American Naturalist*, **112**, 637–9.

Zohary, M. (1950). Evolutionary trends in the fruiting head of Compositae. *Evolution*, **4**, 103–9.

Zohary, M. & Fahn, A. (1950). On the heterocarpy of *Aethionema*. *Palestine Journal of Botany*, **5**, 28–31.

12

Homage to the null weasel

P. H. HARVEY AND K. RALLS

Introduction

Ecology flourished in North America during the 1960s. Largely under the influence of G. Evelyn Hutchinson and Robert H. MacArthur, the link between process and pattern became a dominant theme: patterns of species variation were used to identify processes and, in turn, processes were used to predict patterns. For example, one of the most influential ecological papers of this century is Hutchinson's (1959) 'Homage to Santa Rosalia' in which he claims that (i) when size-adjacent close competitors are found in sympatry, the size ratio of the larger to the smaller is about 1.3 in linear dimensions and (ii) this figure 'may tentatively be used as an indication of the kind of difference necessary to permit two species to co-occur in different niches but at the same level of a food web'. There followed two decades of mathematical models aimed at identifying processes likely to cause this ratio together with numerous examples of sympatric species pairs differing by the required amount. Hutchinson's ratio gained wide acceptance as an important ecological rule, as did character displacement, latitudinal diversity gradients, and particular forms of species incidence curves and species–area relationships.

More recently the euphoria has been dampened by gloom resulting from the use of 'null models' which ask whether such patterns exist and, if they do, whether we need to invoke biotic interactions to explain them. The answers are often surprisingly elusive. Returning to Hutchinson's ratio as an example: it seems that it has been variously interpreted as a modal, mean, optimum or minimum figure, and most fantastically as 'a value that some proper summation of differences along all *n* axes of niche space should attain' (see Roth, 1981). As the ratio became more widely accepted, investigators who noted it in their own work often inferred the action of competition and sometimes went on to claim this as confirmatory evidence of competition causing the ratio. Of course, if the ratio was not quite 1.3 (Hutchinson had allowed 1.1–1.4), it was often near enough. And if ratios were greater (they could hardly be less), this was often cited as evidence that competition was not important in a particular case. But what if we examine *randomly constructed* communities of animals belonging to the same ecological guild? How similar would size-adjacent species be? Would the distribution of ratios differ significantly from those found in real communities? And would they cluster around 1.3? Recent papers (Roth, 1981; Simberloff & Boecklen, 1981) have started to ask such questions and, although their analyses leave much to be desired (see Harvey *et al.*, 1983; Schoener, 1984), ecological faith in the ubiquity of Hutchinson's ratio has been shaken. We could equally well have chosen other 'ecological rules' whose efficacy has recently been challenged. While some phenomena are now more or less bereft of reliable examples (e.g. character displacement) and some await improved statistical treatment (e.g. size ratios), others have been corroborated (e.g. species–area relationships).

Ecological pattern can be both predicted and recognized as a correlate of environmental variation at a variety of levels (changes in morphology, physiology or behaviour both within individuals or among guilds and taxonomic groupings). It has become commonplace to relate species differences in behaviour and morphology to variation in ecology (Clutton-Brock & Harvey, 1984). However, data on intraspecific variation are less commonly available. This is unfortunate because many potentially confounding variables, such as life-history characteristics and physiology, are more tightly constrained within species than between them.

In this chapter, we use the condensed results of a survey across North America to test hypotheses that might explain intraspecific variation in size and sexual dimorphism of three species of congeneric weasels: the ermine *Mustela erminea*, the long-tailed weasel *Mustela frenata* and the least weasel *Mustela nivalis*. A more detailed discussion of these data and our statistical analyses of them will be presented elsewhere (Ralls & Harvey, 1985).

The three species comprise a 'hunting set' (Rosenzweig, 1966) and have similar ecology and morphology. They provide a rare opportunity for testing ecological hypotheses since, because they are so similar, most hypotheses about differences in size and possibly sexual dimorphism predict covariation among the species. In addition, because the species have overlapping but not identical ranges, we can test for the effects of interactions between the species on differences in body size by comparing areas of allopatry and sympatry between size-adjacent pairs of species. Finally, because climatic data are available from throughout the ranges of the three species, we can test for changes in size and dimorphism in response to climatic differences.

These three species have been reasonably well studied in the past and a variety of authors have interpreted their results in the light of contemporary ecological theory. We have the opportunity to test some of the predictions of their theories.

The data

Two of the three weasel species (*erminea* and *nivalis*) have holarctic distributions, while the third (*frenata*) is limited to the American continent. The distributions of the three species in North America are shown in Fig. 1. The three species have been trapped throughout their ranges and, although

Fig. 1. Distribution of *Mustela erminea* (vertical lines), *Mustela frenata* (horizontal lines) and *Mustela nivalis* (circles) in North America (after Hall, 1981).

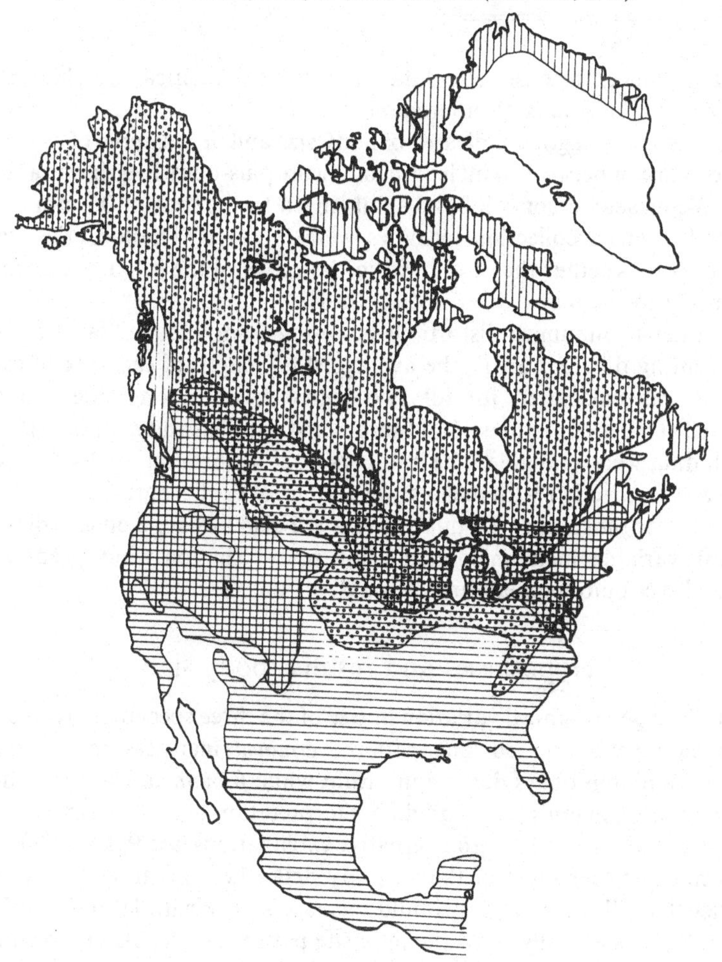

Table 1. *Descriptive statistics for skull length (in mm) of samples of weasels from North America*

Species	Sex	Mean	Variance	Coefficient of variation	Spearman's r (skull length and latitude)	Number of 2° squares
erminea	M	42.5	10.3	0.24	0.63	253
	F	36.8	6.3	0.17	0.69	181
frenata	M	49.0	5.2	0.11	0.04	183
	F	43.6	9.2	0.21	0.16	141
nivalis	M	32.0	2.9	0.09	0.11	50
	F	29.9	1.5	0.05	−0.38	43

we shall concentrate on our data from North America, we shall also incorporate the results of other surveys.

Skull length is a good indicator of body size and, in a sample of *erminea*, the correlation between skull length and head-plus-body length was above 0.97. We measured condylobasal skull length to the nearest 0.1 mm from adults for which collection localities were reported. The localities were scored as to whether they were within the range of the other two species according to the maps given in Hall (1981).

In order to minimize distortions resulting from unequal sample sizes representing different areas, the average skull length of each sex of each species was calculated for all two-degree latitude–longitude squares (centred at the intersections of odd-numbered parallels and meridians) from which data were available. The numbers of 2° squares represented by sex and species, together with mean and variance of skull lengths, are given in Table 1. For those same 2° squares, data are also available on a variety of climatic variables including precipitation, solar radiation, wind speeds, and dry and wet bulb temperatures throughout the year.

Hypotheses and results: body size

Analysis of geographical variation in size of the three species across North America reveals one obvious pattern: *erminea* increases in size with latitude, and thus obeys Bergmann's Rule, while *frenata* and *nivalis* do not. Although Bergmann's (1847) original interpretation of his rule is no longer accepted, there are at least 16 alternative explanations (see Pyke, 1978). We need not consider most of these in any detail here because they would predict that all three species should increase in size with latitude; in fact *nivalis* females actually show the opposite pattern (see Table 1). For three such similar species to exhibit such different patterns of size variation with latitude is, perhaps, surprising. We interpret this finding to mean that the balance of selective forces that influence a large part of the geographical size variation in *erminea* is different for the other two species. Earlier but less extensive analyses by Hall (1951), Rosenzweig (1968) and McNab (1971) revealed similar patterns.

Character displacement and character release

McNab (1971) argued that consideration of the relative distributions of the three species could be used to explain the increase in size with latitude in *erminea*. This is potentially an attractive explanation because the relevant selective forces acting on *erminea* relate to the presence or absence of the other two species. His interpretation was based on character displacement: *erminea* is large in the north of its range because that is where the larger *frenata* is absent and *erminea* is small in parts of the south of its range because that is where the smaller *nivalis* is absent.

McNab's study was criticized by Grant (1972), who wrote that 'The evidence for character displacement is no more than suggestive . . . the data are inadequate . . . In the graphs relating body size to latitude . . . the lines drawn are biassed towards showing [expected] change in body size. More detailed study is required to substantiate the character displacement hypothesis'. We have used three approaches to make more detailed tests of the hypothesis.

Analysis of variance

Our data on *erminea* and *frenata* are sufficiently extensive to allow a four-way analysis of variance to test for changes in skull size with (i) sex, (ii) longitude classified as west or east of the 105° W meridian, (iii) presence or absence of *nivalis*, and (iv) presence or absence of *frenata*/*erminea*. There were significant effects with sex (males are larger than females in each species), and with the presence or absence (+/−) of the other species: *erminea* was larger in the absence of *frenata* and smaller in the absence of *nivalis*.

At first sight these findings seem to lend support to the character-displacement hypothesis. However, *erminea* increases in size with latitude in 15 of the 16 groups of samples defined by particular combinations of sex, longitude, +/− *nivalis* and +/− *frenata*. This means that *erminea* follows Bergmann's Rule *within* areas in which both *nivalis* and *frenata* are consistently present or absent. In other words, McNab's explanation for Bergmann's Rule as a consequence of character displacement is not likely to be the complete story since *erminea* increases in size with latitude irrespective of the occurrence of the other two species. Indeed, the

Table 2. *The average length of* erminea *skulls (in mm) from localities in several areas near the limits of the range of* (*a*) nivalis *and* (*b*) frenata

(*a*) +/− *nivalis*

Area	*nivalis*	Mean	Variance	Number of localities
SE Canada	+	42.4	2.4	16
	−	44.5	2.4	24
Northwest Territories	+	44.1	4.4	13
	−	44.0	2.4	12
SE Alaska and W Canada	+	43.1	2.5	11
	−	41.5	1.7	4

(*b*) +/− *frenata*

Area	*frenata*	Mean	Variance	Number of localities
SE Canada and NE United States	+	41.7	2.9	31
	−	42.9	2.6	17
SW Canada and NW United States	+	40.3	2.5	21
	−	41.8	3.2	5

significant effects of the presence and absence of *frenata* and *nivalis* are more simply interpreted as a consequence of the facts that *frenata* is absent in the north (where *erminea* is large) while *nivalis* is absent in parts of the south (where *erminea* is small).

Allopatry and sympatry

We have reasonable samples of *erminea* from five areas at the limits of the distribution of one of the other two species. In those areas we have been able to test whether *erminea* skull size changes from an area of sympatry with another species to one of allopatry. In three of the areas (southeastern Canada, the Northwest Territories, and southeastern Alaska and western Canada) *erminea* occurs both alone and in the presence of *nivalis*, and in another two locations (southeastern Canada and the northeastern United States, and southwestern Canada and the northwestern United States) *erminea* is found with and without *frenata*. The predictions are that *erminea* should be smaller when *nivalis* is absent and larger when *frenata* is absent. The data used for this analysis are not the average sizes for 2° latitude–longitude squares, but sizes of skulls averaged by collection localities. The results are summarized in Table 2 and provide no statistical support for the character-displacement hypothesis.

McNab (1971) thought that an area of very small *erminea* in the southwest of the species range was a result of the absence of the smaller *nivalis*. However, examination of our more extensive samples of *erminea* reveal that the major size transitions in this area do not coincide with the borders of *nivalis*' range.

Covariation between species

If character displacement or character release (*sensu* Grant 1972, 1975) is an important influence on body size in the three species, we might also expect correlated changes in body size to occur when size-adjacent species coexist. For example, when *frenata* is particularly large, *erminea* might also be expected to increase in size. Interspecific size differences are the least when the female of the larger species is compared with the male of the smaller. We have looked for covariation in size across areas of sympatry between male *nivalis* and female *erminea*, and between male *erminea* and female *frenata*. In the first case the correlation coefficient does not differ significantly from zero, and the small correlation which does exist between male *erminea* and female *frenata* (Spearman's $r=0.20$, $n=73$, $p=0.05$) was found to be a consequence of two different geographical patterns: *erminea* is very small in the southwest of its range and *frenata* is small in the west of its range. Several 2° squares are represented from the area of very small *erminea* and, when they are removed from the analysis, the correlation vanishes (Spearman's $r=-0.02$, $n=58$, not significant).

We conclude that there is no evidence for character displacement between these weasel species in North America. This leads us to be wary of statements about weasel size that invoke character displacement and character release either in North America or elsewhere. For example, in his 'Homage to Santa Rosalia' Hutchinson (1959) points out that *erminea* is smaller in Ireland, where it occurs alone, than on the British mainland, where *nivalis* is also present. He cites this as a possible example of character displacement and, following this lead, Williamson (1972) writes that 'it is certainly reasonable to suppose that the size differences in the skulls results from natural selection caused by competition'. However, it is now apparent that *erminea* from the south of Ireland are similar in size to those on the British mainland – it is only in the north of Ireland that *erminea* is so small (Fairley, 1981). In similar vein, King & Moors (1979) argue that *erminea* is usually smaller in western North America than in Britain because the larger *frenata* is present in western North America and that, since *nivalis* is absent, *erminea* occupies its niche. However, we can identify other areas in North America where large *erminea* are found in the presence of *frenata* (e.g. parts of British Columbia and Alberta) and other areas where small *erminea* are found with *nivalis* (e.g. in northern Saskatchewan).

Prey size

The characteristic elongated cylindrical shape of weasels may be an adaptation for following prey into restricted areas, such as burrows. It is known that the larger weasel species, and the males within a species, often feed on larger prey (reviewed by Anonymous, 1976; Simms, 1979). Therefore, it seems reasonable to suggest that geographical size variation *within* species may result from selection for weasels that are successful at hunting local prey species. This hypothesis would be contradicted if latitudinal variation in skull length of *erminea* did not relate to prey size. One indirect way of testing this is suggested by Barnett's (1977) analysis of size variation in grey squirrels (*Sciurus carolinensis*) which showed that, although skull length increased with latitude, various structures related to feeding did not. This suggests that skull-size variation with latitude in grey squirrels may not be related to food size.

We carried out an analysis similar to Barnett's on 48 male *erminea* skulls from a wide range of latitudes. Five 'feeding measures' (e.g. lower jaw length, canine size) and five 'non-feeding measures' (e.g. length of posterior half of skull, width of the skull at the mastoid process) were taken from each skull. The feeding measures correlated with latitude at least as strongly as the non-feeding measures (0.79–0.83 compared with 0.42–0.80 respectively) as predicted by the prey-size hypothesis.

Rosenzweig (1968) and Simms (1979) looked for a relationship between prey- and weasel-size variation in North America. Rosenzweig (1968) calculated two measures of prey size at various localities. His first measure was the average size of prey likely to be eaten by *any* weasel species. Since the species do not covary in size, perhaps it is not surprising that this measure did not correlate significantly with variation in weasel size. Rosenzweig's second measure was the average size of prey likely to be eaten by *each* weasel species, thus taking into account dietary differences among the species, and these results were equally non-significant. However, despite these findings, prey size may still be important. As Rosenzweig, himself, pointed out, the lack of relationships could be a consequence of the 'somewhat arbitrary and artificial nature' of his prey-size estimates. For example, the relevant measures might be that of a species, or combination of species, which is eaten only at some critical season of the year, when the usual prey are scarce or inactive.

Simms (1979) found that, in southern Ontario, *erminea* may be a vole specialist: *erminea* was absent from sites where *frenata* was present feeding on rabbits and birds but which lacked voles, and at sites where *erminea* was present it fed primarily on voles although *Peromyscus* was common. Simms carried out a series of 'tunnel trials' which indicated that female *erminea* were small enough to follow voles in their subnivean and subterranean burrows. He then compared weasel sizes in eastern and central North America with those of 'the commonest suitable-sized small mammals in their ranges' and claimed to show that female *erminea* covary with prey size. Simms' data are from only five areas and are far from convincing: the prey species at two of the sites are substantially larger than the female *erminea* (in terms of 'minimum passable tunnel diameter') and cannot be considered a reasonable constraint on weasel size. More data of this sort are clearly desirable but, if the results prove negative, we could again argue that the size of the relevant prey species at some or all sites had not been measured.

Simms uses his conclusion that *erminea* are optimally sized vole predators to explain why *erminea* is generally larger in Europe than in North America. It is, he argues, because European *erminea* feed on the 'ubiquitous' water vole (*Arvicola terrestris*). This may be partly true, but it is unlikely to be the whole story: many feeding and gut-content analyses from various parts of western Europe, including the British mainland where *erminea* is particularly large, reveal that it is not a vole specialist, even less a water vole specialist (Day, 1968; Anonymous, 1976; Erlinge, 1979; King & Moors, 1979). However, recent work in southern Sweden reveals a clear preference by male (but not female) *erminea* for water voles (Erlinge, 1981) and Klimov (1940) states that water voles are also a major food supply for *erminea* in Western Siberia.

Although male and female weasels differ in diet (see Anonymous, 1976; Simms, 1979; Erlinge, 1981), there is also considerable dietary overlap between the sexes (see Day, 1968; Fitzgerald, 1977; and especially Moors, 1980). Therefore, the geographical covariation in size between the sexes of each species, which we shall show below, cannot be used as evidence against the prey-size hypothesis. Indeed, it might be cited in its favour.

Seasonality

One explanation of Bergmann's Rule has been discussed by Boyce (1979), who pointed out that environments become more seasonal at higher latitudes. He suggested that larger animals are better able to survive periods of food scarcity in seasonal environments. Similar arguments have been made by other authors (reviewed in Clutton-Brock & Harvey, 1983). This hypothesis might explain other environmental correlates of large size in a variety of animal species (high versus low altitude, continental versus coastal locations, and savanna versus rain forest at the same latitude – see Boyce, 1979). For this hypothesis to explain latitudinal size variation in *erminea*, we would have to assume that *erminea* responded differently from the other two species to changes in seasonality. This seems unlikely because *erminea* is intermediate in size between the other two species and, *a priori*, factors which affect *frenata* and *nivalis* are also likely to affect *erminea*. Nevertheless, since some climatic data were available, we have attempted to test the seasonality hypothesis. We calculated various measures of seasonality as the annual coefficient of variation of our various climatic measures (see above) for different 2° squares and found no evidence to support the hypothesis for any measure of seasonality with any species. Even for *erminea* in North America, latitude is a better predictor of size than seasonality in any climatic variable.

Hypotheses and results: sexual dimorphism

The reasons for the evolution of sexual dimorphism in weasels have been the subject of considerable speculation (Brown & Lasiewski, 1972; Erlinge, 1979; Simms, 1979; Powell, 1979; Moors, 1980). Three main explanations have been proposed: (i) intrasexual selection favours large over small males in competition for mates, (ii) feeding competition between the sexes selects for divergent body sizes, and (iii) selection for channelling energy into reproduction favours small females which have lower metabolic maintenance costs.

Intrasexual selection

Many papers have demonstrated a positive correlation between polygyny and sexual dimorphism across species in various vertebrate taxa (Clutton-Brock & Harvey, 1977; Clutton-Brock, Harvey & Rudder, 1977; Shine, 1978, 1979; Berry & Shine, 1980; Alexander *et al.*, 1979). As predicted, males are usually the larger sex. The order Carnivora is no exception with the polygynous species being more dimorphic than the monogamous ones (Gittleman, 1983), and in *erminea* larger males are dominant over smaller ones (Erlinge, 1977). Weasels are known to be polygynous and intrasexual selection is thus a reasonable explanation of the fact that males are larger than females. The extent to which this will help to explain *geographical variation* in sexual dimorphism within weasel species is not known.

Intersexual competition

Brown & Lasiewski (1972) suggested that the metabolic costs of being long and thin led to increased intraspecific feeding competition among weasels, resulting in the evolution of feeding niche differences between the sexes and hence sexual dimorphism. Powell (1979) found a positive correlation between an index of body elongation and sexual dimorphism across 19 mustelid species, as would be predicted by the hypothesis.

There are several problems with this explanation. First, it is not clear that increased metabolic needs inevitably lead to increased intraspecific competition. If a population is food limited, then increased hunting efficiency will be favoured by natural selection as long as the increased food intake more than matches the additional metabolic costs, but the population will *remain* food limited (albeit possibly at a different population density) and increased intraspecific competition is not a necessary consequence. Second, we have repeated Powell's elongation analysis using a larger sample (26 species, data extracted from the literature by Gittleman (1983)). There is no significant correlation between elongation and sexual dimorphism within the genus *Mustela* ($r = 0.28$, $n = 8$) or among mustelid genera ($r = 0.17$, $n = 14$). Powell's significant correlation resulted from treating all species as independent points: in his sample there were seven *Mustela* species which are generally more dimorphic and more elongated than other mustelids. His relationship was, therefore, simply a restatement of Brown & Lasiewski's (1972) original observation that weasels are both more dimorphic and more elongated than other mustelids.

However, even if an elongate body shape does not lead to increased intraspecific competition, sexual dimorphism could still have evolved through intraspecific competition for food. This leads to our third criticism of the hypothesis: monogamous species in which pairs occupy a joint territory should be more dimorphic than polygynous species. This is not the case (Gittleman, 1983). Our fourth criticism (see also Moors, 1980) is that the feeding-niche hypothesis *considered alone* would not predict that males are invariably the larger sex. But, given a single common mustelid ancestor in which males were larger, such a difference might be retained through successive speciation events. Alternatively, a small degree of intrasexual selection could swing the balance in favour of males being the larger sex. We discount one other criticism of the feeding-niche hypothesis because it depends on the type of group selection envisaged by Wynne-Edwards (1962) which has now been discredited (e.g. see Maynard Smith, 1964, 1976): 'the benefits of avoiding or at least strictly limiting intersexual competition for a resource such as food are obvious, because the continued survival of the species depends on one sex not being able to outcompete the other' (Moors, 1980).

Small females

Moors (1980) has calculated that a typical male-sized female *nivalis* would need to catch an extra 0.5 vole per day merely to supply her own increased metabolic needs. He sees this as evidence for selection favouring small females which are able to channel more energy into reproduction. If the hunting efficiencies of males and females were the same, we suggest that a more profitable way of viewing the size difference would be to seek the selective pressures (such as intrasexual selection) which have operated to make males so large and energetically inefficient. Since nothing is known about the relationship between weasel size and hunting efficiency, we shall not speculate further on this point.

Moors goes on to argue that, while females are selected to be small, males are selected to be large for reasons of intrasexual competition (see above). He concludes that 'the optimum sizes of males and females are likely to vary independently'. This prediction does not hold (see Fig. 2).

Fig. 2. The relationship between male and female skull length in samples of (*a*) *Mustela erminea*, (*b*) *Mustela frenata*, and (*c*) *Mustela nivalis*. Individual points for the North American data (triangles) are means of 2° squares (see text). For *erminea* and *nivalis* the points from the USSR (triangles) and Europe (squares) refer to different sampling localities referenced in Ralls & Harvey (1985). *Frenata* is not found in the USSR or Europe.

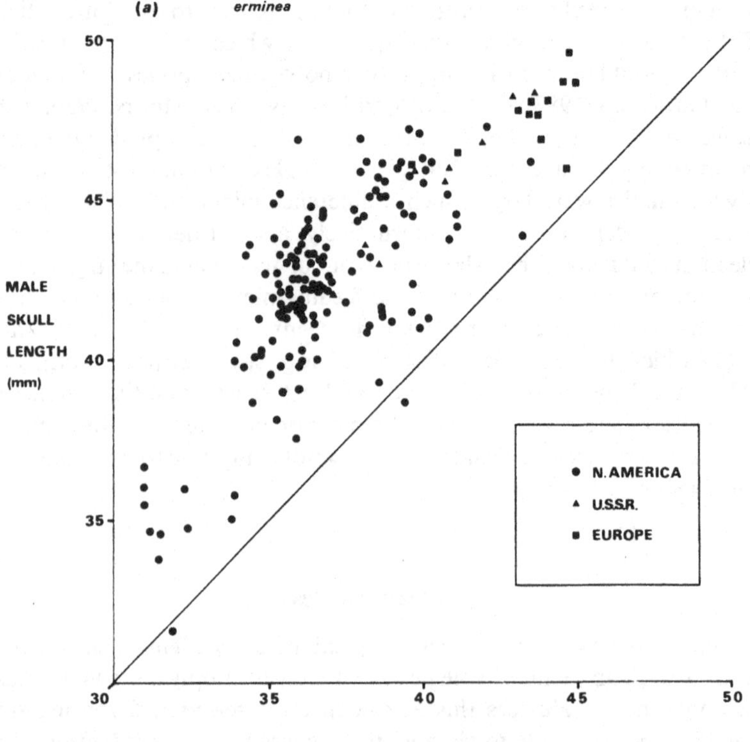

Summary of hypotheses tested

The main hypotheses that we have been able to test, together with our main conclusions, are listed below:

(i) *Character displacement.* It had been suggested that size variation in *erminea* resulted from character displacement with *frenata* and *nivalis.* The data do not support this hypothesis.

(ii) *Prey size.* Geographic variation in skull size might be related to variation in the size of available prey. In *erminea*, variation in the size of skull features that are likely to be related to food size correlates well with variation in the size of other skull features that are more likely to be correlated with overall body size. The data are in accord with the prey-size hypothesis, which needs further testing.

(iii) *Seasonality.* Geographic variation in a species size might be related to seasonality because larger species are able to survive longer periods without food. The absence of covariation in size between the species, and statistical tests relating various measures of climatic seasonality to variation in the size of all three species do not support this hypothesis.

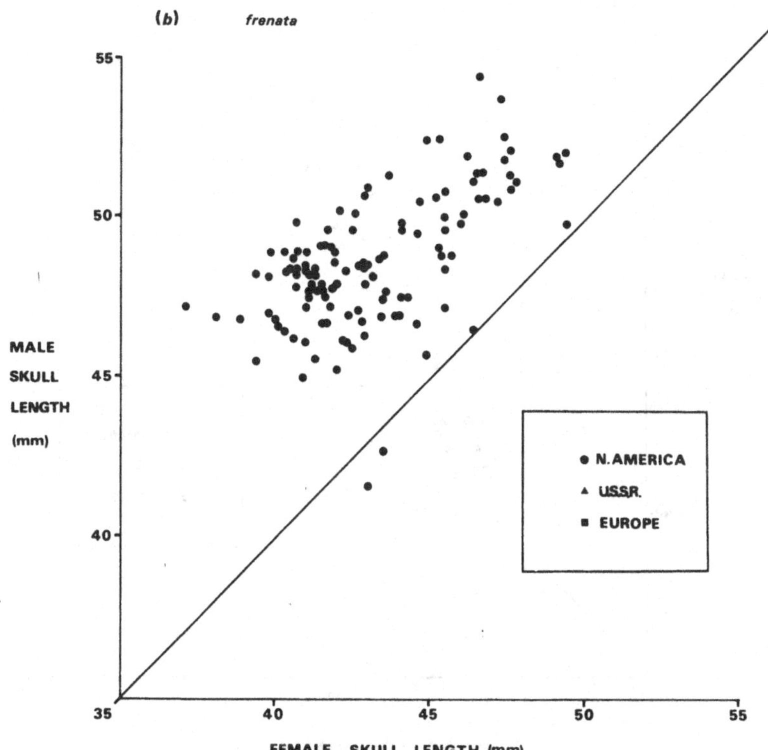

(iv) *Elongation.* Weasels are both more elongated and more sexually dimorphic than other mustelids. It has been suggested that the increased metabolic needs of elongated species lead to increased intraspecific feeding competition, and hence sexual dimorphism. Various lines of evidence lead us to reject this hypothesis.

(v) *Independent size variation in the sexes.* It has been argued that different selective forces act on male and female body size and therefore that, within a species, there should be no geographical covariation in body size. This hypothesis is refuted by the data.

Concluding remarks

The literature on size and sexual dimorphism in weasels gives the impression that we have a reasonably clear understanding of the selective factors influencing both. To the contrary, as this chapter demonstrates, we have a very limited set of facts on which to base an extensive body of relevant theory.

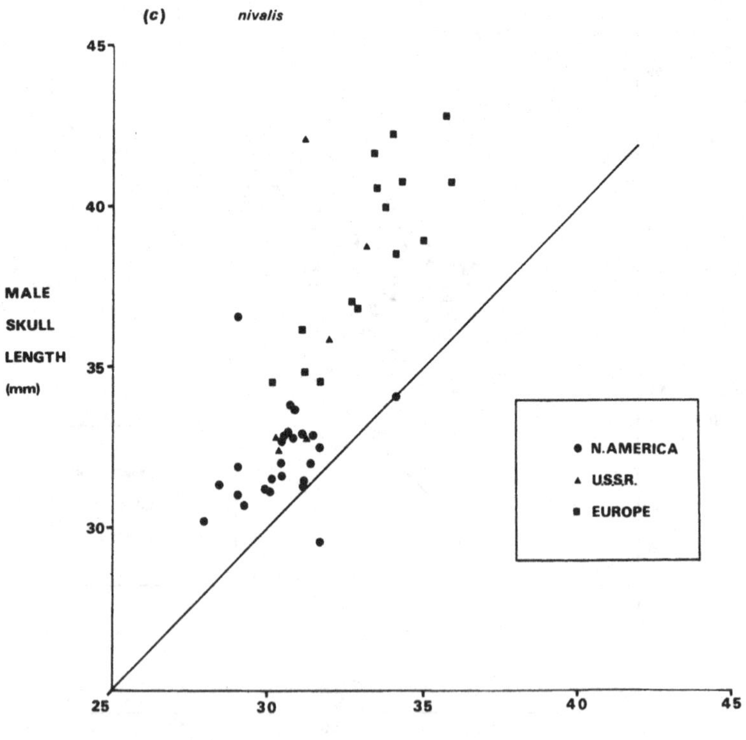

Comparisons relating crude species-specific measures of morphology (e.g. body size, weapon size, organ size) to equally crude measures of behaviour (e.g. timing of activity, breeding system, home-range size, day-range length) have been reasonably successful for both forming and testing adaptationist hypotheses (see Harvey & Mace, 1982; Clutton-Brock & Harvey, 1984). But patterns are rarely strong, and the extent to which scatter does not result from statistical error should give an indication of the dangers of interpreting fine-level intrageneric or intraspecies variation in terms of hypotheses which have been useful for explaining gross variation at higher taxonomic levels. For example, differences in sexual dimorphism between species within mammalian orders can often be related to differences in breeding systems. But there is only a weak correlation (at best) between measures of the degree of polygyny and the extent of sexual dimorphism across polygynous species (Ralls, 1977).

It would be foolhardy to conclude that patterns demonstrated at higher taxonomic levels which imply the importance of particular processes mean that similar patterns and processes explain variation at lower taxonomic levels. Indeed, the absence of geographical covariation in body size among North American weasels suggests that different factors may be influencing patterns in closely related species. Pattern recognition is a useful tool for inferring process in evolutionary ecology, but the taxonomic level of study is an important qualification. At the moment, we are only too well aware that when discussing intraspecific variation in size and sexual dimorphism there is little distinction between preliminary speculations and plausible hypotheses.

This chapter is dedicated to John Maynard Smith in the hope that it is never too late to learn the value of a statistical approach to population biology.

References

Alexander, R. D., Hoogland, J. L., Howard, R. D., Noonan, K. M. & Sherman, P. W. (1979). Sexual dimorphisms and breeding systems in pinnipeds, ungulates, primates and humans. In *Evolutionary Biology and Human Social Behavior*, ed. N. A. Chagnon & W. Irons, pp. 402–35. North Scituate, Mass.: Dewsbury.

Anonymous. (1976). Pop goes the weasel. *Nature*, **260**, 194–5.

Barnett, R. J. (1977). Bergmann's rule and variation in structures related to feeding in the grey squirrel. *Evolution*, **31**, 538–45.

Bergmann, C. (1847). Uber die Verhaltenisse der Warmeokonomie der Thiere zu ihrer Grosse. *Gottinger Studien*, **1**, 595–708.

Berry, J. F. & Shine, R. (1980). Sexual size dimorphism and sexual selection in turtles (order Testudines). *Oecologia*, **44**, 185–91.

Boyce, M. S. (1979). Seasonality and patterns of natural selection for life histories. *American Naturalist*, **114**, 569–83.

Brown, J. C. & Lasiewski, R. C. (1972). Metabolism of weasels: the cost of being long and thin. *Ecology*, **53**, 939–43.

Clutton-Brock, T. H. & Harvey, P. H. (1977). Primate ecology and social organisation. *Journal of Zoology (London)*, **183**, 1–39.

Clutton-Brock, T. H. & Harvey, P. H. (1983). The functional significance of variation in body size among mammals. In *Recent Advances in the Study of Mammalian Behavior*, special publication of the American Society of Mammalogists No. 7, ed. J. F. Eisenberg & D. G. Kleiman, pp. 632–63.

Clutton-Brock, T. H. & Harvey, P. H. (1984). Comparative approaches to investigating adaptation. In *Behavioural Ecology: an Evolutionary Approach*, 2nd edn, ed. J. R. Krebs & N. B. Davies, pp. 7–29. Oxford: Blackwell.

Clutton-Brock, T. H., Harvey, P. H. & Rudder, B. (1977). Sexual dimorphism, socionomic sex ratio and body weight in primates. *Nature*, **269**, 797–800.

Day, M. G. (1968). Food habits of British stoats (*Mustela erminea*) and weasels (*Mustela nivalis*). *Journal of Zoology (London)*, **155**, 485–97.

Erlinge, S. (1977). Agonistic behaviour and dominance in stoats (*Mustela erminea*, L.). *Zeitschrift für Tierpsychologie*, **44**, 375–88.

Erlinge, S. (1979). Adaptive significance of sexual dimorphism in weasels. *Oikos*, **33**, 233–45.

Erlinge, S. (1981). Food preference, optimal diet and reproductive output in stoats (*Mustela erminea*). *Oikos*, **36**, 305–15.

Fairley, J. S. (1981). A north–south cline in the size of the Irish stoat. *Proceedings of the Republic of Ireland Academy*, **81B**, 5–10.

Fitzgerald, B. M. (1977). Weasel predation on a cyclic population of the montane vole (*Microtus montanus*) in California. *Journal of Animal Ecology*, **46**, 367–97.

Gittleman, J. L. (1983). The behavioural ecology of carnivores, D.Phil. Thesis, University of Sussex.

Grant, P. R. (1972). Convergent and divergent character displacement. *Biological Journal of the Linnean Society*, **4**, 39–68.

Grant, P. R. (1975). The classical case of character displacement. *Evolutionary Biology*, **8**, 237–337.

Hall, E. R. (1951). *American Weasels*. University of Kansas Publications No. 4.

Hall, E. R. (1981). *The Mammals of North America*, 2nd edn. New York: John Wiley & Sons.

Harvey, P. H., Colwell, R. K., Silvertown, J. W. & May, R. M. (1983). Null models in ecology. *Annual Reviews of Ecology and Systematics*, **14**, 189–211.

Harvey, P. H. & Mace, G. M. (1982). Comparisons between taxa and adaptive trends: problems of methodology. In *Current Problems in Sociobiology*, ed. King's College Sociobiology Group, pp. 343–61. Cambridge University Press.

Hutchinson, G. E. (1959). Homage to Santa Rosalia or why are there so many kinds of animals? *American Naturalist*, **93**, 145–59.

King, C. M. & Moors, P. J. (1979). On co-existence, foraging strategy and the biogeography of weasels and stoats (*Mustela nivalis* and *M. erminea*) in Britain. *Oecologia*, **39**, 129–50.

Klimov, Y. N. (1940). Data on the biology of the ermine. *Trudy Biologicheskogo Instituta*, **7**, 80–8.

McNab, B. K. (1971). On the ecological significance of Bergmann's Rule. *Ecology*, **52**, 845–54.

Maynard Smith, J. (1964). Group selection and kin selection. *Nature*, **201**, 1145–7.

Maynard Smith, J. (1976). Group selection. *Quarterly Review of Biology*, **51**, 277–83.

Moors, P. J. (1980). Sexual dimorphism in the body size of mustelids (Mammalia: Carnivora): the role of food habits and breeding systems. *Oikos*, **34**, 147–58.

Powell, R. A. (1979). Mustelid spacing patterns: variations on a theme by *Mustela*. *Zeitschrift für Tierpsychologie*, **50**, 153–65.

Pyke, G. H. (1978). Optimal body size in bumble bees. *Oecologia*, **34**, 255–66.

Ralls, K. (1977). Sexual dimorphism in mammals: avian models and unanswered questions. *American Naturalist*, **111**, 917–38.

Ralls, K. & Harvey, P. H. (1985). Geographic variation in size and sexual dimorphism of North American weasels. *Biological Journal of the Linnean Society* (in press).

Rosenzweig, M. L. (1966). Community structure in sympatric carnivora. *Journal of Mammalogy*, **47**, 602–12.

Rosenzweig, M. L. (1968). The strategy of body size in mammalian carnivores. *American Midland Naturalist*, **80**, 299–315.

Roth, V. L. (1981). Constancy in the size ratios of sympatric species. *American Naturalist*, **118**, 394–404.

Schoener, T. W. (1984). Size differences among sympatric, bird-eating hawks: a worldwide survey. In *Ecological Communities: Conceptual Issues and the Evidence*, ed. D. R. Strong, D. Simberloff & L. G. Abele, pp. 254–8. Princeton: Princeton University Press.

Shine, R. (1978). Sexual size dimorphism and male combat in snakes. *Oecologia*, **33**, 269–77.

Shine, R. (1979). Sexual selection and sexual dimorphism in the Amphibia. *Copeia*, 1979(**2**), 297–306.

Simberloff, D. & Boecklen, W. (1981). Santa Rosalia reconsidered: size ratios and competition. *Evolution*, **35**, 1206–28.

Simms, D. A. (1979). North American weasels: resource utilisation and distribution. *Canadian Journal of Zoology*, **57**, 504–20.

Williamson, M. (1972). *The Analysis of Biological Populations*. London: Arnold.

Wynne-Edwards, V. C. (1962). *Animal Dispersion in Relation to Social Behaviour*. Edinburgh: Oliver & Boyd.

13

Asymmetric contests in social mammals: respect, manipulation and age-specific aspects

C. PACKER AND A. PUSEY

Introduction

The application of game theory to the study of fighting behaviour has led to an important distinction being drawn between contests where the opponents have never met each other before, and contests where the opponents have interacted previously. In the former, each contestant may not be able to assess its chances of defeating its opponent; whereas in the latter, each contestant has prior information about the relative size, age, or motivation of the opponent. Thus in a stable social group, animals will be more likely to utilize asymmetries to settle contests without engaging in costly fighting (Maynard Smith, 1982). In this chapter, we will focus on the nature of asymmetries that are commonly found within groups of social mammals and we will show how these affect fighting behaviour.

First, we will describe how asymmetries in 'resource-holding power' (RHP, Parker, 1974) and in 'ownership' are utilized, and how one asymmetry may take priority over another in different mammalian species. Second, we describe how individuals may 'abuse' their opponents' respect of an asymmetry and may utilize complex behaviours in order to cancel an asymmetry. Finally, we will discuss how in mammals age-specific asymmetries will typically include differences in RHP, payoffs and/or costs.

Asymmetries in RHP v. ownership

No two individuals have exactly the same competitive ability and most behavioural biologists are generally measuring differences in resource-holding power (RHP) whenever they measure social dominance. Dominance 'hierarchies' have been detected within age–sex classes of most group-living species, and considerable effort has been devoted to testing whether high dominance rank correlates with reproductive success. For our purposes, however, it is sufficient only to show that a subordinate animal will consistently defer to a dominant animal in a pairwise competition for a valuable resource, such as food or a resting site. Where such dominance relationships are conspicuous, by definition interactions between individuals are affected by an asymmetry in RHP. Contests are usually settled without costly fighting – the outcome instead depends on the recognition of the asymmetry.

However, in a number of species it has been found that individuals respect 'ownership' of a resource, so that an animal with a resource usually will not be supplanted from the resource by another. Such 'respect' of ownership is mutual: the 'intruder' may be an 'owner' during a subsequent interaction. A common example of respect of ownership among territorial animals is that an individual is dominant to a neighbouring conspecific while on his own territory, but he is subordinate while on his neighbour's territory (see Maynard Smith, 1982). When two individuals simultaneously become owners of the same territory, there is no asymmetry of ownership and both 'owners' will fight intensely (e.g. speckled wood butterflies, Davies, 1978).

Respect of ownership extends to other resources besides territories; particularly to receptive females or to certain kinds of feeding sites. In these cases, a dominant individual does not supplant an otherwise subordinate individual from the resource. In some cases, possession of the resource may of itself confer a competitive advantage, either because it is difficult to separate the animal from the resource (e.g. copulating male dung flies, Parker, 1974) or because the resource is highly portable (e.g. subordinate male baboons can carry away very small carcasses when confronted by a dominant male whereas dominant males supplant subordinates from large carcasses (Hamilton & Busse, 1982)). However, 'respect' of ownership can evolve even when ownership does not confer a competitive advantage or involve an asymmetry in payoff (see Maynard Smith & Parker, 1976).

Male Hamadryas baboons restrict the movements of females and thus form stable one-male 'harems'. Experiments by Kummer and colleagues (Kummer, Gotz & Angst, 1974; Bachmann & Kummer, 1980) have shown that (i) male–male fighting for females is common only when two males are simultaneously presented with an unpaired female, (ii) if one male comes into contact with the female only a few minutes before a second male, the second male will respect the first male's ownership of the female, (iii) respect is mutual within pairs of males; one may be an owner on one occasion and a rival on another and (iv) dominant males respect ownership of subordinate males, *except when differences in RHP are extreme.*

More interesting, however, than the respect of what would ordinarily be long-lasting or permanent relationships where females may be expected to maintain the status quo (Clutton-Brock & Harvey, 1976), is the respect of 'temporary ownership' of oestrous females within social groups. In many species living in groups with more than one breeding male, one male will monopolize an oestrous female by maintaining a close proximity to her, preventing her from moving towards other males, and threatening other males if they come close to the female. These 'consortships' last from a few hours to days (sometimes over the entire oestrous period), but a number of different males may consort with the same female in the same or subsequent oestrous periods. In a study of olive baboons, Packer (1979*b*) found that dominant males would defer to subordinate males while the subordinates were consorting. Consequently the mating success of subordinate males was greater than would be expected if RHP were the only variable affecting mating success. However, respect of ownership did not completely overcome differences in RHP, since low-ranking males started consorting earliest in the female's cycle but the highest ranking male had usually gained control of the female by the time she was most likely to conceive. In this study, 'dominance' was based on each male's ability to supplant other males from feeding sites and hence reflected 'RHP' of food.

In Table 1 we contrast the relative importance of asymmetries in RHP and respect of ownership in competition for key resources in three mammalian species: the African lion, the olive baboon, and the chimpanzee. We compare males in each species during competition for access to oestrous females, and we compare females in competition for food. All three species live in stable social groups with multiple breeding females and males. In all three, males form temporary consortships with oestrous females, and no one male is able to monopolize all of the matings with oestrous females. In both lions and baboons, females often feed together and competition for access to food is common. Female chimpanzees only infrequently feed together, but competition for food has been observed.

The three species vary in the extent to which ownership is respected. Lion males are conspicuous in the extent to which they respect their companions' temporary ownership of females (Packer & Pusey, 1982). Fighting only occurs when ownership is unclear or when two consorting males come into close proximity. In contrast, a dominant male chimpanzee can apparently displace another male from a female and successful consortships by subordinate males require both that the female cooperates with the male and that the pair remains hidden from more dominant males (Nishida, 1979; Tutin, 1979; Tutin & McGinnis, 1981). Olive baboons show some respect of ownership (see above), but in most groups high-ranking males are disproportionately successful. Therefore, male lions apparently show the 'Bourgeois strategy' (Maynard Smith, 1976), whereas male chimps rely more exclusively on RHP assessment, and male baboons are intermediate between the other two species.

For the Bourgeois strategy to be an 'evolutionarily stable strategy' in a population of 'hawks' and 'doves', the average costs of injury during an escalated fight should exceed the average payoff from winning (Maynard

Table 1. *Factors affecting respect of ownership in three species*

	Lions	Baboons	Chimpanzees
Males			
A. Extent to which temporary ownership of oestrous females is respected (see text)	High (1)[a]	Moderate (2)	Low (3–5)
B. Chances of being injured during one-on-one fight	High (1, 6)	Moderate (2, 7)	Low (8)
C. Percentage of oestrous periods leading to conception	20–28% (6, 9)	16% (2)	13–20% (5, 10, 11)
D. Percentage of pregnancies leading to young surviving to breeding age	14–44% (12, 13)	*c.* 60% (14)	*c.* 40% (5, 8)
E. Average litter size	2.5 (6)	1 (14)	1 (8)
F. Value of oestrous female $= C \times D \times E$	0.07–0.32	0.10	0.05–0.08
G. Differences in age/size of males in same group	Low: typically of similar age (1)	High: range 8–25 y (2, 14)	High: range 16–34+ y (8)
Females			
A. Extent to which temporary ownership of feeding site is respected (see text)	High (6, 15, 16)	Low (17)	Low (4, 11, 18)
B. Chances of being injured during one-on-one fight	High (6)	Low (17)	Low (8, 19)
C. Value of specific feeding site (see text)	Moderate	Low	Low?
D. Differences in age/size of females within group	Moderate: some groups all same age, others show range of 4–16 y (16)	High: range 5–25 y (14)	High: range 13–34+ y (8)

[a] References: 1: Packer & Pusey, 1982; 2: Packer, 1979*a*; 3: Tutin, 1979; 4: Nishida, 1979; 5: Tutin & McGinnis, 1981; 6: Schaller, 1972; 7: Packer, 1977*a*; 8: Goodall, 1983; 9: Packer & Pusey, 1983; 10: Teleki, Hunt & Pfifferling, 1976; 11: Pusey, 1978; 12: Bertram, 1975; 13: Hanby & Bygott, 1979; 14: Packer, Goodall & Sindimwo, unpublished; 15: Bertram, 1978; 16: Packer & Pusey, unpublished; 17: Hausfater, 1975; 18: Pusey, 1983; 19: Goodall, 1977.

Smith, 1976). It would therefore be expected that the extent to which each species shows Bourgeois behaviour varies either because the costs of fighting are higher in some species than in others or because the 'value' of an oestrous female is lower. Although precise data on costs of injury are unavailable, there are obvious differences in the three species in the degree of weaponry. Lions have well-developed biting dentition and powerful claws, and male baboons have enlarged canine teeth which they keep razor-sharp by honing them against their premolars (Simons, 1972). A lion can easily kill another lion, and male baboons can inflict serious wounds in one-on-one fights (see Table 1). For their body size, male chimps have less-well-developed canines than do cercopithecine monkeys (Harvey, Kavanagh & Clutton-Brock, 1978), and serious wounds are mostly inflicted during gang attacks rather than one-on-one fights (Goodall *et al.*, 1979).

We have estimated the 'value' of an oestrous female in each species given the probability of conception and survivorship of young. The likelihood that a given oestrous period will lead to a surviving offspring is quite similar in all three species (Table 1). Therefore, the difference in the extent of Bourgeois behaviour in the three species is more likely to be due to differences in costs of fighting than to differences in payoffs.

An additional factor, however, might be the different age structure typical of males in the three species (Table 1). Male lions are often closely matched in age and size, whereas male baboons and chimps are often of quite different ages. As a result differences in RHP will typically be smaller in lions and therefore respect of ownership will be more likely to occur (Hammerstein, 1981). Where male lions are not equally matched, males of apparently lower competitive ability also have lower mating success (Packer & Pusey, 1982). However, we do not yet know if this is because inferior males are less likely to gain access to females in the first place, or because their superior companions displace them from their consort partners. Male baboons are not typically matched in age, yet they are intermediate between the other two species in the extent to which they show respect of ownership. This suggests that high costs of fighting as well as age/size discrepancies must be involved.

Respect of ownership is often taken by field workers as evidence that there is an asymmetry in payoff or RHP biased in favour of the owner. However, we doubt that being the temporary owner of an oestrous female in these three species involves either a higher payoff or a lower risk of injury to the owner than to the rival. Packer (1979*b*) previously suggested that a consorting male would always have a higher payoff than a rival, assuming that the nth male to consort with the female has a $1/n$ chance of fathering the offspring, whereas the $(n+1)$th male has only a $1/(n+1)$ chance. This analysis is probably incorrect because both males are competing for exactly the same opportunity to fertilize the female over the next few hours or days

(G. A. Parker & J. Maynard Smith, personal communication). Bachmann & Kummer (1980) showed that the preferences of females have some effect on the rivals' respect of ownership. A female that does not cooperate with the 'owner' can be taken over more easily by a rival. However, of the species listed in Table 1, pairs of male lions and baboons showed mutual respect when each consorted with the same female on different occasions (Schaller, 1972; Packer & Pusey, 1982, and unpublished data) and in chimpanzees, where females' preferences seem to be *most* important to a consortship being successful, ownership is *least* important. Therefore, differences in the behaviour of oestrous females of the three species probably do not account for the varying extent to which ownership is respected.

The lower half of Table 1 compares the extent to which females of the three species respect another female's 'ownership' of a feeding site. Female lions are as Bourgeois at the dinner table as their husbands are in the bedroom. We have never seen an adult female supplant another female from a kill in over 100 h observation of lions feeding together (Packer & Pusey, unpublished). In contrast, female baboons habitually supplant subordinate females from feeding sites and the few available data on female chimps show no evidence of respect of ownership. It is much more difficult to estimate the value of a feeding site except that it is probably highest to a female lion since she can less easily find an alternative feeding site and could starve if excluded by more dominant females from only a few consecutive kills. In fact, dominance orders at kills are undetectable in female lions (Schaller, 1972; Bertram, 1975; Packer & Pusey, unpublished). Again, costs of fighting will be highest in female lions and low both in female baboons and chimps (Table 1). Even a relatively small female lion could inflict a fatal wound on a female conspecific – whereas this is very unlikely in baboons or chimps.

One difficulty in comparing female lions with the two primate species is that a lion's claws enable it to hold on to a carcass as well as injure a competitor and thus a positional advantage of ownership may be involved. However, this can not be the whole story since a female lion will relinquish a kill to an adult male without a fight (Schaller, 1972; personal observation); probably because a male lion is 40% heavier than a female.

The degree to which ownership is respected in these three species does not appear to depend on the degree of kinship between the opponents: individuals within each sex are typically related, except for male baboons (Packer, 1977*a*, 1979*a*) and female chimpanzees (Pusey, 1979). Furthermore, temporary ownership of oestrous females is respected by male lions both when the males are close relatives and when they are unrelated (Packer & Pusey, 1982).

We feel that these comparisons are valid, even though we have largely contrasted a carnivore with two primates. Other species of social carnivore show little if any respect of ownership. There is anecdotal evidence from both wild dogs and spotted hyenas that dominant individuals can supplant subordinates from receptive females and from food (van Lawick & van Lawick-Goodall, 1971; Kruuk, 1972; Frame *et al.*, 1979). These two species differ from lions in that they lack vicious claws though they do have sharp teeth.

In summary, there are striking differences across species of mammals in the extent to which ownership of a resource is respected. These differences are probably due to variation both in the risks of injury from fighting and in the extent to which differences in RHP exist within social groups.

Manipulating asymmetries

Having seen that respect of ownership can sometimes override differences in RHP, we will now describe how individuals may manipulate their opponents' respect of ownership or other asymmetry, and how cooperative behaviour may often evolve in order to overcome ownership.

Asymmetry 'abuse'

An individual is taking advantage of an opponent's respect of ownership whenever he behaves like an 'owner' in order to avoid a fight, but the 'resource' he is holding has no immediate value. For example, when a male baboon is being threatened by another male he will sometimes seek out a female that is not at the height of oestrus and 'consort' with her, and his opponent will stop threatening him (Packer, 1979*a*,*b*). The 'consorting' male's 'ownership' of the female is respected, even though the female has no immediate value; and the 'consorting' male ceases to show further interest in the female once his opponent has moved off. Thus, respect of ownership extends to situations where the aggressive interaction had not initially involved that specific resource, and a male may manipulate an opponent's respect of ownership to his own advantage.

Another asymmetry that apparently does not directly involve a resource is the carrying of infants by male baboons. While a male is carrying an infant in an encounter with another male, he is less likely to be threatened by and is more able to displace the other male, than when he is not carrying an infant (Packer, 1980). This behaviour probably originates from the protection of an infant by its father against a potentially infanticidal male (Busse & Hamilton, 1981). However, such 'protection' is an insufficient explanation of the behaviour in contemporary populations. Males seek out infants while being threatened by other males (Packer, 1980; Stein, 1981) and in one population the carrying male was more likely to have been the infant's father than was the opponent in only 39.7% of cases ($N = 161$ cases of infant carrying, where data were available on the males' mating activity with the infant's mother at the time of conception (Packer, personal observation)). More important than the pattern of kinship between infant, carrying male and opponent is the fact that males generally carry infants during encounters with more dominant males (Ransom & Ransom, 1971; Popp, 1978; Packer, 1980; Collins, 1981; Stein, 1981). Thus, carrying an infant temporarily overrides differences in RHP in the same way as does respect of 'ownership' of an oestrous female.

Therefore, carrying of infants may be another example of an individual manipulating an asymmetry in order to take advantage of his opponent's recognition of that asymmetry. Initially the behaviour may have functioned exclusively to protect the infant from infanticide. The usual explanation for infanticide by males is that by killing a female's young she will return to sexual receptivity more quickly (Hrdy, 1974). While protecting his infant, the father would be fighting for a high payoff – survival of his offspring – whereas the opponent would only be fighting for a chance to mate with its mother. We suggest that, once widespread, the respect of the asymmetry in payoff as signalled by carrying an infant was 'exploited' by subordinates. Males actually compete for access to infants and the carrying of infants has become the source of considerable infant mortality in some populations (Collins *et al.*, 1984).

Both of these examples apparently involve exploiting an opponent's respect for asymmetry. In neither case, however, does it seem that the 'abuse' is specifically used to gain access to a resource nor does it result in a permanent rise in dominance ranking (Packer, 1979*b*), but it is instead used to prevent the male from being threatened further. Therefore, if two males were having a dispute for some other reason, the male that suddenly consorts with an infertile female or that picks up an infant does so to 'turn off' his opponent's attack. By creating a situation that has no worthwhile payoff to the opponent, he has made it no longer worthwhile for the opponent to risk injury. The 'abusive' male benefits only by avoiding injury. As long as no resource is at stake, these behaviours need not be viewed as bluff. However, if the 'abusive' male did occasionally use them to gain a resource it might be expected that they would be less likely to persist in the population (Andersson, 1980). These ideas are obviously speculative, but suggest the possibility that quite complex behaviours involving the creation of 'asymmetries' may be used to settle contests in ways not previously appreciated.

Overwhelming ownership

In section 1 we showed that the respect of ownership can overcome differences in RHP, *unless differences in RHP are extreme.* An obvious technique whereby an individual may create an extreme difference in RHP is to form an alliance with a partner and to challenge an opponent cooperatively. Cooperative pairs of subordinate male baboons can easily displace a single, highly dominant individual (DeVore, 1962) and pairs of male baboons can also separate a consorting male from an oestrous female (Packer, 1977*a*). Hence, cooperating individuals overcome the asymmetry of ownership by greatly increasing their combined RHP in comparison with an owner. There are a number of examples of cooperation enabling individuals to overcome ownership: nomadic bands of male langurs can cooperatively evict a breeding male from his troop (Sugiyama, 1965; Hrdy, 1974), large coalitions of male lions can evict smaller resident coalitions from prides (Bygott, Bertram & Hanby, 1979), and large groups of primates can expand the size of their territories at the expense of smaller neighbouring groups (see Wrangham, 1980).

Age-specific asymmetries

The life history of an animal as long-lived as most mammals is obviously complex, and is especially so in species living in groups comprised of individuals of different ages. Contests with different opponents may involve different asymmetries in RHP, different effects of injury on fitness, and different values of the resource.

It is already widely recognized that RHP will change with age in species where RHP is related to body size. In consequence, young individuals may adopt alternative strategies until they are large enough to become successful competitors. For example, some young male elephant seals successfully avoid direct competition with larger adult males for access to females by 'sneaking' into harems (Le Boeuf, 1974). Because of their small size and lack of secondary sex characteristics, they are apparently difficult to distinguish from adult females.

It would also be expected that ageing animals will be more willing to engage in a serious fight since they are nearing the end of their lifespan anyway and thus have less to lose from sustaining a permanent injury (e.g. Popp & DeVore, 1979). Therefore, animals may show an increased 'irascibility' when nearing senescence in the same way that females are expected to invest most heavily in their final offspring (e.g. Pianka & Parker, 1975). Meat is an unusual, but highly prized food item for chimpanzees, and ageing males will compete vigorously to gain access to meat (Wrangham, 1975). Older males thus gain or maintain access to meat more often than would be predicted from their usual dominance relations.

Recent studies have also suggested that the value of a resource will depend on the age of the contestant. For example, in many vertebrates young animals must rapidly gain weight in order to survive their first winter (e.g. yellow bellied marmots, Armitage & Downhower, 1974). Thus, access to a particular food item may be more critical to the survival of a yearling than to an adult (Barkan, Craig & Brown, 1982). In wild dogs, immatures have priority of access to meat and will aggressively displace adults from kills (van Lawick & van Lawick-Goodall, 1971; Schaller, 1972). While this must largely reflect the provisioning strategies of the adults, priority of access by yearlings persists even when subsequent litters are born (Malcolm & Marten, 1982). Since yearling dogs are apparently more subject to starvation than are adults, priority of access by yearlings may result from yearling 'dominance' due to an asymmetry in payoff (Barkan *et al.*, 1982).

However, higher payoffs to younger animals have to be countered by the fact that wounds received while young may lead to a permanent loss in competitive ability. Thus the precise effects of fighting on lifetime reproductive success need to be considered. We will discuss the following example in detail in order to emphasize the importance of considering precisely how the costs of fighting will affect lifetime reproductive success (see also Caryl, 1980).

Males in most mammalian species disperse from their natal group or areas (Packer, 1979*a*; Greenwood, 1980) and in some, dispersal often does not occur until after males reach full size. In such exogamous species, males that bred with close genetic relatives would have offspring of lower average viability due to inbreeding depression (see Ralls, Brugger & Ballou, 1979; Ralls & Ballou, 1982). Therefore, when competing for access to mates, a male closely related to an oestrous female would have a lower expected payoff than a male less closely related to the female.

Full-sized male baboons still residing in their natal troop will not compete aggressively against immigrant males for access to oestrous females even though such 'natal males' often compete successfully against immigrants for access to feeding sites (Packer, 1979*a*). This disparity in competitive behaviour for access to the two resources is not simply due to a higher value of food to the younger male (as suggested above) because males of similar age compete vigorously for females after they have emigrated and males that return briefly to their natal troops cease competing for females until they leave again. Although males might mate surreptitiously with females in their natal troop, they never defend their exclusive access to the female in the way typical of immigrant males.

Could the payoff asymmetry account for this behaviour? Packer (1977*b*) modelled the lifetime reproductive success of males as follows. A male baboon was assumed to be at the peak of his competitive ability when he reached full size, but he did not emigrate until a fixed time after this age. Competitive ability was assumed to determine the probability of winning fights against all opponents, but declined linearly as a function of the number of fights the male had engaged in. A continuous decline in competitive ability is expected due to the steady accumulation of scar tissue and loss of mobility due to wounding. Each male has a finite lifespan and, after the onset of senescence, the probability of fighting success declines rapidly to zero irrespective of the subsequent frequency of fighting. Each individual was viewed as having two phases to its lifetime – before or after emigration – and the payoff to gaining access to females before emigration is lower than after emigration. Each male could either behave as a 'hawk' (competing aggressively for access to females) or a 'dove' (deferring to a 'hawk', and only gaining access to females in competition with other 'doves'). Only hawks suffer a loss in competitive ability from fights and only when fighting other hawks. A male could show the same or a different strategy before and after emigration.

It was shown that the strategy observed in the wild, 'play dove while in natal troop, play hawk after emigrating', could in fact be an evolutionarily stable *lifetime* strategy, but only when (i) the rate of decline in fighting ability was sufficiently high that a male that waited until emigration to play hawk would still be likely to 'use up' most of his fighting potential before reaching senility, or (ii) when levels of inbreeding depression were severe (Packer, 1977*b*).

Otherwise the male should always play hawk both before and after emigrating. In any case, the best strategy for a male would obviously be to emigrate as soon as he reached full-size, as is often the case (Packer, 1979*a*). Males that remained longer in the natal troop apparently did so because they were subject to unusually high levels of aggression while attempting to join nearby troops.

Conclusions

There are striking differences across mammalian species in the extent to which 'respect of ownership' influences competitive encounters. In broad agreement with predictions generated by Maynard Smith's analyses of the Bourgeois strategy, respect of ownership appears to be of greatest importance in contests where fighting is most costly. It is possible that animals sometimes 'abuse' their opponents' respect of an asymmetry such as ownership, and the advantage of ownership is often overcome by opponents challenging an owner cooperatively. Complex life-history strategies typical of mammals and other vertebrates may be profitably interpreted by considering age-specific changes in RHP, potential costs of injury, and payoff from gaining access to a particular resource.

References

Andersson, M. (1980). Why are there so many threat displays? *Journal of Theoretical Biology*, **86**, 773–81.

Armitage, K. & Downhower, J. (1974). Demography of yellow-bellied marmot populations. *Ecology*, **55**, 1233–45.

Bachmann, C. & Kummer, H. (1980). Male assessment of female choice in hamadryas baboons. *Behavioral Ecology and Sociobiology*, **6**, 315–21.

Barkan, C. P. L., Craig, J. L. & Brown, J. L. (1982). Dominance, aggression and payoff asymmetries in Mexican jays. Paper given at American Ornithological Union Meeting, Chicago.

Bertram, B. C. R. (1975). Social factors influencing reproduction in wild lions. *Journal of the Zoological Society of London*, **177**, 463–82.

Bertram, B. C. R. (1978). *Pride of Lions*. London: Dent.

Busse, C. & Hamilton, W. J., III. (1981). Infant carrying by male chacma baboons. *Science*, **212**, 1282–3.

Bygott, J. D., Bertram, B. C. R. & Hanby, J. P. (1979). Male lions in large coalitions gain reproductive advantages. *Nature*, **282**, 839–41.

Caryl, P. G. (1980). Escalated fighting and the war of nerves: games theory and animal combat. In *Perspectives in Ethology*, ed. P. P. G. Bateson & P. H. Klopfer. New York: Plenum Press.

Clutton-Brock, T. H. & Harvey, P. H. (1976). Evolutionary rules and primate societies. In *Growing Points in Ethology*, ed. P. P. G. Bateson & R. A. Hinde, pp. 557–84. London: Academic Press.

Collins, D. A. (1981). Behaviour and patterns of mating among adult yellow baboons (*Papio c. cynocephalus*). Ph.D. thesis, University of Edinburgh.

Collins, D. A., Busse, C. & Goodall, J. (1984). Infanticide in two populations of savanna baboons. In *Infanticide: Comparative and Evolutionary Perspectives*, ed. G. Hausfater & S. B. Hrdy. New York: Aldine Publishing Company.

Davies, N. B. (1978). Territorial defence in the speckled wood butterfly (*Pararge aegeria*): the resident always wins. *Animal Behaviour*, **26**, 138–47.

DeVore, I. (1962). The social behavior and organization of baboon troops. Ph.D. thesis, University of Chicago.

Frame, L. H., Malcolm, J. R., Frame, G. W. & van Lawick, H. (1979). Social organization of African wild dogs (*Lycaon pictus*) on the Serengeti Plains, Tanzania, 1964–1978. *Zeitschrift für Tierpsychologie*, **50**, 225–49.

Goodall, J. (1977). Infant killing and cannibalism in free-living chimpanzees. *Folia primatologica*, **28**, 259–82.

Goodall, J. (1983). Population dynamics during a 15 year period in one community of free-living chimpanzees in the Gombe National Park, Tanzania. *Zeitschrift für Tierpsychologie*, **61**, 1–60.

Goodall, J., Bandora, A., Bergmann, E., Busse, C., Matama, H., Mpongo, E., Pierce, A. & Riss, D. (1979). Intercommunity interactions in the chimpanzee population of the Gombe National Park. In *The Great Apes*, ed. D. A. Hamburg & E. R. McCown, pp. 13–53. Menlo Park: Benjamin/Cummings.

Greenwood, P. J. (1980). Mating systems, philopatry and dispersal in birds and mammals. *Animal Behaviour*, **28**, 1140–62.

Hamilton, W. J., III & Busse, C. (1982). Social dominance and predatory behavior of chacma baboons. *Journal of Human Evolution*, **11**, 567–73.

Hammerstein, P. (1981). The role of asymmetries in animal contests. *Animal Behavior*, **29**, 193–205.

Hanby, J. P. & Bygott, J. D. (1979). Population changes in lions and other predators. In *Serengeti: Dynamics of an Ecosystem*, ed. A. R. E. Sinclair & M. Norton-Griffiths, pp. 249–62. Chicago: University of Chicago Press.

Harvey, P. H., Kavanagh, M. & Clutton-Brock, T. H. (1978). Sexual dimorphism in primate teeth. *J. Zool. Lond.*, **186**, 475–86.

Hausfater, G. (1975). Dominance and reproduction in baboons (*Papio cynocephalus*). *Contributions to Primatology*, **7**. Basil: Karger.

Hrdy, S. B. (1974). Male–male competition and infanticide among langurs (*Presbytis entellus*) of Abu, Rajastan. *Folia primatologica*, **22**, 19–58.

Kruuk, H. (1972). *The Spotted Hyena*. Chicago: University of Chicago Press.

Kummer, H., Gotz, W. & Angst, W. (1974). Triadic differentiation: an inhibitory process protecting pair bonds in baboons. *Behaviour*, **49**, 62–87.

Le Boeuf, B. J. (1974). Male–male competition and reproductive success in elephant seals. *American Zoologist*, **14**, 163–76.

Malcolm, J. R. & Marten, K. (1982). Natural selection and the communal rearing of pups in African wild dogs (*Lycaon pictus*). *Behavioral Ecology and Sociobiology*, **10**, 1–13.

Maynard Smith, J. (1976). Evolution and the theory of games. *American Scientist*, **64**, 41–5.

Maynard Smith, J. (1982). *Evolution and the Theory of Games*. Cambridge University Press.

Maynard Smith, J. & Parker, G. A. (1976). The logic of asymmetric contests. *Animal Behaviour*, **24**, 159–75.

Nishida, T. (1979). The social structure of chimpanzees of the Mahale mountains. In *The Great Apes*, ed. D. A. Hamburg & E. R. McCown, pp. 73–121. Menlo Park: Benjamin/Cummings.

Packer, C. (1977*a*). Reciprocal altruism in *Papio anubis*. *Nature*, **265**, 441–3.

Packer, C. (1977*b*). Inter-troop transfer and inbreeding avoidance in *Papio anubis* in Tanzania. Ph.D. thesis, University of Sussex.

Packer, C. (1979*a*). Inter-troop transfer and inbreeding avoidance in *Papio anubis*. *Animal Behaviour*, **27**, 1–36.

Packer, C. (1979*b*). Male dominance and reproduction activity in *Papio anubis*. *Animal Behaviour*, **27**, 37–45.

Packer, C. (1980). Male care and exploitation of infants in *Papio anubis*. *Animal Behaviour*, **28**, 512–20.

Packer, C. & Pusey, A. E. (1982). Cooperation and competition within coalitions of male lions: kin selection or game theory? *Nature*, **296**, 740–2.

Packer, C. & Pusey, A. E. (1983). Male takeovers and female reproductive parameters: a simulation of oestrus synchrony in lions. *Animal Behaviour*, **31**, 334–40.

Parker, G. A. (1974). Assessment strategy and the evolution of animal conflicts. *Journal of Theoretical Biology*, **47**, 223–43.

Pianka, E. R. & Parker, W. S. (1975). Age-specific reproductive tactics. *American Naturalist*, **109**, 453–64.

Popp, J. L. (1978). Male baboons and evolutionary principles. Ph.D. thesis, Harvard University.

Popp, J. L. & DeVore, I. (1979). Aggressive competition and social dominance theory: synopsis. In *The Great Apes*, ed. D. A. Hamburg & E. R. McCown, pp. 317–38. Menlo Park: Benjamin/Cummings.

Pusey, A. E. (1978). The physical and social development of wild adolescent chimpanzees (*Pan troglodytes schweinfurthii*). Ph.D. thesis, Stanford University.

Pusey, A. E. (1979). Intercommunity transfer of chimpanzees in Gombe National

Park. In *The Great Apes*, ed. D. A. Hamburg & E. R. McCown, pp. 465–79. Menlo Park: Benjamin/Cummings.

Pusey, A. E. (1983). Mother–offspring relationships in chimpanzees after weaning. *Animal Behaviour*, **31**, 363–77.

Ralls, K., Brugger, K. & Ballou, J. (1979). Inbreeding and juvenile mortality in small populations of ungulates. *Science*, **206**, 1101–3.

Ralls, K. & Ballou, J. (1982). Effects of inbreeding on infant mortality in captive primates. *International Journal of Primatology*, **3**, 491–505.

Ransom, T. W. & Ransom, B. S. (1971). Adult male–infant relations among baboons (*Papio anubis*). *Folia primatologica*, **16**, 179–95.

Schaller, G. B. (1972). *The Serengeti Lion*. Chicago: University of Chicago Press.

Simons, E. L. (1972). *Primate Evolution*. New York: Macmillan.

Stein, D. M. (1981). The nature and function of social interactions between infant and adult male yellow baboons. Ph.D. thesis, University of Chicago.

Sugiyama, Y. (1965). Behavioural development and social structure in two troops of hanuman langurs (*Presbytis entellus*). *Primates*, **6**, 381–418.

Teleki, G., Hunt, E. E., Jr & Pfifferling, J. H. (1976). Demographic observations (1963–1973) on the chimpanzees of Gombe National Park, Tanzania. *Journal of Human Evolution*, **5**, 559–98.

Tutin, C. E. G. (1979). Mating patterns and reproductive strategies in a community of wild chimpanzees (*Pan troglodytes schweinfurthii*). *Behavioral Ecology and Sociobiology*, **6**, 29–38.

Tutin, C. E. G. & McGinnis, P. R. (1981). Chimpanzee reproduction in the wild. In *Reproductive Biology of the Great Apes*, ed. C. E. Graham, pp. 239–64. New York: Academic Press.

van Lawick, H. & van Lawick-Goodall, J. (1971). *Innocent Killers*. London: Collins.

Wrangham, R. W. (1975). The behavioural ecology of chimpanzees in Gombe National Park, Tanzania. Ph.D. thesis, Cambridge University.

Wrangham, R. W. (1980). An ecological model of female-bonded primate groups. *Behaviour*, **75**, 262–300.

14

Functions of communal care in mammals

J. L. GITTLEMAN

Introduction

Mammals display three forms of parental care: maternal, biparental and communal. Although not mutually exclusive, each form represents a different type of parental investment. Maternal care is, of course, most common and is represented by female placentation, lactation and general provisioning of the young. Biparental care is characterized by the establishment of a monogamous pair-bond between the male and female, with parental duties divided between each sex. Both maternal and biparental care are, in most cases, well-explained in terms of individuals increasing their genetic contribution to future generations (Trivers, 1972; Maynard Smith, 1977). However, communal care (where individuals of one or both sexes assist in the rearing of other individuals' offspring, and sometimes relinquish breeding to do so) appears to contradict this view of natural selection theory.

Functional explanations for communal care in mammals have not received as much attention as similar forms of parental care in cooperative-breeding birds (e.g. Brown, 1978, 1983; Emlen, 1978) and eusocial insects (e.g. Hamilton, 1972; West Eberhard, 1975), though reviews of allo-parenting and adoption in mammals are available (Spencer-Booth, 1970; Hrdy, 1976; Eisenberg, 1981; Riedman, 1982). At this early stage it is important to identify which classes of communal care in mammals require description and those selective forces critical to examining current evolutionary models of their functional significance. In this paper I will: (i) define the various forms of communal care in mammals and outline general characteristics associated with them; (ii) describe ecological and behavioural advantages promoting communal care; and (iii) discuss various disadvantages or costs accruing from communal rearing systems. This chapter differs from Riedman's (1982) recent review by dealing exclusively with mammalian trends and focussing on species which have been studied in the wild. This may remove spurious behavioural and developmental trends in captive animals (see Kaufmann & Kaufmann, 1963; Jones, 1967; Rood, 1980) and pinpoint important ecological constraints on communal care.

Characteristics of mammalian communal care

Generally, communal care (CC) in mammals may be defined as individuals as well as the parents assisting in the feeding, guarding and indirect provisioning of the young. This definition does not specify the effects of helping upon the fitnesses of the individuals involved. As with biparental care (see Kleiman & Malcolm, 1981), CC may occur directly or indirectly. Direct forms are acts that are not expressed in everyday activities but which occur in response to the presence of young. These acts include feeding, carrying, grooming, playing and actively protecting young. Indirect forms include behaviours performed independently of the presence of young which inadvertently increase survivorship of young. Examples of indirect CC include territorial defense of critical resources, scent marking, herding movements, or long-distance vocalizations that facilitate the spacing of individuals or groups.

Communal-caring species tend to live in small (3–10 individuals), relatively stable groups in well-defended territories, although exceptions will be mentioned below. Four distinct types of CC are observed. These involve (see Table 1): (i) nuclear family groups with a reproductive pair and offspring from previous breeding seasons (*Canis lupus*, *Saguinus oedipus*) – common names are listed in Table 1; (ii) a matriarchy with a reproductive female or reproductive females, daughters and sisters (*Elephas maximus*, *Nasua narica*); (iii) a harem consisting of a male and related or unrelated females (*Alouatta palliata*, *Cynomys ludovicianus*); and (iv) multi-male/multi-female groups containing both related and unrelated individuals (*Mungos mungo*, *Hydrochoerus hydrochaeris*).

Helpers range in age from juveniles to mature breeding adults. Characteristic of nuclear family or matriarchal groups are non-breeding helpers, and therefore the number of helpers is determined by the number of non-breeders in the group. A common feature among communal-caring species is protracted development and the delay of sexual maturation until the age of two or more years. As in birds (Brown, 1974, 1982; Gaston, 1978; Wiley, 1981), delayed breeding in mammals may be correlated with risks in premature breeding and energetic costs of reproduction, such as establishment of territories and mate competition. The precise causal relationship between delayed breeding and CC has yet to be thoroughly studied; nevertheless, inter-specific differences do support this trend. For example, after removing the effects of size, communal species (e.g. *Panthera leo*, *Canis lupus*, *Nasua narica*) in the order Carnivora reach sexual maturity – defined

Table 1. *Types of communal care in mammals*

Taxon	Common Name	Type (synchrony)[a]	Reference
Rodentia			
Cynomys ludovicianus	Black-tailed prairie dog	3 (S)	Smith *et al.*, 1973; Hoogland, 1981
Spermophilus beldingi	Belding's ground squirrel	2	Sherman, 1980
Castor canadensis	Beaver	1	Tevis, 1950; Wilsson, 1971
Hydrochoerus hydrochaeris	Capybara	4 (S)	Macdonald, 1981
Carnivora			
Canis lupus	Wolf	1	Rabb, Woolpy & Ginsburg, 1967; Mech, 1970
Canis latrans	Coyote	1	Camenzind, 1978
Lycaon pictus	African wild dog	4	Malcolm & Marten, 1982
Vulpes vulpes	Red fox	1	Macdonald, 1979
Cuon alpinus	Dhole	4	Johnsingh, 1982
Nasua narica	Coati	2 (S)	Kaufmann, 1962; Janzen, 1970; Russell, 1979
Helogale parvula	Dwarf mongoose	1	Rasa, 1977; Rood, 1978
Mungos mungo	Banded mongoose	4 (S)	Rood, 1974, 1975
Hyaena brunnea	Brown hyaena	1	Owens & Owens, 1978; Mills, 1982
Crocuta crocuta	Spotted hyaena	4	Kruuk, 1972
Panthera leo	African lion	4 (S)	Schaller, 1972; Bertram, 1975; Packer & Pusey, 1983*a,b*
Pinnipedia			
Mirounga angustirostris	Northern elephant seal	3 (S)	Le Boeuf, Whiting & Gantt, 1972
Chiroptera			
Tadarida brasiliensis	Mexican free-tailed bat	2 (S)	Davis, Herreid & Short, 1962
Myotis thysanodes	Little brown bat	2 (S)	O'Farrell & Studier, 1973
Primates			
Lemur catta	Ring-tailed lemur	3 (S)	Jolly, 1966; Sussman, 1977; Klopfer & Boskoff, 1979
Microcebus murinus	Lesser mouse lemur	2 (S)	Martin, 1972, 1973
Alouatta villosa	Mantled howler	3	Jolly, 1972
Saguinus oedipus	Cotton-head marmoset	1	Dawson, 1977
Cercopithecus aethiops	Vervet	3	Struhsaker, 1967
Cercopithecus cambelli	Lowe's guenon	3	Bourlière, Hunkler & Bertrand, 1970
Cercopithecus ascanius	Redtail monkey	3	Struhsaker, 1977
Macaca mulatta	Rhesus macaque	3	Kaufmann, 1966

Table 1 (*cont.*)

Taxon	Common name	Type (synchrony)[a]	Reference
Macaca sinica	Toque monkey	3	Dittus, 1979
Presbytis entellus	Hanuman langur	2	Hrdy, 1977; McKenna, 1981
Presbytis cristatus	Silvered langur	3	Bernstein, 1968
Colobus guereza	Black-and-white colobus	3	Oates, 1977
Pan troglodytes	Chimpanzee	4	Pusey, 1978
Gorilla gorilla	Mountain gorilla	2	Schaller, 1963; Harcourt, 1979
Artiodactyla			
Connochaetes taurinus	Wildebeest	2 (S)	Estes & Estes, 1979
Proboscidea			
Elephas maximus	Indian elephant	2	McKay, 1973

[a] Species and types of communal care in Eutherian mammals. Types are represented by: (1) nuclear family (reproductive pair, offspring from previous breeding seasons); (2) matriarchy (reproductive female, daughters, sisters); (3) harem (male, allied females); (4) multi-male/multi-female group. Breeding synchrony, if known, is indicated in parentheses.

Fig. 1. Age at sexual maturity plotted against generic body weight for different forms of parental care in Carnivora. ● Communal care, ○ biparental care, △ female care. Data taken from Gittleman (1984).

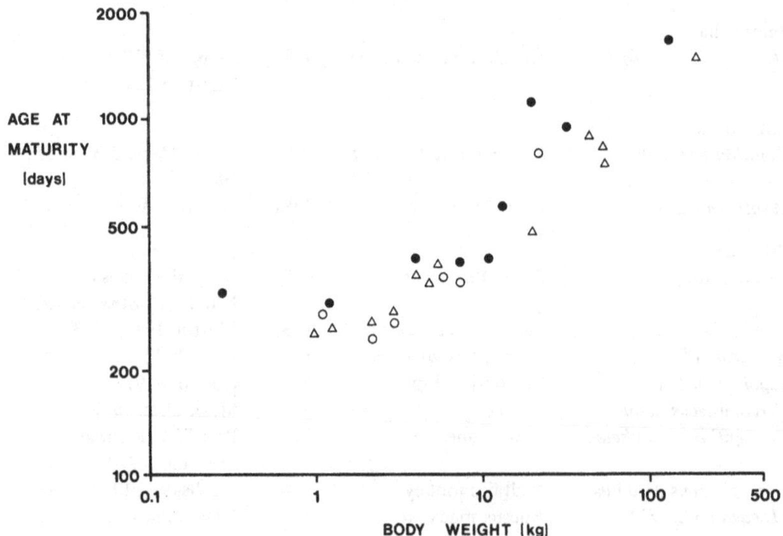

as average age of females at first reproduction – later than maternal (e.g. *Panthera tigris*) or biparental (e.g. *Hyaena brunnea*) species (see Fig. 1). This pattern is also found in closely related taxa. In the genus *Cynomys*, the communal breeding black-tailed prairie dog (*C. ludovicianus*) does not reproduce until at least two years of age whereas the white-tailed prairie dog (*C. leucurus*), which lives in loose aggregations, usually breeds in the spring of its first year (Hoogland, 1981).

Among communal species where more than one female reproduces in a given breeding season, females often breed synchronously (see Table 1 for species where synchronized births have been reported). This would be essential for communal suckling (e.g. *Panthera leo*, *Hydrochoerus hydrochaeris*), except where the duration of lactation is long enough for lactation periods to overlap (Vehrencamp, 1980). Although breeding synchronization is found in many communal mammals, it is not a unique characteristic of CC, for numerous non-communal species synchronize birthing (e.g. *Nyctalus nuctula*: Kleiman, 1969), and it may be induced by a variety of reproductive parameters (see Packer & Pusey, 1983*a*).

As in avian communal systems (Brown, 1978; Emlen, 1978; Koenig & Pitelka, 1981), the taxonomic, ecological and geographic variation among communal mammals is extensive. CC is found in at least seven orders (Carnivora, Primates, Chiroptera, Proboscidea, Artiodactyla, Pinnipedia and Rodentia), roughly 15% of all mammalian species (Eisenberg, 1981). This percentage is noticeably greater when data from captive animals are included (Spencer-Booth, 1970; Riedman, 1982). It should be borne in mind that most of these species are group living in general and thus intensively studied, perhaps inflating the occurrence of CC among eutherian mammals (Eisenberg, 1977, 1981). Dietetic and foraging patterns are diverse; species displaying CC include strict carnivores (*Panthera leo*, *Canis lupus*), omnivores (*Vulpes vulpes*), folivores (*Presbytis entellus*), frugivores (*Macaca mulatta*, *Nasua narica*) and insectivores (*Helogale parvula*). The vegetation, zonation and geography in which communal species reside are equally varied and suggest no common characteristics. In summary, CC is associated with a variety of factors in different mammals and, as in birds (Brown, 1978, 1983; Emlen, 1978), probably evolved independently under a number of conditions. In the remainder of this chapter I will discuss some of the benefits and costs specific to those conditions which gave rise to CC in mammals.

Benefits and costs of communal care in mammals: ecological constraints

Benefits and costs accrued from CC can be measured in relation to the reproductive adults, offspring or helpers. The following discussion deals

Table 2. *Benefits and costs accrued by communal helpers to the breeding adults and young*

Benefits	Costs
1. Predator defence	1. Attract predators
a. increased vigilance	2. Increased aggression
b. 'selfish herd'	3. Mix-up of litters
2. Suitable breeding territory	4. Disease and ectoparasites
3. Defend territory	5. Displace breeder
4. Communal suckling	6. Infanticide
5. Maternal learning	7. Increased competition (for food, mates, denning areas, etc.)
6. Feed lactating female(s)	
7. Acquisition of food	
8. Allopreen or thermoregulate	

with effects that helpers confer on the breeding adults and their offspring, although in some cases (e.g. conspicuousness to predators, suitable breeding areas) these are spread across the communal group. The behavioural and ecological characteristics which represent the benefits and costs discussed herein are listed in Table 2.

Benefits

The two most common functional explanations of CC in mammals relate to benefits gained in antipredator defence and food acquisition. Antipredator defence by helpers may be directed to inter-specific or conspecific threats, and operate by increased vigilance (including improved fighting ability) or a 'selfish herd' effect (Hamilton, 1971). Increased vigilance occurs in two forms, either in small groups where one or two helpers remain with the young in a denning (natal) area or in large herds where many pairs of eyes (or ears) are better than one in detecting predators. In the former type, it is known that many social canids (*Canis lupus*, *C. latrans*, *Lycaon pictus*) and viverrids (*Mungos mungo*, *Helogale parvula*) fend off potential predators as a communal group but are unable to do so solitarily. In specific cases, where young have been abandoned or left unattended, the young are often killed (e.g. *Lycaon pictus*; Malcolm & Marten, 1982). In smaller species (ground squirrels, prairie dogs) increased vigilance may lead to earlier predator detection and alarm calls, thereby increasing the chances of avoiding predation and protecting young. However, this would predict that larger communal groups would improve predator detection and increase survivorship of young, a prediction that remains untested (see Hoogland, 1979*a*, 1981). Species that cannot directly (e.g. by fighting) or indirectly (e.g. by escaping or giving alarm calls) avoid predation by increased vigilance must rely on a 'selfish herd' or dilution effect, particularly when the level of predation is intense. Over a 5-year period Estes & Estes (1979) found that wildebeest (*Connochaetes taurinus*) suffered around 50% calf mortality, predominantly from spotted hyaenas (*Crocuta crocuta*). Calves reared in large matriarchal groups had a survival rate of 75% compared with 41% in smaller herds, even though a larger group cannot escape an attacking hyaena clan. Thus a benefit of communal rearing in wildebeest, as perhaps in other communal ungulates (e.g. impala, Grant's and Thomson's gazelles), is related to a dilution effect more than the ability to prevent a kill.

Food acquisition is also an important element of CC. Direct feeding of young or regurgitating food to the lactating mother by helpers aids considerably in the maintenance and rearing of young. Such a benefit is particularly critical in species relying on scarce food resources (e.g. Kleiman & Eisenberg, 1973; Macdonald & Moehlman, 1982). Generally the breeding female remains with the young at all times until weaning, while helpers provision food for her during lactation and occasionally feed the young (*Vulpes vulpes*: Macdonald, 1979; *Canis mesomelas*: Moehlman, 1979; *Lycaon pictus*: Malcolm & Marten, 1979; *Hyaena brunnea*: Mills, 1982; *Helogale parvula*: Rood, 1978). Communal suckling in some species (e.g. *Panthera leo*, *Hydrochoerus hydrochaeris*) also eases energetic demands of lactation. Because CC may decrease the costs of lactation and probably also the difficulty in finding food, this may give rise to larger and heavier litters: in the order Carnivora, maternal and biparental species have lighter litter weights (birth weight multiplied by litter size) than communal species (see Fig. 2). The precise causal factors (e.g. amount of food, quality of food) influencing this relationship remain unknown; furthermore, study of such factors may disentangle positive correlations between number of helpers and survival rate recently found for black-backed and golden jackals (Moehlman, 1979; Macdonald & Moehlman, 1982), African lions (Bygott, Bertram & Hanby, 1979), coyotes (Bekoff & Wells, 1982) and possibly African wild dogs (Malcolm & Marten, 1982).

Predator defence and food acquisition are beneficial characteristics accrued from communal rearing in many, if not most, mammals. However, these factors may be relatively unimportant in other species. Discussions of cooperative breeding in birds have emphasized two environmental pressures in the evolution of cooperative breeding (Emlen, 1982*a*): (i) patchy, stable habitat where competition for suitable breeding territories is severe (see Gaston, 1978; Koenig & Pitelka, 1981); and (ii) unpredictable environments, often due to rainfall cycles, which promote dramatic changes in the availability of food and suitable breeding habitats. (Brown (1982, personal communication) criticizes this ecological dichotomy and suggests that territoriality and energy dynamics that prevent an individual from breeding, and thus select for helping, are more salient factors contributing to the origin of avian communal-breeding systems. Although such factors may apply to mammalian CC, examination of payoffs from helping in particular species in relation to ecological stability may be of heuristic value.) Two examples illustrate the importance of these factors in mammalian CC. Mexican free-tailed bats (*Tadarida brasiliensis*) are migratory in North America, and in the summer females establish breeding colonies in which birthing is synchronized (Davis, Herreid & Short, 1962). Newborn young are naked and flightless, requiring close contact to maintain body temperatures. The young are clustered in such large colonies that it is impossible for females to identify their own offspring and young are suckled indiscriminately, unlike in most other Chiroptera (Eisenberg, 1981). Roosting of female creches in caves that reach peak population levels in the hottest months may be advantageous because of the constant air and rock temperatures. Variation in clustering is an effective means of regulating heat loss (Herreid, 1963): experimental studies show that solitary-roosting bat species have significantly higher oxygen levels and weight losses than species roosting in colonies (Trune & Slobodchikoff, 1976). Competition for cave roosts is intense, involving considerable

Fig. 2. Litter weight (litter size × birth weight) plotted against generic female body weight for different forms of parental care in Carnivora. ● Communal care, ○ biparental care, △ female care. Data taken from Gittleman (1984).

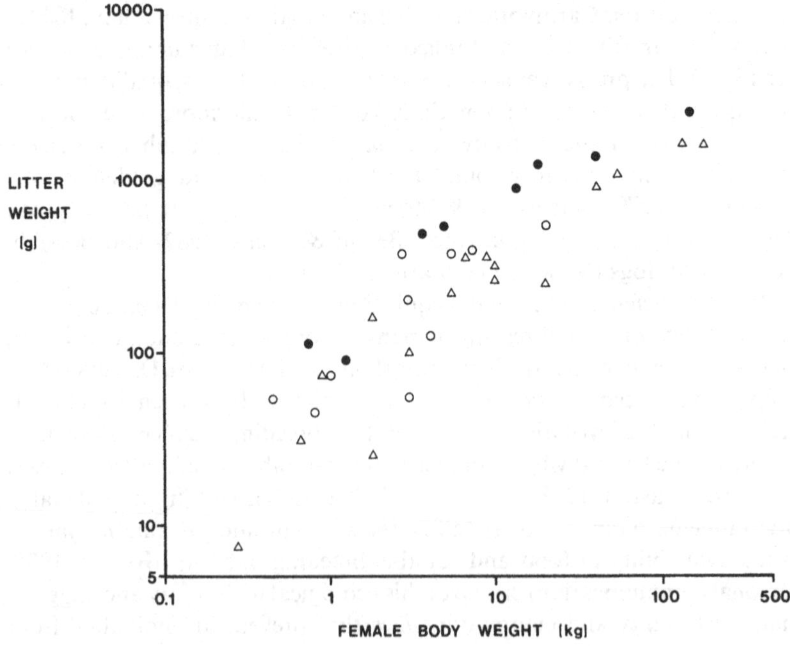

pushing, nudging and biting, and often resulting in injury (Davis *et al.*, 1962). At the same time, predation on Mexican free-tailed bats is low and foraging behaviour is unselective (Ross, 1961). Thus, this species and perhaps other Chiroptera (Eisenberg, 1981) form CC more in response to selection for suitable breeding sites (caves) which are particularly advantageous for homeothermy (McCracken, 1984).

Variable and unpredictable environments may also promote helping behaviour in mammals (see Emlen, 1982*a*). As an example, the capybara (*Hydrochoerus hydrochaeris*) lives in large, mixed-sex groups numbering over 50 individuals, which congregate at the edge of rapidly fluctuating water sources (Macdonald, 1981). The young are protected and suckled indiscriminately by females in the group. Group sizes increase during the dry season when water holes are patchily distributed. Thus group sizes and the number of communal helpers are primarily determined by the availability of water resources. However, as Macdonald (1981) points out, this relationship may be mediated through other factors: water may provide an escape from predation, a suitable habitat for mating (all matings occur in water), plant foods and a medium for thermoregulation. Nevertheless, the capybara serves as an example of how an unpredictable environmental factor, not immediately related to predation or food acquisition, may influence mammalian CC.

The above discussion indicates that predator defence, food acquisition and suitable breeding sites are ecological factors promoting CC. In the context of these factors I have mentioned additional payoffs in communal suckling, feeding lactating females and thermoregulation. Two other benefits, though not critically important for the evolution of CC, deserve mention. In larger mammals, because females produce a limited number of offspring which require prolonged and intensive care, the cost of maternal errors may be great (Riedman, 1982). Therefore CC may serve as necessary parental practice while caring for another individual's offspring. Three characteristics are relevant to this 'learning hypothesis' (Hrdy, 1976): (i) differences in maternal ability between primiparous and multiparous females should be apparent from mothering experiences; (ii) inexperienced mothers (nulliparous) should be primarily involved in helping; and (iii) reproductive success should increase with maternal experience. In general, mammalian studies, particularly on primates, support this hypothesis (see Hrdy, 1976, 1977; Riedman, 1982); however, the relationship between maternal experience and reproductive success is unknown, and in some cases 'experienced' (multiparous) non-lactating or non-pregnant mothers are involved in helping (Hanuman langurs: Hrdy, 1977). The fact that experienced mothers are involved in helping is unsurprising: females resting from recent motherhood or non-reproductive old females would be expected to assist in terms of energetic costs in rearing. Neither case refutes the learning hypothesis; rather, they merely specify some conditions which are exceptions and suggest that the learning hypothesis alone does not supply a sufficient function for the evolution of CC.

CC probably increases territorial defence in response to scarcity of food, suitable breeding areas and mates. Territorial benefits from communality depend on the degree of individual mobility (ranging areas) and seasonal fluctuations of scarce resources (Eisenberg, 1981). In the case of African wild dogs (*Lycaon pictus*) which rely on migratory ungulate herds as a primary food source (Schaller, 1972), defensibility is impossible. In contrast, territorial boundaries of wolves (*Canis lupus*) on Isle Royale, where prey (mainly caribou) are less mobile, reveal that larger packs have greater sized territories than smaller groups (Mech, 1970). Territoriality has also been shown to increase where suitable breeding areas are patchily distributed or availability of mates is low (Waser & Wiley, 1980; Eisenberg, 1981). Communal helpers may increase such territoriality and in turn the breeding adults and young may benefit.

Thus far I have discussed various forms of CC as if they arise as rigid characteristics to particular ecological or behavioural factors. This is not the case, however. Just as selective factors have promoted CC, they may also relax it (see Macdonald & Moehlman, 1982). If food availability, breeding sites, or other ecological factors produce risks to individuals when helping then the establishment of CC may decline (Emlen, 1982*a*,*b*). For example, in wolves (Harrington & Mech, 1980), coyotes (Bekoff & Wells, 1982), red foxes (Macdonald, 1981), African wild dogs (Malcolm & Marten, 1982) and golden jackals (Macdonald & Moehlman, 1982), potential helpers disperse from communal groups during seasons of low food availability and return when resources are more plentiful. It is through such 'natural experiments' that, as in birds (e.g. Emlen, 1981), specific ecological constraints promoting or relaxing CC in mammals will be isolated.

Costs

Previous reviews have emphasized the benefits and played down the costs of CC in mammals (Vehrencamp, 1980; Riedman, 1982; but see Hrdy, 1976). As with grouping in general (Bertram, 1978; Harvey & Greene, 1981), many of the harmful characteristics resulting from CC are subtle and difficult to measure. Furthermore, it is usually *assumed* that costs rarely outweigh benefits (otherwise CC would not be observed). Even so, detailed field studies are now showing that behavioural and ecological disadvantages of grouping in general are significant and that many of these may impose constraints on CC. Generally, three costs arise from communality (see Table 2): increased conspicuousness to predators, increased competition, and increased transmission of disease or ectoparasites. Other costs, such as infanticide, potential mix-up of litters or increased aggression, either occur less frequently or are directly related to these three characteristics.

The level of attraction to potential predators from CC is dependent on vegetation cover and defensive ability. Although many of the social carnivores mentioned above suffer occasional mortality from other predators (Schaller, 1972; van Lawick, 1973), CC probably does not significantly increase its likelihood. In contrast, ungulate herds in large matriarchal groups and small mammal colonies in large mixed-sex groups, both of which reside in open vegetation, are potentially more conspicuous to predators. For example, large aggregations of female wildebeest with young are attacked more frequently by spotted hyaenas than smaller groups (Estes & Estes, 1979). However, aside from communality *per se*, it is difficult to remove the additional factor that wildebeest females in the final stages of pregnancy, particularly after the foetus has descended to the lower abdomen, are more vulnerable to predation. In small mammals, CC may produce vocal conspicuousness, visibility of denning areas or mounds, and visibility of individuals: in black-tailed prairie dogs (Hoogland, 1979*b*) all of these characteristics were correlated with ward size (number of individuals in a group relative to other groups).

The frequency and intensity of aggressive behaviours correlate with group size in many communal species (see Zimen, 1976; Hoogland, 1979*b*; Macdonald, 1981). In those species possessing horns, large canines or claws, increased aggression may lead to serious injury and possibly to death in both adults and young. Most cases of increased aggression with group size are associated with resource limitations (food, mates, etc.), thus promoting competition. In the context of aggression in communal groups, considerable attention has been given to the evolution and significance of infanticide (Hrdy, 1979; Sherman, 1981). Examples of parents killing offspring are commonly explained by ecological stress, deformed young and/or young of an inappropriate sex. Infanticide by helpers may have other causes, such as competition for food, mates or denning areas, cannibalism, or accidental occurrence. Such predation may incur substantial mortality: at least 25% of all African lion cubs, 17% of spotted hyaena young and 8% of Belding ground squirrel pups are eaten by helpers (Kruuk, 1972; Bertram, 1975; Sherman, 1981). Because of the many possible reasons for infanticide by helpers, it is difficult to isolate specific causes. Sherman (1981) found that in Belding ground squirrels, one-year-old males probably kill to obtain food which increases overwinter survivorship and the possibility of reproductive success in the following spring, while females probably kill to remove current and future competitors from preferred nest sites. In larger species where individuals are more dispersed and have larger home ranges, infanticide may also be caused by other factors. Infanticide in African lions most often occurs after new males immigrate into a pride. The death of an infant induces oestrus in lionesses and increases the chances of matings by the new males (Packer & Pusey, 1983*b*), a pattern also found in some primates (Hrdy, 1979). Therefore, infanticide among communal mammals may be caused by many individuals for various reasons, and its cause should always be distinguished from the causes of grouping in general.

Other possible costs resulting from CC are misdirected parental care, transmission of disease or ectoparasites, or displacement of breeders. Evaluation of each of these costs requires detailed information on relatedness among individuals in a communal group, observations over successive breeding seasons, and close inspection of individual physical condition. Few studies have achieved these standards. The likelihood of misdirected parental care or transmission of disease is greater in large, transient populations (e.g. communal ungulates) or in those species that remain in tight clusters (e.g. communal bats, small mammals). An extreme example of the latter is found in the Mexican free-tailed bat and the little brown bat where females frequently lose their young within a mass of hundreds and suckle so indiscriminately that milk has been found in the stomachs of adults (Davis *et al.*, 1962), but this may occur less often than originally reported (McCracken, 1984). Mix-up of litters is probably rare in mammals, however, because mothers have numerous (behavioural and hormonal) protection mechanisms (Holmes & Sherman, 1983) to guard their substantial prenatal investment.

Transmission of disease or ectoparasites is more common and may occur in many forms and degrees. Hoogland (1979*b*) states that, in small mammals, 'Individuals of colonial species probably contract diseases and ectoparasites more often than do individuals of closely related solitary species, and that disease and ectoparasites are probably more troublesome in large colonies than small colonies'. Flea infestation correlates positively with colony size in black-tailed and white-tailed prairie dogs (Hoogland, 1979*b*), and transmission of mites, beetle larvae and ticks is more extensive in large colonies of Mexican free-tailed bats (Davis *et al.*, 1962). Although the effects of parasitism on offspring survival and reproductive success in natural populations are unknown, laboratory studies indicate that parasitism probably reduces fitnesses of mammalian hosts (Smith, 1977).

In summary, costs resulting from CC will depend on the number of helpers in a group, frequency of contact between them, potentially harmful morphological characters and various ecological factors. Future studies examining the benefits of CC in mammals should include these costs so that a more balanced assessment of those selective factors will emerge.

Conclusions

The above discussion suggests that many diverse behavioural and ecological conditions have selected for communal care in mammals. Although treated at a general level, partly because of insufficient data, the benefits and costs operating on communal systems are almost certain to be unique to each species. A similar conclusion, following much more detailed study of avian communal breeding systems, was drawn by Brown (1978): '. . . species termed "communal" are often fundamentally different from each other, sharing only a few traits, such as various kinds of helping behaviour. The selection pressures operating in these species may also be extremely different'. The trends and conclusions set forth in this chapter are tentative and are intended as a general framework upon which to place a variety of environmental factors that influence mammalian communal care. I finish by suggesting some problem areas which might help in refining our questions.

(i) What are the interactive relationships between specific benefits and costs in particular communal systems? As an example of this problem, helpers in black-backed jackals (*Canis mesomelas*) defend young against potential predators and feed young both indirectly (by feeding lactating mother) and directly (by regurgitating to young). As these benefits increase so does the likelihood of aggression, attraction to predators, competition or even infanticide. Therefore study must be made of the balance struck between costs and benefits in particular communal systems.

(ii) Is protracted development and delayed sexual maturity a causal or consequential factor of communal care? Many species develop slowly in relation to complexity of food acquisition, habitat structure, litter size, and metabolic rate, to name only a few influential factors (see Eisenberg, 1981). Communal care, however, is not consistently associated with these factors. The study of within-species variation, focussing on ecological fluctuations that promote delayed maturity, should be helpful in answering this question.

(iii) Given that an individual delays breeding and remains in its natal group, what are the factors influencing it to provide care? Some of these have been mentioned above in the context of energetic constraints on CC. However, this discussion ignored the fact that helping involves different decisions at various stages in an individual's ontogeny. Indeed, perhaps *the* only generality among communal mammals, as in cooperative breeding birds (Brown, 1978, 1983), may be flexibility. Obviously, resource availability, age and mate competition will weigh heavily in this decision-making process.

(iv) Lastly, further attention should be given to the characteristics that are required for a mammalian species to be labelled 'communal'. This chapter has identified varying degrees of communality, from little communal effort expended in large ungulate herds to alarm calls in prairie dogs to extensive guarding and provisioning in social carnivores and primates. Thus, communal-rearing systems in mammals appear to exist along a continuum rather than as an all-or-none phenomenon (see Michener & Murie, 1983; Hoogland, 1983).

I thank Drs R. H. Bixler, J. L. Brown, P. H. Harvey and D. G. Kleiman for comments on a previous draft of this chapter. This work was partially funded by a Smithsonian Postdoctoral Fellowship Grant.

References

Bekoff, M. & Wells, M. C. (1982). Behavioral ecology of coyotes: social organization, rearing patterns, space use, and resource defense. *Zeitschrift für Tierpsychologie*, **60**, 281–305.

Bernstein, I. S. (1968). The lutong of Kuala Selangor, *Behaviour*, **32**, 1–16.

Bertram, B. C. R. (1975). Social factors influencing reproduction in wild lions. *Journal of Zoology*, **177**, 463–82.

Bertram, B. C. R. (1978). Living in groups: predators and prey. In *Behavioural Ecology: An Evolutionary Approach*, ed. J. R. Krebs & N. B. Davies, pp. 64–96. Oxford: Blackwells.

Bourlière, F., Hunkler, C. & Bertrand, M. (1970). Ecology and behavior of Lowe's Guenon (*Cercopithecus cambelli lowei*) in the Ivory Coast. In *Old World Monkeys*, ed. J. R. Napier & P. H. Napier, pp. 297–350. New York: Academic Press.

Brown, J. L. (1974). Alternative routes to sociality in jays – with a theory for the evolution of altruism and communal breeding. *American Zoologist*, **14**, 63–80.

Brown, J. L. (1978). Avian communal breeding systems. *Annual Reviews of Ecology and Systematics*, **9**, 123–55.

Brown, J. L. (1982). Optimal group size in terrestrial animals. *Journal of Theoretical Biology*, **95**, 793–810.

Brown, J. L. (1983). The evolution of helping behavior – an ontogenetic and comparative perspective. In *Evolution of Adaptive Skills: Comparative and Ontogenetic Approaches*, ed. E. S. Gollin. New York: Academic Press.

Bygott, J. D., Bertram, B. C. R. & Hanby, J. P. (1979). Male lions in large coalitions gain reproductive advantages. *Nature*, **282**, 839–41.

Camenzind, F. J. (1978). Behavioral ecology of coyotes on the National Elk Refuge, Jackson, Wyoming. In *Coyotes: Biology, Behavior, Management*, ed. M. Bekoff, pp. 267–94. New York: Academic Press.

Davis, R. B., Herreid, C. F. & Short, H. L. (1962). Mexican free-tailed bats in Texas. *Ecological Monographs*, **32**, 11–46.

Dawson, G. A. (1977). Composition and stability of social groups of the Tamarin, *Saguinus oedipus geoffroyi*, in Panama: ecological and behavioral implications. In *Biology and Conservation of Callitrichidae*, ed, D. G. Kleiman, pp. 23–39. Washington, D.C.: Smithsonian.

Dittus, W. P. J. (1979). The evolution of behaviors regulating density and age-specific sex ratios in a primate population. *Behaviour*, **69**, 265–302.

Eisenberg, J. F. (1977). The evolution of the reproductive unit in the class

Mammalia. In *Reproductive Behavior and Evolution*, ed. J. S. Rosenblatt & B. R. Komisaruk, pp. 39–71. New York: Plenum Press.

Eisenberg, J. F. (1981). *The Mammalian Radiations*. Chicago: University of Chicago Press.

Emlen, S. T. (1978). The evolution of cooperative breeding in birds. In *Behavioural Ecology: An Evolutionary Approach*, ed. J. R. Krebs & N. B. Davies, pp. 245–81. Oxford: Blackwells.

Emlen, S. T. (1981). Altruism, kinship, and reciprocity in the white-fronted bee-eater. In *Natural Selection and Social Behavior*, ed. R. D. Alexander & D. W. Tinkle, pp. 217–30. New York: Chiron Press.

Emlen, S. T. (1982*a*). The evolution of helping behavior. I. An ecological constraints model. *American Naturalist*, **119**, 29–39.

Emlen, S. T. (1982*b*). The evolution of helping behavior. II. The role of behavioral conflict. *American Naturalist*, **119**, 40–53.

Estes, R. D. & Estes, R. K. (1979). The birth and survival of wildebeest calves. *Zeitschrift für Tierpsychologie*, **50**, 45–95.

Gaston, A. J. (1978). The evolution of group territorial behavior and cooperative breeding. *American Naturalist*, **112**, 1091–100.

Gittleman, J. L. (1984). The behavioural ecology of carnivores. Ph.D. thesis, University of Sussex.

Hamilton, W. D. (1971). Geometry for the selfish herd. *Journal of Theoretical Biology*, **31**, 295–311.

Hamilton, W. D. (1972). Altruism and related phenomena, mainly in social insects. *Annual Reviews of Ecology and Systematics*, **3**, 193–232.

Harcourt, A. C. (1979). Social relationships among adult female mountain gorillas. *Animal Behaviour*, **27**, 251–64.

Harrington, F. H. & Mech, L. D. (1980). Wolf helpers at the den? Paper presented at the Animal Behavior Society Meeting. Fort Collins, Col. (June, 1980).

Harvey, P. H. & Greene, P. J. (1981). Group composition: an evolutionary perspective. In *Group Composition: Theoretical and Clinical Perspectives*, ed. H. Kellerman, pp. 149–69. New York: Grune & Stratton.

Herreid, C. F., II (1963). Temperature regulation of Mexican free-tailed bats in cave habitats. *Journal of Mammalogy*, **44**, 560–73.

Holmes, W. G. & Sherman, P. W. (1983). Kin recognition in animals. *American Scientist*, **71**, 46–55.

Hoogland, J. L. (1979*a*). The effect of colony size on individual alertness of prairie dogs (Sciuridae: *Cynomys* spp.). *Animal Behaviour*, **27**, 394–407.

Hoogland, J. L. (1979*b*). Aggression, ectoparasitism, and other possible costs of prairie dog (Sciuridae, *Cynomys* spp.) coloniality. *Behaviour*, **69**, 1–35.

Hoogland, J. L. (1981). The evolution of coloniality in white-tailed and black-tailed prairie dogs (Sciuridae: *Cynomys leucurus* and *C. ludovicianus*). *Ecology*, **62**, 252–72.

Hoogland, J. L. (1983). Black-tailed prairie dog coteries are cooperatively breeding units. *American Naturalist*, **121**, 275–80.

Hrdy, S. B. (1976). Care and exploitation of nonhuman primate infants by conspecifics other than the mother. *Advances in the Study of Behaviour*, **6**, 101–58.

Hrdy, S. B. (1977). *The Langurs of Abu: Female and Male Strategies of Reproduction*. Cambridge: Harvard University Press.

Hrdy, S. B. (1979). Infanticide among animals: a review, classification, and examination of the implications for reproductive strategies of females. *Ethology and Sociobiology*, **1**, 13–40.

Janzen, D. H. (1970). Altruism by coatis in the face of predation by a *Boa constrictor*. *Journal of Mammalogy*, **51**, 387–9.

Johnsingh, A. J. T. (1982). Reproductive and social behaviour of the dhole, *Cuon alpinus* (Canidae). *Journal of Zoology, London*, **198**, 443–63.

Jolly, A. (1966). *Lemur Behavior: A Madagascar Field Study*. Chicago: University of Chicago Press.

Jolly, A. (1972). *The Evolution of Primate Behavior*. New York: Macmillan.

Jones, C. (1967). Growth, development, and wing loading in the evening bat, *Nyctileus humeralis*. *Journal of Mammalogy*, **48**, 1–19.

Kaufmann, J. H. (1962). Ecology and social behavior of the coati, *Nasua narica*, on Barro Colorado Island, Panama. *University of California Publications in Zoology*, **60**, 95–222.

Kaufmann, J. H. (1966). Behavior of infant rhesus monkeys and their mothers in a free-ranging band. *Zoologica*, **60**, 17–28.

Kaufmann, J. H. & Kaufmann, A. (1963). Some comments on the relationship between field and laboratory studies of behavior, with special reference to coatis. *Animal Behaviour*, **11**, 464–9.

Kleiman, D. G. (1969). Maternal care, growth rate, and development in the noctule, pipistrelle and serotine bats. *Journal of Zoology, London*, **157**, 187–212.

Kleiman, D. G. & Eisenberg, J. F. (1973). Comparisons of canid and felid social systems from an evolutionary perspective. *Animal Behaviour*, **21**, 637–59.

Kleiman, D. G. & Malcolm, J. R. (1981). The evolution of male parental investment in mammals. In *Parental Care in Mammals*, ed. D. J. Gubernick & P. H. Klopfer, pp. 347–87. New York: Plenum Press.

Klopfer, P. H. & Boskoff, K. J. (1979). Maternal behavior in prosimians. In *The Study of Prosimian Behavior*, ed. G. A. Doyle & R. D. Martin, pp. 123–54. New York: Academic Press.

Koenig, W. D. & Pitelka, F. A. (1981). Ecological factors and kin selection in the evolution of cooperative breeding in birds. In *Natural Selection and Social Behavior*, ed. R. D. Alexander & D. W. Tinkle, pp. 261–80. New York: Chiron Press.

Kruuk, H. (1972). *The Spotted Hyena: A Study of Predation and Social Behavior*. Chicago: University of Chicago Press.

Le Boeuf, B. J., Whiting, R. J. & Gantt, R. F. (1972). Perinatal behavior of northern elephant seal females and their young. *Behaviour*, **43**, 121–57.

McCracken, G. F. (1984). Communal nursing in Mexican free-tailed bat maternity colonies. *Science*, **223**, 1090–1.

Macdonald, D. W. (1979). Helpers in fox society. *Nature*, **282**, 69–71.

Macdonald, D. W. (1981). Dwindling resources and the social behaviour of capybaras (*Hydrochoerus hydrochaeris*) (Mammalia). *Journal of Zoology, London*, **194**, 371–91.

Macdonald, D. W. & Moehlman, P. D. (1982). Cooperation, altruism, and restraint in the reproduction of carnivores. In *Perspectives in Ethology, vol. 5, Ontogeny*, ed. P. P. G. Bateson & P. H. Klopfer, pp. 433–67. New York: Plenum Press.

McKay, G. M. (1973). The ecology and behavior of the Asiatic elephant in southeastern Ceylon. *Smithsonian Contributions in Zoology*, **125**, 1–113.

McKenna, J. J. (1981). Primate infant caregiving: origins, consequences, and variability with emphasis on the common langur monkey. In *Parental Care in Mammals*, ed. D. J. Gubernick & P. H. Klopfer, pp. 389–416. New York: Plenum Press.

Malcolm, J. R. & Marten, K. (1982). Natural selection and the communal rearing of pups in African wild dogs (*Lycaon pictus*). *Behavioral Ecology and Sociobiology*, **10**, 1–13.

Martin, R. D. (1972). A preliminary study of the lesser mouse lemur (*Microcebus murinus* J. F. Miller 1777). *Zeitschrift für Tierpsychologie*, **9**, 43–89.

Martin, R. D. (1973). A review of the behaviour and ecology of the lesser mouse lemur. In *Comparative Ecology and Behaviour of Primates*, ed. R. P. Michael & J. H. Crook, pp. 1–68. London: Academic Press.

Maynard Smith, J. (1977). Parental investment: a prospective analysis. *Animal Behaviour*, **25**, 1–9.

Mech, L. D. (1970). *The Wolf: The Ecology of an Endangered Species.* New York: American Museum of Natural History.

Michener, G. R. & Murie, J. O. (1983). Black-tailed prairie dog cotteries: are they cooperatively breeding units? *American Naturalist*, **121**, 266–74.

Mills, M. G. L. (1982). The mating system of the brown hyaena (*Hyaena brunnea*, Thundberg, 1820) in the southern Kalahari. *Behavioral Ecology and Sociobiology*, **10**, 131–5.

Moehlman, P. D. (1979). Jackal helpers and pup survival. *Nature*, **277**, 382–3.

Oates, J. F. (1977). The social life of a black-and-white colobus monkey. *Colobus guereza. Zeitschrift für Tierpsychologie*, **45**, 1–60.

O'Farrell, M. J. & Studier, E. H. (1973). Reproduction, growth, and development in *Myotis thysanodes* and *M. lucifugus. Ecology*, **54**, 18–30.

Owens, M. J. & Owens, D. D. (1978). Feeding ecology and its influence on social organisation of brown hyaenas (*Hyaena brunnea*, Thunberg) of the central Kalahari desert. *East African Wildlife Journal*, **16**, 113–35.

Packer, C. & Pusey, A. E. (1983*a*). Male takeovers and female reproductive parameters: a simulation of oestrous synchrony in lions (*Panthera leo*). *Animal Behaviour*, **31**, 334–40.

Packer, C. & Pusey, A. E. (1983*b*). Adaptations of female lions to infanticide by incoming males. *American Naturalist*, **121**, 716–28.

Pusey, A. E. (1978). The physical and social development of wild adolescent chimpanzees (*Pan troglodytes schweinfurthi*). Ph.D. thesis, Stanford University.

Rabb, G. R., Woolpy, J. H. & Ginsburg, B. E. (1967). Social relationships in a captive group of wolves. *American Zoologist*, **7**, 305–11.

Rasa, O. A. E. (1977). The ethology and sociology of the dwarf mongoose. *Zeitschrift für Tierpsychologie*, **43**, 337–406.

Riedman, M. L. (1982). The evolution of alloparental care and adoption in mammals and birds. *Quarterly Review of Biology*, **57**, 405–35.

Rood, J. P. (1974). Banded mongoose males guard young. *Nature*, **248**, 176.

Rood, J. P. (1975). Population dynamics and food habits of the banded mongoose. *East African Wildlife Journal*, **13**, 89–111.

Rood, J. P. (1978). Dwarf mongoose helpers at the den. *Zeitschrift für Tierpsychologie*, **48**, 277–87.

Rood, J. P. (1980). Mating relationships and breeding suppression in the dwarf mongoose. *Animal Behaviour*, **28**, 143–50.

Ross, A. (1961). Notes on food habits of bats. *Journal of Mammalogy*, **42**, 66–71.

Russell, J. K. (1979). Reciprocity in coati (*Nasau narica*) bands. Ph.D. thesis, University of North Carolina.

Schaller, G. B. (1963). *The Mountain Gorilla: Ecology and Behavior.* Chicago: University of Chicago Press.

Schaller, G. B. (1972). *The Serengeti Lion: A Study of Predator–Prey Relations.* Chicago: University of Chicago Press.

Sherman, P. W. (1980). The limits of ground squirrel nepotism. In *Sociobiology: Beyond Nature/Nurture?*, ed. G. W. Barlow & J. Silverberg, pp. 504–44. Boulder: Westview Press.

Sherman, P. W. (1981). Reproductive competition and infanticide in Belding's ground squirrels and other animals. In *Natural Selection and Social Behavior*, ed. R. D. Alexander & D. W. Tinkle, pp. 311–31. New York: Chiron Press.

Smith, D. H. (1977). Effects of experimental bot fly parasitism on gonad weights of *Peromyscus maniculatus. Journal of Mammalogy*, **58**, 679–81.

Smith, W. J., Smith, S. L., Oppenheimer, E. C., de Villa, J. G. & Ulmer, F. A. (1973). Behavior of a captive population of black-tailed prairie dogs: annual cycle of social behavior. *Behaviour*, **46**, 189–220.

Spencer-Booth, Y. (1970). The relationship between mammalian young and conspecifics other than the mother and peers. *Advances in the Study of Behavior*, **3**, 120–80.

Struhsaker, T. T. (1967). Social structure among vervet monkeys (*Cercopithecus aethiops*). *Behaviour*, **29**, 83–121.

Struhsaker, T. T. (1977). Infanticide and social organization in the redtail monkey (*Cercopithecus ascanus schmidti*) in the Kibale Forest, Uganda. *Zeitschrift für Tierpsychologie*, **4**, 75–84.

Sussman, R. W. (1977). Feeding behaviour of *Lemur catta* and *Lemur fulvus*. In *Primate Ecology*, ed. T. H. Clutton-Brock, pp. 1–36. London: Academic Press.

Tevis, L. Jr (1950). Summer behavior of a family of beavers in New York State. *Journal of Mammalogy*, **31**, 40–65.

Trivers, R. L. (1972). Parental investment and sexual selection. In *Sexual Selection and the Descent of Man, 1871–1971*, ed. B. Campbell, pp. 136–79. Chicago: Aldine.

Trune, D. R. & Slobodchikoff, C. N. (1976). Social effects of roosting on the metabolism of the pallid bat (*Antrozous pallidus*). *Journal of Mammalogy*, **57**, 656–63.

van Lawick, H. (1973). *Solo.* London: Collins.

Vehrencamp, S. L. (1980). The roles of individual, kin, and group selection in the evolution of sociality. In *Handbook of Neurobiology*, vol. 3, *Social Behavior and Communication*, ed. P. Marler & J. G. Vandenbergh, pp. 351–94. New York: Plenum Press.

Waser, P. M. & Wiley, R. H. (1980). Mechanisms and evolution of spacing in animals. In *Handbook of Behavioral Neurobiology*, vol. 3, *Social Behavior and Communication*, ed. P. Marler & J. G. Vandenbergh, pp. 159–223. New York: Plenum Press.

West Eberhard, M. J. (1975). The evolution of social behavior by kin selection. *Quarterly Review of Biology*, **50**, 1–33.

Wiley, R. H. (1981). Social structure and individual ontogenies: problems of description, mechanism and evolution. In *Perspectives in Ethology*, vol. 4, ed.

P. P. G. Bateson & P. H. Klopfer, pp. 105–33. New York: Plenum Press.
Wilsson, L. (1971). Observations and experiments on the ethology of the European beaver (*Castor fiber* L.). *Viltrevy Swedish Wildlife*, **8**, 115–266.
Zimen, E. (1976). On the regulation of pack size in wolves. *Zeitschrift für Tierpsychologie*, **40**, 300–41.

The evolutionary ecology of sex

15

Recombination and sex: is Maynard Smith necessary?

J. FELSENSTEIN

In the years before he discovered the comforts of life as an FRS, John Maynard Smith did much to popularize work on evolutionary explanations of reproductive strategies and genetic systems. W. D. Hamilton and he can fairly be called the leaders of this movement, which has had such a strong impact on population biology during the last 20 years. The contributions to the present volume include a number, primarily in this section, that illustrate the breadth and depth of the revival of interest in these subjects in the last two decades, particularly the papers of León, Charlesworth, Charnov, and Clutton-Brock.

Any review of the status of our understanding of the evolution of genetic systems and reproductive strategies would probably conclude that some phenomena, such as the sex ratio, are relatively well explained by evolutionary theories (but see Maynard Smith, 1978*a*, pp. 178–9). Others such as dioecy and the degeneration of Y chromosomes are arguably understood, while we can scarcely claim to have adequate evolutionary explanations for the number of chromosomes, the order of genes on chromosomes, diploidy, or the rates of mutation.

In this continuum of increasing ignorance, where does the phenomenon of recombination fall? Theories of the evolution of recombination have existed since the work of R. A. Fisher and H. J. Muller: we are scarcely short of models. But Maynard Smith (1978*a*, p. 10), on reviewing all models available, wrote that he 'fear[ed] that the reader may find these models insubstantial and unsatisfactory. But they are the best we have'. I wish here to review briefly the status of this problem, and will conclude that we probably possess the elements of a sufficient explanation for the evolution of recombination, and that to have hoped for more was unrealistic. The problem has had a fascination for biologists, all the more so if it is described as 'the evolution of sex' rather than 'the evolution of recombination'. Williams (1975, p. *v*) felt that, given the failure to solve it, 'there is a kind of crisis at hand in evolutionary biology'.

The first genetic explanation for the evolution of recombination was that it serves as a source of variation. East (1918, p. 284) wrote that

> if N variations occur in the germplasm of an asexually reproducing organism, only N types can be formed to offer raw material to selective agencies. But if N variations occur in the germplasm of a sexually reproducing organism 2^n types can be formed. The advantage is almost incalculable. Ten variations in an asexual species mean simply 10 types, 10 variations in a sexual species mean the possibility of 1,024 types. Twenty variations in the one case is again only 20 types to survive or perish in the struggle for existence; 20 variations, in the other case, may present 1,032,576 [*sic*] types to compete in the struggle.

East's argument is stated in terms of benefit to the population, and his 'types' are multilocus genotypes. If the loci determined a phenotype additively and did not interact, then the amount of phenotypic variation should be essentially the same in the two cases. Even if the loci do interact, then we must still take into account that it will be much more difficult to preserve a favourable combination in the presence of recombination. The same phenomenon – recombination – that forms the favourable combination continually breaks it down. This has rarely been mentioned when East's argument is used.

If East's argument shows the dangers of intuitive argument unchecked by any algebra or computation, the theories of R. A. Fisher and of H. J. Muller show the benefits of intuition. Fisher (1930; 1958, p. 136) pointed out that asexual reproduction had the problem that '... the beneficial mutations which occur will have only the minutest chance of not appearing in types of organisms so inferior to some of their competitors, that their offspring will certainly be supplanted by some of the latter'. Muller (1932, p. 121) wrote that 'without sexual reproduction, the various favorable mutations that occur must simply compete with each other, and either divide the field among themselves or crowd each other out till the best adapted for the given conditions remains'.

Between 1932 and the revival of work on this subject in the mid-1960s, there was only one other major contribution. Sturtevant & Mather (1938) argued that if natural selection in some generations favoured coupling linkage disequilibrium and in others repulsion, there would be an advantage to having recombination. A form of this argument had also been given by Muller (1932). We shall see that this argument and that of Fisher and Muller between them encompass all models for the evolution of recombination. Neither argument invokes individual selection.

There the matter lay for some time. Muller returned to the subject 20 years later (1958) with a note giving a formula that was an attempt to calculate the amount of advantage conferred by recombination under the Fisher–Muller model. In 1964 he raised a new argument. If deleterious mutations continually occurred in a population, and were held in check by selection, each individual would contain a variable number of deleterious mutant alleles. If the average number of deleterious mutants carried per individual is not small and the population size is not infinite, it will occasionally happen that every haploid genome in the population contains at least one mutant.

Such a condition is not irreversible if there is recombination, which can form a mutant-free haploid genome from two haploid genomes that contain different mutants. But in asexual reproduction, Muller pointed out, the mutant-free genome could not arise. He wrote (1964, p. 8):

that an asexual population incorporates a kind of ratchet mechanism, such that it can never get to contain, in any of its lines, a load of mutations smaller than that already existing in its at presently least-loaded lines.

Muller's 'ratchet mechanism' appears, at first, to be an entirely new argument but we will see that it and the original Muller–Fisher argument are closely related.

Following Muller's 1964 paper, interest in the subject increased. One reason for this revival of interest is straightforward: it was only in the 1960s that population geneticists began to feel confident in their understanding of the way natural selection and linkage interacted, as a result of the work of Kimura (1956) and Lewontin & Kojima (1960). It will not be useful to review this more recent literature historically; instead we will examine it by subject matter.

Examinations of the Fisher–Muller theory

A number of attempts have been made to calculate the advantage that recombination confers on a population. Muller (1958) gave a formula based on a simple heuristic argument. Crow & Kimura (1965) improved on Muller's formulae by taking into account that an advantageous mutant will increase in frequency along a logistic curve rather than an exponential curve.

Maynard Smith (1971) pointed out that some of the advantageous mutants would in fact become lost in the early generations. He presented a numerical computation and partial computer simulation to show the effect this would have on the relative rates of incorporation of favourable mutants with and without recombination. I further developed Crow & Kimura's formula (Felsenstein, 1974) by allowing for this loss of advantageous new mutations by genetic drift while still rare.

The result is that if we have a haploid population size of N, if the number of favourable mutants per individual per generation is u, summed over all loci, where all are at different loci, and if the selection coefficient in their favour is s, the ratio of the rate of incorporation of new mutants in the presence of recombination to that in its absence is

$$R = \frac{NuU(s)}{s} \ln \{NU(s)\, e^{s/[NuU(s)]} - NU(s) + 1\}, \quad (1)$$

where $U(s)$ is the fixation probability of a new mutant when present alone in a haploid population, as calculated by Kimura (1962):

$$U(s) = \frac{1 - \exp(-2s)}{1 - \exp(-2Ns)}. \quad (2)$$

The resulting formula is perhaps not particularly simple, but it is easily computed. With $N = 10^5$ and $u = s = 10^{-3}$, we can calculate from (1) and (2) that we should have $R = 60.185$. Crow & Kimura's formula gives $R = 69.314$, Muller's formula $R = 69.500$. I have performed simulations of the substitution process in the asexual population, and combining these with the use of Kimura's formula under the assumption of independent substitution at different loci in the presence of free recombination found than all of the predictions, particularly those of Muller and Crow & Kimura, but definitely much higher than 1. The original Fisher–Muller argument stands qualitatively confirmed by simulation, even if none of the formulas for R works very well. The simulations also confirmed the prediction that R would increase with N, u, and s.

All of the above arguments are posed in terms of advantage to the population, and even Fisher concluded (1958, p. 50) that sexual recombination was evolved for the good of the species rather than that of the individual. But there is nothing in the argument to suggest that individual selection must act to eliminate recombination, so that we are free to ask whether the Fisher–Muller mechanism could operate at the level of individual selection. Strobeck, Maynard Smith & Charlesworth (1976) have simulated, both deterministically and stochastically, a model in which one locus has a favourable mutant and another an overdominant polymorphism. The favourable mutant initially occurs in a chromosome with one allele at the polymorphic locus. To get into the chromosomes carrying the other allele requires recombination. A modifier causing recombination exists, and it turns out that when the favoured allele succeeds in spreading by this recombination, it to some extent carries the modifier allele with it, causing an increase of the frequency of the modifier allele. This phenomenon is the hitchhiking effect previously discussed by Maynard Smith and Haigh (1974; Haigh & Maynard Smith, 1976). The hitchhiking effect, as we shall see, is closely related to the effects envisaged by Fisher and Muller.

Felsenstein & Yokoyama (1976) examined selection for a modifier of recombination in a model with favourable mutants occurring and substituting in a finite population. In the case where the modifier is recessive, so that those haploid genomes recombining must be the ones that carry the modifier, there is no distinction between individual and group selection. Fisher's and Muller's arguments predict that multiple mutant genotypes will arise more readily in the subpopulation bearing the recombination-causing modifier allele, and that subpopulation will therefore tend to achieve higher fitness and outcompete the other, taking the recombination modifier with it to fixation. This can be viewed either as a process of hitchhiking or as a process of group selection. The simulations also showed that when the recombination-producing modifier allele was dominant, it also increased in frequency, presumably also by a hitchhiking effect. There is thus no need to go as far as Fisher did in invoking group selection.

Muller's ratchet

The 'ratchet effect' discussed by Muller has also been subjected to quantitative treatment. I have discussed it and simulated its operation (Felsenstein, 1974), showing that it was real and suggesting that it would be seen when Nu was large and Ns intermediate in size. Felsenstein & Yokoyama (1976) simulated the behaviour of a modifier of recombination when deleterious mutants were occurring, and found that there would be individual selection favouring recombination, which could be detected operating in the simulations under some parameter combinations. Haigh (1978) made an approximate analysis of the behaviour of the ratchet, concluding that it would operate when $u/s>1$, u being the rate of deleterious mutation summed over all loci. Maynard Smith (1978) has discussed Haigh's results, which imply that the ratchet may not always be an important effect. Heller & Maynard Smith (1978) have pointed out that Muller's ratchet can operate in self-fertilizing populations as well. Charlesworth (1978) has used Muller's ratchet to explain the progressive inactivation of Y chromosomes in X–Y sex determination systems and the origin of dosage compensation.

The source of the effects

There has been considerable uncertainty as to whether the Fisher–Muller phenomenon, the hitchhiking effects, and the Muller ratchet are real, and, if so, what evolutionary forces are responsible for them. The computer simulations have demonstrated the reality of both effects, but this has not helped us much in understanding them, as the difficulty of intuiting the behaviour of a series of linked loci with mutation, selection and genetic drift has been considerable. Maynard Smith (1968, 1971*a*) and Crow & Kimura (1969) debated the reality of the Fisher–Muller phenomenon and, in particular, whether it was an artifact of the assumption in the models and simulations that each mutation occurred at a new locus.

My own contributions to this clarification (Felsenstein, 1974) were to point out the critical importance of genetic drift to the argument, and to suggest that one phenomenon, that found by Hill & Robertson (1966) when they investigated linkage effects on selection, was the common phenomenon underlying the Fisher–Muller, hitchhiking, and ratchet effects.

Hill & Robertson (1966) were attempting to explain why simulation studies had repeatedly shown that with directional selection on linked genes that determined the same phenotype but did not interact, tighter linkage retarded the response to selection. I had showed (Felsenstein, 1965) that if the fitnesses were multiplicative across loci, no linkage disequilibrium would arise, but that truncation selection would be expected to create repulsion disequilibria which were in the right direction to cause a retardation by linkage. But this could not have caused the large effects seen, and these turned up in Hill and Robertson's simulations even when the loci were made to interact multiplicatively!

Hill & Robertson's solution was to point out an interaction between genetic drift and natural selection. The effect occurs because, in a finite population, favoured alleles at different loci do not remain at linkage equilibrium. Genetic drift causes them sometimes to be in coupling, sometimes in repulsion. Those associations are maintained for longer periods, the tighter is the linkage. This randomly generated linkage disequilibrium would not, in itself, seem to cause any effect on the fixation probability of either mutant. After all, the coupling disequilibria should increase, and the repulsion disequilibria decrease, selection response at each locus. Would these not be expected to cancel on average?

In fact, they do not. Hill & Robertson showed that the helpful effect of coupling is not quite as great as the harmful effect of repulsion: the net result is that tight linkage slows down response to selection by increasing the duration and size of these random disequilibrium effects. One way to see that there will be a net negative effect of these disequilibria is to consider each locus as if it were an environmental effect on the fitnesses of genotypes at the other locus. Variation in fitness in the genetic background owing to the different alleles at the background locus will, in effect, cause an increase in the variation of the number of offspring between individuals. That will have the effect of producing a reduction in effective population size, which will reduce expected response to selection. Linkage, by keeping background genotypes in association with a given locus for longer, increases the effect of this variation in fitness and thereby decreases effective population size.

This argument applies qualitatively whether the favoured alleles at the background locus are rare, as in the Fisher–Muller argument, or common, as in the ratchet. The same phenomenon thus explains both, although, of course, its strength will differ in different cases.

The patch theory: is it different?

Williams & Mitton (1973) and Williams (1975) have introduced a theory of the evolution of recombination which seems very different. Maynard Smith (1976, 1977*a*, 1977*b*, 1978) has characterized this as a theory giving a short-term advantage to recombination, as contrasted with the Fisher–Muller theory, a long-term theory. In Williams' theory, patches of environment are invaded by a small number of propagules, and there is subsequently intense competition among the offspring of these propagules. Without recombination (in haploids) all offspring of a propagule are identical to the parent, save for new mutants. With recombination, they differ much more.

The effect is on the partitioning of variation into within- and between-family components. In an infinitely large species with n loci, all 2^n types will be present; there is no less variability at the species level – it is just that, when the species has no recombination, this variability is mostly between sibships rather than within sibships. Natural selection being intense, the winner of competition in each patch is likely to be the best genotype, and that is much more likely to appear among the sexuals within the patch. There is an advantage to recombination, and simulations by Maynard Smith (1976) have verified this.

The theory could not appear more different from the Fisher–Muller theory. Appearances are, however, deceiving. Let us modify this sib-competition theory a bit. Assume that there is only one patch, and that all the survivors in it constitute the 'propagules' that 'colonize' it in the next generation. There are finitely many such survivors, so that the chance of finding the best genotype among the asexuals continues in this case to be less than among the sexuals. Further, let us assume that competition is not infinitely fierce. We now have, without ever leaving the sib-competition model, arrived at a Fisher–Muller model. The two are the same in their essence. In both linkage disequilibrium owing to random genetic drift is at work: in Williams's model linkage equilibrium in both sexual and asexual subpopulations would imply equal frequencies of the favoured genotype in both, which would abolish the advantage of recombination. In both models it is hitchhiking that is responsible for fixation of the allele for recombination.

It is the fierceness of the selection, not the basic structure of the model, that makes the sib-competition model appear to be a 'short-term' model. The version we have constructed has less selection and is longer-term, without ceasing to be a sib-competition model. In fact, sib-competition is present in the Fisher–Muller argument as well. There, the number of 'propagules' is always finite: if it were not, there would be an infinite population size and all genotypes would appear in both subpopulations at equal frequencies.

The varying selection theory

The argument of Sturtevant & Mather (1938) has been discussed by Maynard Smith (1971*a*, 1977*a*, 1978), who expressed some doubts as to how often its preconditions could be met. It requires that the combinations *AB* and *ab* be favoured in one generation, but the combinations *Ab* and *aB* soon after, so that the presence of recombination makes the switch in the sign of linkage disequilibrium faster. Such a rapid change in the direction of selection, combined with the proper interaction of loci, seems implausible. Charlesworth (1976) simulated this varying selection model and found that the conditions for evolution of recombination were quite restrictive. Maynard Smith (1977*a*, 1978) also discusses work by Slatkin (1975) and Charlesworth & Charlesworth (1979) in which the variation of the pattern of natural selection occurs in different spatial regions rather than through time. This scheme is a bit more plausible, but would still be expected to select for reduced as often as for increased recombination.

A less implausible scheme was discovered by Maynard Smith (1980). Natural selection for an intermediate optimum in a phenotype will produce an excess of the *Ab* and *aB* gametes, if capital letters stand for alleles increasing the phenotype. Response to a shift in the optimum requires a change of gene frequencies, and this would occur most quickly if the gamete types *AB* and *ab* were preponderant. The result is that there can be natural selection for recombination. The effect is not large, but the selection scheme is likely to be fairly common.

Parasite theories

One circumstance has been suggested as producing the kind of change in the pattern of epistasis that the Sturtevant–Mather theory requires; this is parasitism. Jaenike (1978), Bremermann (1979), Glesener (1979) and Hamilton (1980) have suggested that if a parasite is sensitive to multilocus genotypes but is able to adapt quickly to cope with the most frequent genotypes then a generation having mostly *AB* and *ab* hosts will breed parasites adapted to those genotypes, so that in the next generation *Ab* and *aB* will be favoured. Price & Waser (1982) have presented some further theory relevant to such a frequency-dependent scheme.

There can be no doubt that the Sturtevant–Mather theory is the major competitor to the Fisher–Muller theory as the explanation for the origin and maintenance of recombination. There is a pleasing symmetry about the pair. Both have recombination at an advantage as a result of its action in reducing linkage disequilibrium. The theories differ primarily in the forces that cause the disequilibrium (natural selection or genetic drift).

Selection favouring closer linkage

There would be little anxiety about the status of our explanation of the evolution of recombination were we unable to come up with any evolutionary forces acting to eliminate it. There are two categories of effect acting in this direction. First, it has been widely appreciated, at least since the work of Kimura (1956) and Lewontin & Kojima (1960), that under a very broad variety of circumstances recombination would be expected to be selected against, as a result of its effect in breaking up coadapted genotypes. This requires epistasis, but not of any particular kind. Nei (1967) has argued that modifiers reducing recombination will be favoured whenever there is epistasis of any kind, and a formal proof was provided by Feldman (1972). This should be a fairly powerful effect, given the ease with which its conditions are met.

Second, there is a 'cost of meiosis' which can be eliminated by asexual reproduction. Maynard Smith (1971*b*; 1972, pp. 115ff) showed that in an anisogamous organism there is a two-fold advantage to mutants causing asexual reproduction. This is associated with the demographically useless production of male gametes with half the reproductive effort in anisogamous sexuals. I will not discuss this important effect here, as it does not occur in isogamous organisms (Maynard Smith & Williams, 1976) and cannot be invoked as an argument for gradual reduction of recombination in a sexual population.

Do we understand the evolution of recombination?

I would argue that we are in possession of two classes of phenomena capable of explaining the evolution of recombination. In both linkage disequilibrium reduces adaptedness, and the natural selection for recombination occurs for reasons related to this. What we do not know is how important either effect is compared with the selection for reduced recombination or for asexuality, or how important either is quantitatively compared with the other. We do not have a general picture of how often environments change or how often favourable mutants occur, how great the selection coefficients are in either case, or how common and large are fitness interactions between loci or changes of environment that cause a change in the pattern of interaction.

At the present stage of our understanding of population biology, we simply do not have a sufficient understanding to make a realistic model of the mixture of evolutionary forces: we can hope for qualitative understanding but not quantitative prediction. We have the elements of a theory of the evolution of recombination; to have hoped for more was unrealistic. I see no reason to regard that state of affairs as a crisis in evolutionary biology.

Another kind of reductionism

The career value of wide publicity for one's work, rather than appreciation of its usefulness by one's colleagues, seems to have increased a great deal in evolutionary biology in the last decade. The result has been a strong selection pressure in favour of declaring one's theories to be 'new', and against pointing out their relationship to previous work. The result has been a kind of random noise which often threatens to swamp the signal. What we need is to reduce the number of theories, not by being less imaginative, nor by giving in to the pressure towards the narrow empiricism which is dominant among experimental biologists, but by seeking to understand the relationships between new theories and old ones. In this respect, as in many others, John Maynard Smith has played a role of leadership: he has always been concerned to understand the relationship between ostensibly different theories and to communicate this. For this reason, had he never introduced any new theories John Maynard Smith could still have been judged to have played an essential role in our understanding of the evolution of recombination, and this alone would have been grounds enough for answering the question of this article's title in the affirmative.

This work was supported in part by task agreement no. DE-AT06-76EV71005 of contract no. DE-AM06-76RL02225 between the U.S. Department of Energy and the University of Washington. I wish to thank M. Slatkin and P. H. Harvey for helpful editorial suggestions.

References

Bremermann, H. J. (1979). Theory of spontaneous cell fusion. Sexuality in cell populations as an evolutionarily stable strategy. Applications to immunology and cancer. *Journal of Theoretical Biology*, **76**, 311–34.

Charlesworth, B. (1976). Recombination modification in a fluctuating environment. *Genetics*, **83**, 181–95.

Charlesworth, B. (1978). Model for the evolution of Y chromosomes and dosage compensation. *Proceedings of the National Academy of Sciences, USA*, **75**, 5618–22.

Charlesworth, D. & Charlesworth, B. (1979). Selection on recombination in clines. *Genetics*, **91**, 581–9.

Crow, J. F. & Kimura, M. (1965). Evolution in sexual and asexual populations. *American Naturalist*, **99**, 439–50.

Crow, J. F. & Kimura, M. (1969). Evolution in sexual and asexual populations: a reply. *American Naturalist*, **103**, 89–91.

East, E. M. (1918). The role of reproduction in evolution. *American Naturalist*, **52**, 273–89.

Feldman, M. W. (1972). Selection for linkage modification. I. Random mating populations. *Theoretical Population Biology*, **3**, 324–46.

Felsenstein, J. (1965). The effect of linkage on directional selection. *Genetics*, **52**, 349–63.

Felsenstein, J. (1974). The evolutionary advantage of recombination. *Genetics*, **78**, 737–56.

Felsenstein, J. & Yokoyama, S. (1976). The evolutionary advantage of recombination. II. Individual selection for recombination. *Genetics*, **83**, 845–59 (corrigendum, **85**, p. 372, 1977).

Fisher, R. A. (1930). *The Genetical Theory of Natural Selection.* Oxford: Clarendon Press.

Fisher, R. A. (1958). *The Genetical Theory of Natural Selection*, 2nd edn. New York: Dover Publications.

Glesener, R. R. (1979). Recombination in a simulated predator–prey interaction. *American Zoologist*, **19**, 763–71.

Haigh, J. & Maynard Smith, J. (1976). The hitchhiking effect – a reply. *Genetical Research*, **27**, 85–7.

Haigh, J. (1978). The accumulation of deleterious genes in a population – Muller's ratchet. *Theoretical Population Biology*, **14**, 251–67.

Hamilton, W. D. (1980). Sex versus non-sex versus parasite. *Oikos*, **35**, 282–90.

Heller, R. & Maynard Smith, J. (1978). Does Muller's ratchet work with selfing? *Genetical Research*, **32**, 289–93.

Hill, W. G. & Robertson, A. (1966). The effect of linkage on limits to artificial selection. *Genetical Research*, **8**, 269–94.

Jaenike, J. (1978). An hypothesis to account for the maintenance of sex within populations. *Evolutionary Theory*, **3**, 191–4.

Kimura, M. (1956). A model of a genetic system which leads to closer linkage by natural selection. *Evolution*, **10**, 278–87.

Kimura, M. (1962). On the probability of fixation of mutant genes in a population. *Genetics*, **47**, 713–19.

Lewontin, R. C. & Kojima, K. (1960). The evolutionary dynamics of complex polymorphisms. *Evolution*, **14**, 458–72.

Maynard Smith, J. (1968). Evolution in sexual and asexual populations. *American Naturalist*, **102**, 469–73.

Maynard Smith, J. (1971*a*). What use is sex? *Journal of Theoretical Biology*, **30**, 319–35.

Maynard Smith, J. (1971*b*). The origin and maintenance of sex. In *Group Selection*, ed. G. C. Williams, pp. 163–71. Chicago: Aldine–Atherton.

Maynard Smith, J. (1972). *On Evolution.* Edinburgh: Edinburgh University Press.

Maynard Smith, J. (1976). A short-term advantage for sex and recombination through sib competition. *Journal of Theoretical Biology*, **63**, 245–58.

Maynard Smith, J. (1977*a*). Why the genome does not congeal. *Nature*, **268**, 693–6.

Maynard Smith, J. (1977*b*). The sex habit in plants and animals. In *Measuring Selection in Natural Populations*, ed. F. B. Christiansen & T. M. Fenchel. Lecture Notes in Biomathematics No. 19, pp. 315–31. Berlin, Heidelberg & New York: Springer-Verlag.

Maynard Smith, J. (1978). *The Evolution of Sex.* Cambridge University Press.

Maynard Smith, J. (1980). Selection for recombination in a polygenic model. *Genetical Research*, **35**, 269–77.

Maynard Smith, J. & Haigh, J. (1974). The hitch-hiking effect of a favourable gene. *Genetical Research*, **23**, 23–35.

Maynard Smith, J. & Williams, G. C. (1976). Reply to Barash. *American Naturalist*, **110**, 897.

Muller, H. J. (1932). Some genetic aspects of sex. *American Naturalist*, **66**, 118–38.

Muller, H. J. (1958). Evolution by mutation. *Bulletin of the American Mathematical Society*, **64**, 137–60.

Muller, H. J. (1964). The relation of recombination to mutational advance. *Mutation Research*, **1**, 2–9.

Nei, M. (1967). Modification of linkage intensity by natural selection. *Genetics*, **57**, 625–41.

Price, M. V. & Waser, N. M. (1982). Population structure, frequency-dependent selection and the maintenance of sexual reproduction. *Evolution*, **36**, 35–43.

Slatkin, M. (1975). Gene flow and selection in a two-locus system. *Genetics*, **81**, 787–802.

Strobeck, C., Maynard Smith, J. & Charlesworth, B. (1976). The effect of hitchhiking on a gene for recombination. *Genetics*, **82**, 547–58.

Sturtevant, A. H. & Mather, K. (1938). The interrelations of inversions, heterosis and recombination. *American Naturalist*, **72**, 447–52.

Williams, G. C. (1975). Sex and evolution. *Monographs in Population Biology*, **8**. Princeton: Princeton University Press.

Williams, G. C. & Mitton, J. B. (1973). Why reproduce sexually? *Journal of Theoretical Biology*, **39**, 545–54.

16

Birth sex ratios and the reproductive success of sons and daughters

T. H. CLUTTON-BROCK

Variation in mammalian sex ratios

Among invertebrates, there is extensive evidence that parents manipulate the sex ratio of their progeny at hatching so as to maximize their own fitness (Trivers & Hare, 1976; Charnov, 1982). Similar results might be expected among vertebrates but, despite widespread interest, there is little agreement over whether adaptive variation in sex ratios occurs in animals where sex is chromosomally determined (Clutton-Brock & Albon, 1982). Though some reviews claim that foetal or birth sex ratios vary in a manner consistent with adaptive predictions (Trivers & Willard, 1973; Silk, 1983), several critiques have questioned this conclusion, pointing to the lack of variation in many studies, the dubious statistical basis of many suggested trends, the paucity of evidence for genetic variation in sex ratios among domestic mammals and the absence of any known mechanism by which conception sex ratios can be experimentally manipulated (Beatty, 1970; Myers, 1978; Williams, 1979; Maynard Smith, 1980).

Re-analysis of sex-ratio data for mammals (Clutton-Brock & Iason, unpublished) shows that although a considerable proportion of suggested trends do not approach statistical significance, others are difficult to explain by chance alone. Table 1 lists 12 studies of non-human mammals that found statistically significant trends in foetal or birth sex ratios where the sex ratio produced by at least one category of mothers deviates significantly from parity.

It is not impossible that a proportion of these cases, too, may have arisen by chance – for the published evidence of sex-ratio trends represents a biassed sample since evidence that sex ratios vary is more likely to be printed than evidence that they do not. As the number of biologists that examine sex-ratio data may be large, the null hypothesis that sex ratios do not vary has probably been wrongly rejected in a number of cases and there is no satisfactory way of estimating how large this number might be. However, it is unlikely that this explains all of the trends listed in Table 1

Table 1. *Sex ratio trends in which a significant difference has been demonstrated between two or more categories of parents and the birth sex ratio of at least one category deviates significantly from parity*

	Categories compared	Sex ratio (% males)	Sample size		Reference
Litter size					
White tailed deer	Single births	66.7*[a]	$n=63$	*	Verme, 1969
(prime aged does)	Twins	45.5	$n=213$		
	Triplets	41.7	$n=12$		
Horses	Single births	49.5	$n=2783$	*	Platt, 1978
	Twins	35.5*	$n=62$		
Mother's age					
White tailed deer	Yearling mothers	30.0*	$n=40$	***	Verme, 1969
	2.5–6.5 y old	66.7**	$n=63$		
Maternal nutrition					
Laboratory mice	Low-fat diet	24.4**	$n=41$	**	Rivers & Crawford, 1974
	Control	50.4	$n=119$		
White-tailed deer	High plane	43.2*	$n=220$	***	Verme, 1969
	Low plane	72.1***	$n=68$		
Barbari goats	High-energy diet	46.5	$n=172$	**	Sachdeva *et al.*, 1973
	Medium-energy diet	54.1	$n=159$		
	Low-energy diet	67.9**	$n=84$		
Maternal stress					
Laboratory rats	Stressed (restraint)	35.8**	$n=106$	*	Lane & Hyde, 1973
	Control	50.7	$n=142$		
Laboratory rats	Stressed (cold)	44.0*	$n=448$	*	Moriya & Hiroshige, 1978
	Control	51.7	$n=464$		
Laboratory rats	Stressed (ACTH)	46.1	$n=369$	**	Geiringer, 1961

Maternal dominance					
Rhesus macaques, pre-1972	High-ranking mothers	25.0*	$n=24$	**	Simpson & Simpson, 1982
	Others	68.2	$n=22$		
Rhesus macaques, 1972–9	High-ranking mothers	31.0*	$n=29$	*	Simpson & Simpson, 1982
	Medium-ranking mothers	61.3	$n=31$		
	Low-ranking mothers	60.6	$n=33$		
Date of birth					
Grey seals	October 14–27	58.2	$n=122$	*	Coulson & Hickling, 1961
(juveniles)	October 28–November 10	52.6	$n=671$		
	November 11–24	50.1	$n=459$		
	November 24+	42.5*	$n=181$		
Weddell seals	1st quartile of birth date	55.9*	$n=408$	**	Stirling, 1971
(juveniles)	2nd quartile of birth date	51.0	$n=408$		
	3rd quartile of birth date	44.9*	$n=408$		
	4th quartile of birth date	43.9*	$n=408$		
Timing of insemination					
White tailed deer	Mating – h after onset of oestrus				
	13–24	14.3***	$n=28$	***	Verme & Ozoga, 1982
	25–36	38.7	$n=31$		
	37–48	62.5	$n=40$		
	49–96	80.8**	$n=26$		

From T. Clutton-Brock & G. R. Iason, unpublished.

[a] Asterisks against a percentage indicate a significant deviation from parity, those in the penultimate column denote a significant difference in sex ratio between categories (* <0.05, ** <0.01, *** <0.001), tested using the G test (Sokal & Rohlf, 1969) and two-tailed probability values.

for, within several data sets, trends show considerable internal consistency (e.g. Simpson & Simpson, 1982).

Evidence that birth sex ratios vary does not, of course, indicate that trends are adaptive. The principal adaptive theory of vertebrate sex ratios argues that in cases where parental investment affects the breeding success of offspring of one sex to a greater extent than that of another, parents who can afford to invest heavily should bias the sex ratio of their offspring towards the sex whose breeding success is most variable (Trivers & Willard, 1973). This is usually assumed to be males, for variance in male breeding success generally exceeds variance in female success in polygynous species (Trivers & Willard, 1973). However, predictions are complicated by the fact that it is the effects of parental investment on offspring success and not variance in success itself that will determine which sex should be favoured (Clutton-Brock & Albon, 1982). If parental investment has a greater influence on the eventual breeding success of offspring of the less variable sex, parents that can afford to invest heavily may increase their fitness by biassing the sex ratio of their offspring towards that sex despite any difference in overall variance (Clutton-Brock & Albon, 1982).

Unfortunately, little is known of the effects of parental characteristics on the comparative breeding success of sons and daughters (Clutton-Brock, 1982) and this makes it possible to produce an adaptive explanation for virtually any observed sex-ratio trend. In general, the absence of significant trends in many cases where they might be expected to vary (e.g. Clutton-Brock, Albon & Guinness, 1981) and the heterogeneity of those observed (see Table 1) argue against claims that existing sex-ratio data support either any particular adaptive hypothesis – or even adaptive hypotheses in general. Several alternative, nonadaptive explanations of sex-ratio variation exist (e.g. Myers, 1978; James, 1980*a*,*b*) including the possibility that the greater susceptibility of male zygotes to environmental stress is a nonadaptive consequence of sexual selection favouring faster intra-uterine growth and more rapid development in males (Clutton-Brock, Albon & Guinness, 1982) but none of these provides a satisfactory account of trends, either. It seems likely that more than one process may be responsible for consistent variation in birth sex ratios.

Under these circumstances, advances in understanding the functional significance of variation in birth sex ratios are more likely to stem from investigation of the adaptive significance of particular trends than from further analyses of the overall distribution of sex-ratio biasses. Where sex-ratio trends can be shown to exist, studies need to investigate whether or not the trend is likely to enhance the parent's fitness by comparing the eventual breeding success of male and female offspring and examining whether or not the distribution of the sex ratio accords with an adaptive explanation. Long-term studies of mammals are now approaching a stage at which the effects of parental characteristics on the survival and breeding success of sons and daughters can be compared (Clutton-Brock *et al.*, 1982; Silk, 1983) and two groups of studies have recently attempted to test hypotheses concerning the adaptive significance of sex-ratio trends.

Maternal dominance and the sex ratio in primates

More extensive data on individual differences in reproductive success and their causes is available for non-human primates than for any other order of mammals. The first evidence of a consistent association between the birth sex ratio and social behaviour was provided by Altmann (1980) who demonstrated that the social rank of female yellow baboons (*Papio cynocephalus*) in the Amboseli National Park was consistently related to the sex ratio: dominant mothers produced more daughters than subordinate ones (see Table 2).

Subsequent studies of the sex ratio of offspring born in a British colony of rhesus macaques (*Macaca mulatta*) revealed a similar difference between dominant and subordinate females and in this case the sex ratio produced by dominant mothers differed significantly from parity in both data sets (Table 2). A similar (though non-significant) tendency for subordinates to produce male-biassed sex ratios was found in a Californian colony of bonnet macaques (*M. radiata*) though, in contrast to the other two studies, dominant mothers produced equal numbers of sons and daughters (Silk, 1983).

Dominant female primates commonly may show enhanced reproductive performance (Drickamer, 1974; Dunbar & Dunbar, 1977; Silk *et al.*, 1981) and, on the grounds that male success is generally more variable than female success in polygynous vertebrates (Clutton-Brock, 1983), dominant mothers might initially have been expected to produce more sons. The suggested explanation of the opposite trend (see Altmann, 1980; Simpson & Simpson, 1982; Silk, 1983) appeals to the alternative argument that maternal rank has a greater effect on the breeding success of daughters than sons.

In baboons and macaques, male offspring disperse from their natal group at adolescence and their mothers' rank may have little influence on their adult rank and breeding success (Silk, 1983). In contrast, female offspring remain in their natal troop and associate with their matrilineal relations throughout their lives (Koford, 1963; Dittus, 1979; Hausfater, 1975; Packer, 1975; Sugiyama, 1976). Female rank (which is related to reproductive success) is correlated with the rank of the mother and other relatives (Drickamer, 1974; Silk, 1983) and a mother's rank may thus have a pronounced effect on the reproductive success of her daughters. In addition, there is evidence that in at least two species of macaques (*Macaca sinica* and

Table 2. *Birth sex ratios in relation to female dominance in studies of cercopithecine primates*

Species	Stage of development	Categories	Sex ratio (% males)	Sample size	G	Degrees of freedom	p (two-tailed)[a]	Source
Yellow baboon	Birth	Rank: High	34.5	29	5.80	1	*[a]	Altmann, 1980
(*Papio cynocephalus*)		Rank: Low	68.2	22				
Rhesus macaques	Birth	Rank of kin group:						
(*Macaca mulatta*)		High	57.1	28	1.93	2	n.s	Breuggeman, 1978
		Medium	50.0	58				
		Low	41.5	53				
Bonnet macaques	Birth	High-ranking females	51.8	83	2.30	1	n.s	Silk, 1983
(*Macaca radiata*)		Low-ranking females	62.5	120				
Rhesus macaques	Birth	pre-1972 sample						
		Rank: High	25.0*	24	8.91	1	**	Simpson & Simpson, 1982
		Rank: Other	68.2	22				
Rhesus macaques	Birth	1972–9 sample						
		Rank: High	31.0*	29	7.27	2	*	Simpson & Simpson, 1982
		Rank: Medium	61.3	31				
		Rank: Low	60.6	33				
Rhesus macaques	Birth	High-ranking matrilines	54	Total for combined samples 719		Author's test	***	Meikle *et al.*, 1984
		Low-ranking matrilines	49					

Rhesus macaques	Birth	Rank of mother's troop:				
		1	55	Total sample 719	$r_s = -.78$ $p < 0.05$	Meikle *et al.*, 1984
		1	60			
		2	53			
		2	41			
		3	47			
		3	49			
		4	43			
		5	40			

From T. Clutton-Brock & G. R. Iason, unpublished.

[a] All tests re-calculated using the G test (Sokal & Rohlf, 1969) except where a correlation coefficient is shown.

M. radiata) female juveniles receive more threats from resident adult females than juvenile males receive – possibly because female offspring born into a group represent future competitors for the daughters of resident females (Dittus, 1979, 1980; Silk, 1983). In both species, sex differences in the number of threats received are associated with an increase in mortality among juvenile females. In *M. radiata*, this effect is confined to the daughers of subordinate mothers (Silk, 1983), thus strengthening any relationship between a female's work and the sex ratio of her offspring.

However, additional evidence now complicates both the association between maternal rank and the sex ratio and the suggested explanation of the trend. No significant differences in the sex ratio of offspring born to dominant and subordinate mothers were found in another colony of rhesus macaques (Breuggemann, 1978). In addition, data from rhesus populations on the Puerto Rican islands of La Cueva and Guayacan (Meikle *et al.*, 1984) show that females belonging to high-ranking troops and matrilines produce more *sons* than those in low-ranking troops or matrilines – though the magnitude of the bias is small (Table 2).

In addition, the evidence that maternal rank has little effect on the success of sons is insecure. Of the two studies cited by Silk (1983) as evidence that the ranks of males after puberty and dispersal are not consistently related to their mothers' ranks (Sade, 1972; Kawanaka, 1973) neither contains any systematic analysis of the relationship between the ranks of mothers and their sons. And, in contrast to Silk's suggestion that maternal rank is unlikely to have much influence on the rank and breeding success of sons, analysis of long-term data from the Puerto Rican colonies suggests that the rank of a mother's group has a stronger effect on the reproductive success of sons than on that of daughters (Meikle *et al.*, 1984). Thus, while data from long-term primate studies offer important opportunities to examine sex-ratio trends and to test adaptive explanations, there is currently little agreement between the results of different studies.

Maternal dominance and the sex ratio in red deer

Data that can be used to examine the relationship between maternal rank and the sex ratio in other groups of mammals is uncommon, but a 12-year study of reproduction in red deer (*Cervus elaphus*) on the Isle of Rhum (Inner Hebrides) now offers the possibility of examining these effects in an ungulate (Clutton-Brock *et al.*, 1982).

As in macaques, social rank in red deer was positively associated with measures of reproductive success among red deer hinds (see Table 3). Dominant mothers gave birth to significantly more sons than daughters (Table 4*a*), a trend which was consistent across four different subdivisions of

Table 3. *Measures of breeding performance in hinds above versus below median dominance rank*[a]

	Subordinate		Dominant		
Median age of first breeding (y)	3.90	(60)	3.48	(52)	$G=4.32$, $p<0.05$
Fecundity rate (y^{-1})	0.65	(46)	0.68	(43)	$G=0.01$, n.s.
Per cent conceiving before median conception date	38	(55)	58	(50)	$G=4.91$, $p<0.05$
Median birth weight of calves (kg)	6.4	(55)	6.8	(45)	$G=14.38$, $p<0.001$
Per cent calf mortality	45	(55)	28	(50)	$G=3.45$, $0.1>p>0.05$
$\bar{x}$ Lifespan (y)	9	(14)	11	(14)	Fisher's exact probability $p<0.05$
Median lifetime reproductive success (calves reared to 1 y old)	2.3	(14)	6.0	(15)	Fisher's exact probability $p<0.05$

From Clutton-Brock, Albon & Guinness (1984).

[a] Data from cohorts born between 1967 and 1978. Sample sizes of mothers (shown in brackets) differ between measures, since not all variables were available for all individuals. The analyses control for the effects of age as well as for geographical variation in reproductive performance within the study area. All p values are two-tailed.

the study population (Table 4*b*). The same trend was also found within cohorts of hinds born in different years (Table 4*c*), showing that the association between rank and the sex ratio was not a consequence of a dependence of rank on age.

Comparisons of breeding success over at least nine years for 44 sons and 58 daughters born to 50 different hinds of known dominance rank showed that the sons of dominant hinds were more successful than their daughters, while the daughters of subordinate animals were more successful than their sons (Fig. 1).

These results indicated that the association between rank and the sex ratio was adaptive in this population (Clutton-Brock & Harvey, 1979), but it was still possible that all mothers conceived 60% sons, that subordinate mothers lost a proportion of their offspring during gestation and that foetal mortality rates were higher among males than females. If so, differences in fecundity would have been expected between dominant and subordinate mothers – but none were apparent (Table 3). Clear evidence that sex-ratio variations in invertebrates represent evolved strategies have been derived by testing theoretical predictions concerning the extent of sex-ratio biasses among the offspring of different females (Charnov, 1982). Unfortunately similar tests are inappropriate in mammals since variation in birth sex ratios is apparently constrained by the physiology of sex determination (see Maynard Smith, 1980). For example, if red deer hinds in the Rhum population were able to vary the sex ratio of their offspring freely, dominant mothers (whose sons outperformed their daughters) might have been expected to produce only sons, and subordinates (whose daughters outperformed their sons) only daughters. That this was not the case suggested that sex ratios could not be manipulated freely, obviating adaptive predictions about the extent of the sex-ratio bias.

One other quantitative prediction about the distribution of sex ratios can, however, be made from knowledge of the effects of maternal characteristics on the breeding success of sons and daughters. If sex-ratio variation is adaptive, mothers whose sone are mostly to be more successful than their daughters should exceed the average sex ratio, while those whose daughters are more successful than their sons should produce sex ratios lower than the average. As a result, the point at which the slopes of male and female success intersect when regressed on maternal characteristics should predict the point at which the sex ratio first exceeds the population average. For example, Fig. 1 shows that above point 10 on the scale of maternal rank, sons outperformed daughters while, below this, daughters outperformed sons. Consequently, mothers above Rank 10 should have produced male-biassed sex ratios while mothers below Rank 10 should have produced

Fig. 1. Lifetime reproductive success (LRS) of sons and daughters of red deer hinds in relation to their mother's social rank (from Clutton-Brock, Albon & Guinness, 1984). See text for details (filled circles: males; empty circles: females). In these data, the breeding success of sons is positively correlated with the rank of their mothers ($r=0.542$, $t_{36}=3.870$, $p<0.001$) while that of daughters is not ($r=-0.028$, $N=37$, n.s.). Differences between regression slopes calculated for sons and daughters were significant ($F_{1,71}=11.36$, $p<0.001$).

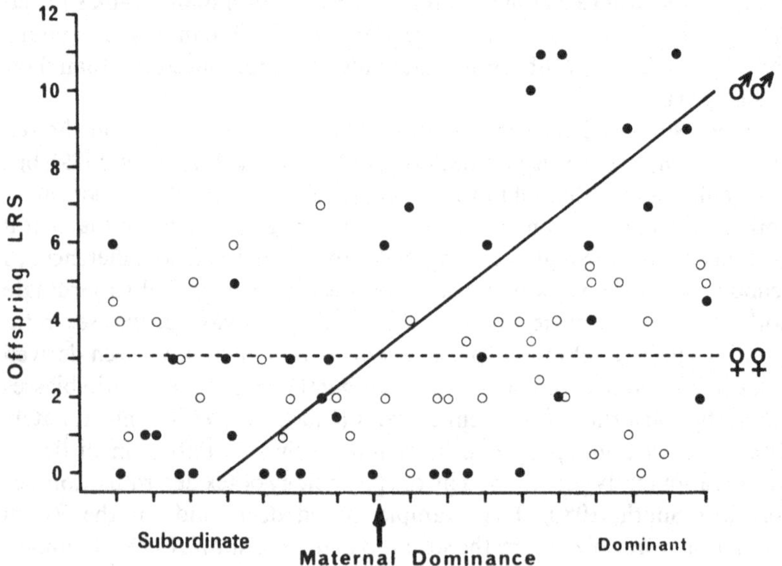

Table 4. *Sex ratio of calves born in relation to mothers' rank*

(a) Analysis of all births 1970–82 from mothers born 1957–74

	Hind dominance rank		
	Low (34)	Medium (29)	High (35)
Males	61	90	149
Females	69	77	97
Males (%)	46.9	53.9	60.6

(b) Analysis by geographical subdivision

	Upper (11)		Kilmory (37)		Intermediate (28)		Samhnsan Insir (22)	
	Subordinate	Dominant	Subordinate	Dominant	Subordinate	Dominant	Subordinate	Dominant
Male	10	24	35	80	26	56	30	39
Female	13	11	34	60	33	37	26	29
Males (%)	43.4	68.6	50.7	57.1	44.1	60.2	53.6	57.4

(c) Analysis by mother's year of birth

	1966 (6)		1967 (8)		1968 (8)		1969 (5)		1970 (8)		1971 (10)		1972 (10)	
	Sub.	Dom.	Sub.	Dom.	Sub.	Dom.	Sub.	Dom.	Sub.	Dom.	Sub.	Dom.	Sub.	Dom.
Males	10	21	14	19	12	13	13	19	7	10	13	17	12	15
Females	10	11	13	12	14	10	15	9	6	2	10	11	12	12
Males (%)	50.0	65.6	51.9	61.3	46.2	56.5	46.4	67.9	53.8	83.3	56.5	60.7	50.0	56.2

From Clutton-Brock, Albon & Guinness, 1984.

[a] Numbers of hinds in each category are shown in brackets.

female-biassed ones. A major discrepancy between the point at which the success of sons exceeded that of daughters and the point at which the sex ratio exceeded the average would be evidence that the association was not unlikely to have evolved as a consequence of selection for sex-ratio manipulation.

Examining correlations between the characteristics of individual animals and the sex ratio of their offspring presents problems in mammals for the total number of young produced during the individual's lifetime is small, and the calculation of percentages for individuals is likely to be subject to considerable random error. To avoid this problem, the red deer study weighted the value for each female by the number of offspring on which it was based. This analysis showed a significant correlation between the ranks of different females and the sex ratio of their offspring ($r=0.291, t_{48}=2.106, p<0.05$). The least squares regression slope for sex ratio on rank exceeded the population average (56% male) close to the rank at which the success of sons began to exceed that of daughters.

The mechanisms underlying the association between a hind's rank, her breeding success and the sex ratio of her offspring were not clear. The correlation between rank and breeding success may have indicated a causal relationship or may have occurred because both rank and breeding success were affected by body size, which was positively correlated with social rank.

Summary

Though a relatively small proportion of studies claiming to demonstrate consistent variation in birth sex ratios in mammals are convincing, there is a growing number of examples where birth sex ratios have differed consistently and significantly. However, there is little indication that the observed trends support any particular adaptive theory and it seems likely that more than one mechanism may be responsible for variation in birth sex ratios.

Two groups of studies have now examined adaptive predictions concerning particular sex-ratio trends. In several primate populations, dominant females produce significantly more daughters than subordinates and some evidence suggests that maternal rank might exert a stronger influence on the breeding success of sons than daughters (Silk, 1983). However, more recent evidence suggests that the relationship between maternal rank and the sex ratio is not as consistent as it appeared at first, and the only direct study of the effects of maternal rank on the breeding success of sons and daughters (Meikle *et al.*, 1984) concluded that the sons of high-ranking mothers were likely to be more successful than their daughters.

In another study of red deer, dominant females produce consistently more sons than subordinates. As an adaptive explanation would predict, the sons of dominant mothers outperformed their daughters while the daughters of subordinates were more successful than their sons. Moreover, there was a reasonably close agreement between the point on the rank scale at which the performance of sons exceeded that of daughters and the point at which the sex ratio began to exceed the population average (Clutton-Brock *et al.*, 1984).

I am grateful to the Director of the Nature Conservancy Council for permission to work on Rhum, which is a National Nature Reserve; to Fiona Guinness, Steve Albon, Glenn Iason and Callan Duck, my collaborators on the Rhum red deer project, for their respective parts in collecting and analysing data on the sex ratio and breeding success in red deer; and to Michael Reiss, Paul Harvey, Jon Seger, John Maynard Smith, Rick Charnov, Nick Davies, Anthony Arak and Dave Harper for comments or discussion. The research is funded by grants from the SERC, NERC, the Leverhulme Trustees and the Royal Society.

References

Altmann, J. (1980). *Baboon Mothers and Infants.* Harvard: Harvard University Press.

Beatty, R. A. (1970). Genetic basis for the determination of sex. *Philosophical Transactions of the Royal Society of London B*, **259**, 3–13.

Breuggeman, J. A. (1978). The function of adult play in free-ranging *Macaca mulatta.* In *Social Play in Primates*, ed. E. O. Smith, pp. 169–91. New York: Academic Press.

Charnov, E. L. (1982). *The Theory of Sex Allocation.* Princeton: Princeton University Press.

Clutton-Brock, T. H. (1982). Sons and daughters. *Nature*, **298**, 11–13.

Clutton-Brock, T. H. (1983). Selection in relation to sex. In *Evolution from Molecules to Men*, ed. D. S. Bendall, pp. 457–81. Cambridge University Press.

Clutton-Brock, T. H. & Albon, S. D. (1982). Parental investment in male and female offspring in mammals. In *Current Problems in Sociobiology*, ed. King's Sociobiology Group, pp. 223–47. Cambridge University Press.

Clutton-Brock, T. H., Albon, S. D. & Guinness, F. E. (1981). Parental investment in male and female offspring in polygynous mammals. *Nature*, **289**, 487–9.

Clutton-Brock, T. H., Albon, S. D. & Guinness, F. E. (1982). Competition between female relatives in a matrilocal mammal. *Nature*, **300**, 178–80.

Clutton-Brock, T. H., Albon, S. D. & Guinness, F. E. (1984). Maternal dominance, breeding success and birth sex ratios in red deer. *Nature*, **308**, 358–60.

Clutton-Brock, T. H. & Harvey, P. H. (1979). Comparison and adaptation. *Proceedings of the Royal Society of London, B*, **205**, 547–65.

Coulson, J. C. & Hickling, G. (1961). Variation in the secondary sex-ratio of the grey seal *Halichoerus grypus* during the breeding season. *Nature*, **190**, 28.

Dittus, W. P. J. (1979). The evolution of behaviors regulating density and age-specific sex ratios in a primate population. *Behaviour*, **69**, 265–302.

Dittus, W. P. J. (1980). The social regulation of primate populations: a synthesis. In *The Macaques: Studies in Ecology, Behavior, and Evolution*, ed. D. G. Lindburg, pp. 263–86. New York: Von Nostrand Reinhold.

Drickamer, L. C. (1974). A ten-year summary of reproductive data for free-ranging *Macaca mulatta. Folia Primatologica*, **21**, 61–80.

Dunbar, R. I. M. & Dunbar, E. P. (1977). Dominance and reproductive success among female gelada baboons. *Nature*, **266**, 351–2.

Fisher, R. A. (1930). *The Genetical Theory of Natural Selection*. Oxford: Clarendon Press.

Geiringer, R. H. (1961). Effect of A.C.T.H. on sex ratio of the albino rat. *Proceedings of the Society for Experimental Biology and Medicine*, **106**, 752–4.

Hausfater, G. (1975). Dominance and reproduction in baboons (*Papio cynocephalus*): a quantitative analysis. *Contributions to Primatology*, **7**, 1–150. Basel: Karger.

James, W. H. (1980*a*). Time of fertilisation and sex of infants. *Lancet*, **112**, 4–6.

James, W. H. (1980*b*). Gonadotrophin and the human secondary sex ratio. *British Medical Journal*, **281**, 711–12.

Kawanaka, K. (1973). Intertroop relationships among Japanese monkeys. *Primates*, **14**, 113–59.

Koford, C. (1963). Ranks of mothers and sons in bands of rhesus monkeys. *Science*, **141**, 356–7.

Lane, E. A. & Hyde, T. S. (1973). The effect of maternal stress on fertility and sex ratio: a pilot study with rats. *Journal of Abnormal Psychology*, **82**, 73–80.

Maynard Smith, J. (1980). A new theory of sexual investment. *Behavioural Ecology and Sociobiology*, **7**, 247–51.

Meikle, D. B., Tilford, B. L. & Vessey, S. H. (1984). Dominance rank, secondary sex ratio and reproduction of offspring in polygamous primates. *American Naturalist*, **124**, 173–88.

Moriya, A. & Hiroshige, T. (1978). Sex ratio of offsprings of rats bred at 5 °C. *International Journal of Biometeorology*, **22**, 312–15.

Myers, P. (1978). Sexual dimorphism in size of vespertilionid bats. *American Naturalist*, **112**, 701–11.

Packer, C. (1975). Male transfer in olive baboons. *Nature*, **255**, 219–20.

Platt, H. (1978). *A Survey of Perinatal Mortality and Disorders in the Thoroughbred*. The Animal Health Trust.

Rivers, J. P. W. & Crawford, M. A. (1974). Maternal nutrition and the sex ratio at birth. *Nature*, **252**, 297–8.

Sachdeva, K. K., Sengar, O. P. S., Singh, S. N. & Lindaht, I. L. (1973). Studies on goats. I. Effect of plane of nutrition on the reproductive performance of does. *Journal of Agricultural Science, Camb.*, **80**, 375–9.

Sade, D. S. (1972). Sociometrics of *Macaca mulatta*. I. Linkages and cliques in grooming matrices. *Folia Primatologica*, **18**, 196–223.

Silk, J. B. (1983). Local resource competition and facultative adjustment of sex ratios in relation to competitive abilities. *American Naturalist*, **121**, 56–66.

Silk, J. B., Clark-Wheatley, C. B., Rodman, P. S. & Samuels, A. (1981). Differential reproductive success and facultative adjustment of sex ratios among captive female bonnet macaques (*Macaca radiata*). *Animal Behaviour*, **29**, 1106–20.

Simpson, M. J. A. & Simpson, A. E. (1982). Birth sex ratios and social rank in rhesus monkey mothers. *Nature*, **300**, 440–1.

Sokal, R. E. & Rohlf, F. J. (1969). *Biometry*. San Francisco: Freeman.

Stirling, I. (1971). Variation in sex ratio of newborn Weddell seals during the pupping season. *Journal of Mammalogy*, **52**, 842–4.

Sugiyama, Y. (1976). Life histories of male Japanese monkeys. In *Advances in the Study of Behaviour*, vol. 7, ed. J. S. Rosenblatt, R. A. Hinde, E. Shaw & C. Beer, pp. 255–84. New York: Academic Press.

Trivers, R. L. & Hare, H. (1976). Haplodiploidy and the evolution of the social insects. *Science*, **191**, 249–63.

Trivers, R. L. & Willard, D. E. (1973). Natural selection of parental ability to vary the sex ratio of offspring. *Science*, **179**, 90–2.

Verme, L. J. (1969). Reproductive patterns of white-tailed deer related to nutritional plane. *Journal of Wildlife Management*, **33**, 881–7.

Verme, L. J. & Ozoga, J. J. (1981). Sex ratio of white-tailed deer and the estrus cycle. *Journal of Wildlife Management*, **45**, 710–15.

Williams, G. C. (1979). The question of adaptive sex ratio in outcrossed vertebrates. *Proceedings of the Royal Society of London*, *B*, **205**, 567–80.

17

Distribution of dioecy and self-incompatibility in angiosperms

D. CHARLESWORTH

Introduction

One way of getting evidence about the selection pressures that have been important in the evolution of breeding systems in plants is to analyse the correlations between breeding systems and other possibly relevant factors. This method was employed by Darwin (1878), who found a correlation between wind-pollination and unisexual flowers (monoecy and dioecy, Darwin, 1878: p. 410) and between dioecy and the tree habit (Darwin, 1878: p. 414). Bawa (1980) has analysed these and other correlations using more extensive recent data. There also appear to be correlations between dioecy and pollination by small bee species (Bawa, 1980), animal dispersal of fruits (Bawa, 1980; Givnish, 1980), hydrophily (Cox, 1983), and dioecy is also commoner in the tropics than in temperate floras (Bawa, 1980). Most recently, Givnish (1982) has examined the correlation between dioecy and self-incompatibility; he concludes that there is no basis for the claim of a negative association between dioecy and self-incompatibility, as has been made by Baker (1959, 1967).

This question is important if we are to decide whether there is a role for the avoidance of inbreeding as a selective factor in the evolution of dioecy, or whether fitness effects directly caused by the male- and female-sterility mutations are solely responsible. Baker (1959) first pointed out that, on the view that dioecy and self-incompatibility are both means of avoiding inbreeding, one would expect that possession of one system would make it unlikely that the other would evolve; he also showed that the facts appeared to agree with this prediction. More recently, theoretical arguments have supported Baker's ideas and have led to the conclusion that the conditions under which dioecy is likely to evolve are most likely to be met in self-compatible populations (Lloyd, 1975; Charlesworth & Charlesworth, 1978). It therefore seems probable that dioecy will be most likely to evolve, or will evolve more often, in families that do not have self-incompatibility systems, or in genera without such systems, when it occurs in families with self-incompatibility, so that a negative association between dioecy and self-incompatibility would be generated; the existence of such an association provides one test of the ideas behind the theoretical models cited above (Thomson & Barrett, 1981). However, it is possible that even if these ideas are fundamentally correct, a negative association would nevertheless not be found, for the following reason. If circumstances that favour outbreeding could cause the evolution of self-incompatibility as often as dioecy, related species might tend to evolve these two different outbreeding systems, causing a positive association. In other words, for a net negative association to be seen, the rate of origin of new self-incompatibility systems should be lower than that of new cases of dioecy. At present, the evidence is that self-incompatibility is generally monophyletic within families (see below) and is probably very unlikely to evolve, compared with dioecy, so that a negative association remains the most likely prediction. If such an association is found, it would certainly support the view that promotion of outbreeding has been an important factor in the evolution of dioecy.

A search for a negative association between dioecy and self-incompatibility is therefore dictated by a prior theoretical prediction and is not purely empirical. A problem with the interpretation of some of the associations that have been discovered is that they are purely empirical and have been individually analysed ignoring other known associations. It is therefore not clear how many factors are likely to be truly causally associated with dioecy, rather than being correlated with other factors that are associated with dioecy. For example, it has been suggested that the association of dioecy with the island habitat is due to a (presumably causal) association of dioecy with fleshy fruits, and to a tendency for islands to be colonized by fleshy-fruited species (Bawa, 1980).

In analysing the possible association between dioecy and self-incompatibility as a means of testing whether avoidance of inbreeding is involved in the evolution of dioecy, it is important to be quite clear about what we wish to test. For example, there are about 108 families with dioecy but no known self-incompatibility system, and between 24 and 57 in which both dioecy and homomorphic self-incompatibility occur, mostly not in close relatives (see below). Even if we exclude wholly dioecious families (about 50, from my tabulations), for which we have no way of knowing whether their dioecy evolved from a self-compatible ancestor, these data make it likely that many cases of evolution of dioecy took place in self-compatible populations (though not necessarily ones with substantial selfing rates, as we have neglected all other means of ensuring outcrossing here). It is therefore reasonable to study the possible ways in which dioecy could evolve, given some degree of selfing in the ancestral species.

However, these data do not permit us to answer the question: is there a lower chance that dioecy will evolve in a self-incompatible (or otherwise outcrossing) species, than in a self-compatible one? To do this, we clearly require some method of calculating the expected number of taxa in which both systems occur, for comparison with the number known. Givnish (1982) has simply done a 2×2 χ^2 analysis for the two pairs of properties: dioecy present in the family versus absent, and incompatibility present versus absent. This implicitly assumes that all families in which neither dioecy nor self-incompatibility is known have been studied as adequately as those in which one or both of these systems has been found. In reality, lack of information is a serious problem and we probably have little or no information about the existence of self-incompatibility (or its absence) in a substantial number of taxa. East (1940) lists a number of families for which no such information exists and, as far as I am aware, this remains true for at least 59 families in all, of which four have known dioecious members. But even when there is some information on a given family, it is in general extremely scanty. The best studied families have not been comprehensively, or even randomly, sampled for the existence of self-incompatibility.

One may also suspect that families with many genera are better known than smaller families. This is supported by the figures in Table 6 (using numbers of genera given in Heywood, 1978). There may also be a bias against strongly outbreeding species in general in the literature, as much of the available information concerns cultivated species which may have been chosen by humans partly because of self-fertility (or, in some cases, selected for this property). Even the use of botanic garden material may lead to such a bias, as self-compatible species are probably easier to maintain (see, for example, Bateman, 1955*a*). It would be more satisfactory to base any study of this question on data from random surveys of floras. At present very few such studies have been done. Bawa (1980) reviews surveys of dioecy, but most surveys have not examined self-incompatibility; the pioneering studies of Bawa (1974), Pojar (1974) and Zapata & Arroyo (1978) are very valuable, but cannot alone provide sufficient data for the present study. Bawa's (1974) study also unfortunately covers a group of species which contains a very high proportion of one family (Leguminosae), and further study of other species sets is therefore desirable.

In this chapter, I present an attempt at a critical tabulation of the existing data. It will be evident that the data do not at present suffice for an adequate statistical analysis of the correlation between dioecy and self-compatibility. The exercise has value chiefly for three reasons. First, it establishes what little is at present reliably known. Second, it suggests the type of data that would be illuminating, if it could be collected. Third, an interesting difference appears between families with homomorphic self-incompatibility, and those without. In what follows, I shall follow the families used by Givnish (1982), though a 'lumping' arrangement (such as this) does not maximize the chance of finding a negative association, as Givnish claims (1982: p. 852), but of course decreases it. This will, however, not matter if we are chiefly interested in data from the genus level, which I shall argue is likely to be the most valuable.

Dioecy

It seems that the presence of dioecy is usually not hard to establish; this is expected from the fact that it can often be detected from morphological criteria. There are problems, however, in a proportion of cases, and this proportion cannot be accurately assessed. First, dioecy may be missed when one or both sexes appears morphologically hermaphrodite. Since all well-studied cases of androdioecy have so far proved to be functionally dioecious (Charlesworth, 1984) one should bear in mind the possibility that other claimed androdioecious species are dioecious. Also apparently gynodioecious species may be functionally dioecious, as in *Charpenteria* (Sohmer, 1972); this has also been suggested for some *Rhus* species (Arroyo & Raven, 1975: p. 507). Second, dioecy can be mistaken for distyly, as males may have short styles and long stamens, while females have short stamens, with non-functional anthers, and long styles. Many examples of this kind of error exist (Table 1 lists some well-known cases, many of which are in families not known to have heterostyly), and the correct interpretation may require detailed and careful studies, as have been done for instance in *Epigaea repens* (Clay & Ellstrand, 1981). Errors of this kind are particularly serious when one wishes to analyse the correlation between dioecy and distyly, within families. A third type of error that may occur is the classification of a species as dioecious merely on the grounds that some plants set no fruit when isolated from pollination; although such an error seems unlikely, Staudt (1952) believes that the claims of dioecy in *Fragaria viridis* were made on this basis, and that the species is really self-incompatible. A fourth type of error, which may be very important, is due to the difficulty of distinguishing between dioecy and sequential monoecy, with whole plants appearing exclusively male or female at any given time, but changing sex during their lifetime, or even during one flowering season. A few cases of this type of breeding system are known, in the Cucurbitaceae (Croat, 1979; Condon & Gilbert, 1982), in *Catasetum* (Gregg, 1975) and in *Arisaema triphyllum* (Policansky, 1981). It is possible that other species classified as dioecious may be similar to these. I have not listed Orchidaceae as having dioecy, despite the claim by Croat (1979).

Despite these problems, the data on dioecy appear quite reliable, and the families and genera in which dioecy is known seem quite well established. Out of the 1303 genera in my tabulations, 920 are listed in Yampolsky & Yampolsky (1922), i.e. 29% are new since that date. This number includes cases of subdioecy and polygamodioecy, but not gynodioecy where there is

Table 1. *Species in which dioecy has the appearance of heterostyly*

Species or genus	Family	Dioecy present elsewhere in family	Reference
Chamissoa	Amaranthaceae	+	East, 1940
Celosia	Amaranthaceae	−	East, 1940
Carissa grandifolia (perhaps gynodioecious)	Apocynaceae	−	Schroeder, 1951; Free, 1970
Amoreuxia sp.	Cochlospermaceae	−	East, 1940
Cleome diandra	Capparaceae	+	Vogel, 1955
Epigaea	Ericaceae	+	Clay & Ellstrand, 1981
Bauhinia reticulata	Leguminosae	+	perhaps dioecious, Urban, 1885; but see Baker, 1983
Erythrina montana	Leguminosae	+	Hernandez, 1982
Passiflora	Passifloraceae	+	East, 1940
Rhamnus catharticus	Rhamnaceae	−	Darwin, 1877
Cinchona	Rubiaceae	+	—
Discospermum	Rubiaceae		Darwin, 1877
Asperula scoparia	Rubiaceae		Darwin, 1877
Carpodetus	Saxifragaraceae	+	Godley, 1979
Solanum appendiculatum	Solanaceae	+	Anderson, 1979
Aegiphila obdurata	Verbenaceae	+	Darwin, 1877

no suspicion that the hermaphrodites are really highly male in function. The list given in Table 2 of Givnish (1982) should probably be augmented by the families in my Table 2, while the Commelinaceae should be deleted (Forman, 1962; R. Faden, personal communication), and so should the Magnoliaceae (as the genera with dioecious members listed in earlier literature are now assigned to different families). The number of families with dioecy but no other known outbreeding system is then 108. A further 16 families have dioecy as well as distyly, and there are about 46 families with dioecy and claimed homomorphic incompatibility, but not distyly. The total number of families with dioecy is thus about 170.

Self-incompatibility

The distribution of heterostyly is rather well known. The three families with tristyly have been known since Darwin (1877), and no well-established cases of dioecy exist in them, though *Dapania* is claimed as androdioecious and *Lepidobotrys* as dioecious (both in Oxalidaceae; Robertson, 1975). Distyly has been reviewed by Ganders (1979), and seems to be present in 22 families.

Table 2. *Families with dioecious members that were not listed by Givnish (1982)*

Family	Species	References
Families without self-incompatibility		
Cymodocaceae	*Thalassodendron, Amphibolis, Halodule, Syringodium*	Cox, 1983
Dilleniaceae	*Tetracera oblongata*	Kubitzki & Baretta-Kuipers, 1969
	Saurauia?	K. S. Bawa (personal communication)
Haloragaceae	*Gunnera*	Yampolsky & Yampolsky, 1922; Godley, 1979
Potomagetonaceae	*Althenia*	Yampolsky & Yampolsky, 1922; McComb, 1966
Nyssaceae	*Nyssa*	East, 1940
Sabiaceae	*Meliosma*	Yampolsky & Yampolsky, 1922
Scheuzeraceae	*Scheuzera*	Yampolsky & Yampolsky, 1922
Smilacaceae	five species of *Smilax*	Croat, 1979
Staphyleaceae	—	Heywood, 1978
Trochodendraceae	*Trochodendron aralioides*	Keng, 1959
Vitaceae	*Vitis*	Westergaard, 1958
Families with homomorphic self-incompatibility		
Malvaceae	*Plagianthus betulinus*	Godley, 1964
	Plagianthus divaricatus	Godley, 1964
	Kydia	Yampolsky & Yampolsky, 1922
	Napea	Yampolsky & Yampolsky, 1922
Myrtaceae	*Pimenta dioica*	Chapman, 1964
	Eugenia	van Wyk, 1982, and personal communication
Families with heterostyly		
Erythrocylaceae	*Erythroxylum rotundifolium*	Schulz, 1970; Opler & Bawa, 1978
Loganiaceae	*Labordia*	Carlquist, 1974
Oxalidaceae	*Lepidobotrys, ?Dapania*	Robertson, 1975

Of these, dioecy is not known in six families, is reasonably well documented in 14, and may perhaps be present in the Iridaceae and Turneraceae also (Table 3). (The Erythroxylaceae and Loganiaceae are not considered by Givnish (1982) to contain dioecious species, but seem to be well-established cases, as far as I can ascertain.) Genera with both dioecy and distyly are known in 6–8 families.

The main difficulties in trying to establish the distribution of breeding systems come when one considers the data on homomorphic self-incompatibility. It is quite a lot of work to demonstrate convincingly that a species is self-incompatible, and rather a high proportion of the claimed cases are based on insufficient evidence. One cannot accept claims that are based only on absence of seed set in flowers isolated from pollinators, as evidence for the existence of a true self-incompatibility system. Nor can one accept as evidence the mere fact that an isolated individual set no seed, particularly when growing away from its normal habitat, as this might in itself affect the ability to fruit, due to changes in the environment, especially the availability of the normal pollinator. In general, to be sure that self-incompatibility exists, one should also prove cross-compatibility with some other plants, in the environment used for the tests. This also rules out claims where the plant is simply extremely poor in fruiting (e.g. *Hevea brasiliensis* (Muzik, 1948), or hermaphrodites of *Rhus ovata* and *integrifolia* (Young, 1972; see also Arroyo & Raven, 1975: pp. 504, 507), or is pollen-sterile, as in hermaphrodites of *Vitis* (Free, 1970). It is also important to show that claimed incompatible crosses do result in seed production, not just fruit; self-incompatible individuals may produce non-seedy fruit, as in *Ananas comosus* (Brewbaker & Gorrez, 1967), *Neoporteria* (Ganders, 1976), or *Citrus grandis* (Li, 1980).

Even when these criteria are satisfied, there may be problems in distinguishing between self-incompatibility and inbreeding depression. It is straightforward conceptually to make a distinction between self-incompatibility and early acting inbreeding depression. The essential point is the distinction between a mechanism operated by the maternal parent plant and controlled by the genotypes of that plant and its pollen donor, and a process acting in the progeny zygote, determined by its own genotype. In practice, however, it is not always easy to distinguish which of these processes is responsible for an observed reduction in seed production after selfing, compared with outcrossing. When self pollen tubes are inhibited on the stigma or in the style, or even in the ovary before fertilization, there is no difficulty in concluding that a true self-incompatibility mechanism exists. But it is possible that delayed-action self-incompatibility systems may exist, in which zygote development is inhibited after fertilization; such an incompatibility system is not necessarily unlikely to evolve in a species in which there is considerable ovule abortion, since maturation of non-selfed ovules could compensate for the loss of selfed ones, so that there need be no net loss in female fertility provided pollinator activity were sufficient to guarantee outcrossing. Since ovule abortion seems to be quite common in angiosperms, with female fertility frequently limited by the resources available for fruit maturation (Stephenson, 1981), this possibility must be taken seriously. A delayed-action type of incompatibility system might also be possible in a species in which effective female fertility is kept to low levels by intense competition among siblings for the chance to establish themselves.

If post-fertilization incompatibility exists, it will be difficult to distinguish from inbreeding depression acting at an early stage of development. The

Table 3. *Presence or absence of dioecy and homomorphic self-incompatibility in families with distyly*

Family	Genus	References	Species in which homomorphic incompatibility is claimed	References
Dioecy not known				
Acanthaceae	—	—	*Thunbergia grandiflora*	East, 1940
			Arrhostoxylum elegans probably gametophytic	Raghuvanshi & Pathak, 1980
			Aphelandra	
Boraginaceae	—	—	Several dubious cases	See text, Philipp & Schou, 1981
Gentianaceae	—	—	—	—
Linaceae	—	—	—	—
Plumbaginaceae	—	—	—	—
Primulaceae	—	—	*Lysimachia nummularia*	Dahlgren, 1922
			Cortusa matthioli	East, 1940
			Trientalis arctica, borealis	Pojar, 1974; Anderson & Beare, 1983
Dioecy well documented				
Connaraceae	*Ellipanthus*	Baker, 1962	—	—
Ehretiaceae	*Cordia*	Opler, Baker & Frankie, 1975	—	—
Erythroxylaceae	*Erythroxylum*	Schulz, 1907; Opler & Bawa, 1978	—	—
Guttiferae	—	Yampolsky & Yampolsky, 1922	*Hypericum calycinum*	East, 1940
Leguminosae	*Bauhinia*?	Urban, 1855; Vogel, 1955; but dioecy not well established, Baker, 1984	Many gametophytic, 1 locus	See Fryxell, 1957
Loganiaceae	—	Carlquist, 1974	—	—
Menyanthaceae	*Nymphoides*	Ornduff, 1966	—	—
Olacaceae	—	Yampolsky & Yampolsky, 1922	*Ximenia americana*	Zapata & Arroyo, 1978

Oleaceae	—	Yampolsky & Yampolsky, 1922; Godley 1979	*Olea europaea*, some varieties perhaps	Fryxell, 1957; but see Baker, 1984
Polygonaceae	—	Yampolsky & Yampolsky, 1922	*Rheum palmatum*	Rosenthal, 1938 (incomplete incompatibility, probably inbreeding depression)
Rubiaceae	Several (5–14)	e.g. Baker, 1958; Beach & Bawa, 1980	*Mussaenda luteola*, *Gardenia thunbergi*	East, 1940
			Coffea	Krug & Carvalho, 1951
Santalaceae	—	Yampolsky & Yampolsky, 1922	—	—
Saxifragaceae	—	Yampolsky & Yampolsky, 1922	Several	Correns, 1928; Sears, 1937; Fryxell, 1957; Gornall & Bohn, 1984
			Ribes aureum?	Goldschmidt-Reischel, 1973
Sterculiaceae	—	Yampolsky & Yampolsky, 1922	*Guazuma tomentosa*	Zapata & Arroyo, 1978
			Sterculia chicha	Taroda & Gibbs, 1982
			Theobroma cacao	Cope, 1962
			Cola nitida	Jacob, 1980
Dioecy uncertain				
Iridaceae	*Nivenia* (androdioecy)	Ornduff, 1983	Several	East, 1929, 1940
Turneraceae	4, all doubtful	Yampolsky & Yampolsky, 1922	—	—

All references to distyly are given in Ganders (1979).

question could be decided by genetic studies, if one could show that the parental genotypes determine failure of the zygotes, rather than the zygotes' own genotypes, but this type of study is extremely difficult to do. Clear evidence of cross-incompatibility among sibs, or parent and offspring, with a limited number of cross-compatible groups, is also satisfactory evidence of a simple genetic basis, not dependent on the progeny plants' own genotypes. In contrast, inbreeding depression would be expected to yield a complex genetic picture, with multiple loci affecting zygote survival. Also, seed set after selfing would be expected to vary from plant to plant, ranging up to the level found after cross-fertilization.

Although it would be surprising if early acting inbreeding depression were extreme enough to make plants seem self-incompatible, the properties of several of the claimed post-fertilization self-incompatibility systems agree with those just outlined, to an extent that makes it necessary to consider this a possible alternative explanation for the low yields of selfed plants, in several species. In addition, the fact that the yield of a cross is inversely correlated with the inbreeding coefficient of the progeny, is also clearly in agreement with this interpretation, but hard to understand on the view that there is a true self-incompatibility system.

For these reasons, the claims of self-incompatibility in *Borago officinalis* (Crowe, 1971), *Myosotis scorpioides* (Varopoulos, 1979), *Medicago* species (Fyfe, 1957) and *Asclepias syriaca* (Sparrow & Pearson, 1948; see also Kephart, 1981) are open to doubt, though *Asclepias verticillata* may be self-incompatible (Kephart, 1981). Other dubious cases, such as *Ribes* and *Acer*, appear in Tables 3–5. The case of *Theobroma cacao*, however, may be a true post-zygotic incompatibility system (Cope, 1962). There appears to be a sporophytic genetic basis (Knight & Rogers, 1955), though the system is not completely clear. Also in this species fruit abortion is known to be high in self-compatible clones, which constitute about one third of those tested (Cope, 1962). *Sterculia chicha* (also in the Sterculiaceae) may have a similar self-incompatibility system (Taroda & Gibbs, 1982), but in *Cola nitida*, inhibition probably occurs much earlier and in this species an incompatibility locus with sporophytic action has also been demonstrated (Jacob, 1980).

Table 3 summarizes the data for families with distyly. None of these families except the Leguminosae (which appears in Table 5 also) has a really well-worked-out case of homomorphic self-incompatibility though 11 families are thought to have self-incompatible members, if we accept the Boraginaceae (see above) and Oleaceae.

Table 4 summarizes the data from families that have no heterostyly and no known dioecious species, while Table 5 deals with families with known, or probable, dioecious species. I have not tried to give the earliest references for any of the information, but have rather preferred to give references that give comprehensive information, or recent references with new information. References for most of the families are given in Fryxell (1957) or de Nettancourt (1977). It will be seen that whenever more than one case of homomorphic self-incompatibility has been studied genetically in the same family, the genetic basis has been found to be the same; the only exceptions are the possible multi-locus control in one Crucifer species (Lewis, 1977), and the finding of two independent loci with gametophytic action in the Solanaceae (Pandey, 1957). In Table 5, I have in addition given references to dioecy and self-incompatibility in those genera where both systems are thought to exist. These occur in perhaps as many as 13 families. Table 4 and Table 5 both illustrate very clearly how uncertain is our information on the existence of self-incompatibility. The ratio of well-established cases to dubious cases is not high, and the suspicion that inbreeding depression, rather than true self-incompatibility, may be the basis for the claims, is noted for several families. It is also clear that new instances of homomorphic self-incompatibility are still being discovered at a significant rate. In my tabulations, there are 415 genera with claimed homomorphic self-incompatibility, compared with 195 claims in Fryxell's (1957) compilation, i.e. 53% of the claims are new since that date.

Correlations between dioecy and homomorphic and heteromorphic self-incompatibility systems

The data tabulated above are summarized in Table 6. The very fragmentary nature of our information about the distribution of homomorphic self-incompatibility, between and within families, means that one can do very little with such a tabulation. I do not believe that a correlation analysis, such as performed by Givnish (1982) is valuable until the data are much more clearly established. It is evidently true, however, that there are no good grounds for concluding that dioecy and self-incompatibility are negatively correlated between families. There is also no significant difference in the incidence of families with dioecy among families with homomorphic self-incompatibility compared with those with distyly. (This conclusion is the same as Givnish's but his statistical test was wrong, as he incorrectly includes families with heterostyly in both the classes 'self-incompatibility present' and 'heterostyly present'.) There is also no significant negative association at the genus level, but this type of analysis is probably not appropriate, as one cannot place all the genera in which neither system is known into the same category, since many have simply not been studied.

As Tables 3 and 5 show, there are claims of both dioecy and homomorphic self-incompatibility in 18 genera. Of these, nine genera (in six families) are reasonably well documented (*Festuca*, *Poa*, *Fragaria*, *Potentilla*, *Rubus*,

Table 4. *Claimed cases of homomorphic self-incompatibility in families with no dioecy*

Family	Number of genera with self-incompatibility	Genetic control	References and notes
Well-established cases			
Amaryllidaceae	8	Not known	—
Apocynaceae	6	Not known	Possible dioecy in *Carissa grandiflora*, Free, 1970
Betulaceae	3	*Corylus avellana*: sporophytic, 1 locus	Thompson, 1979; Germain, Leglise & Delorty, 1981
Bignoniaceae	3	Not known	Bawa, 1974
Bombacaceae	1	Not known	East, 1940; Bawa, 1974
Cactaceae	at least 11	Not known	Ganders, 1976 (also mentions possible dioecy in *Mammillaria*); Lueck & Miller, 1982
Caprifoliaceae	4	Not known	—
Chrysobalanaceae	1	Not known	—
Cochlospermaceae	1	Not known	Bawa, 1974
Commelinaceae	14	*Tradescantia*, *Gibasis* (2 spp.): gametophytic, 1 locus	Owens, 1981
Dipterocarpaceae	1	Not known	Chan, 1981
Fagaceae	1	Not known	Possibly also in *Fagus*, Blinkenberg *et al.*, 1958
Fumariaceae	2	Not known	—
Goodeniaceae	2	Not known	—
Illiciaceae	1	Gametophytic?	Thien, White & Yatsu, 1983
Papaveraceae	6	3 spp. e.g. *P. rhoeas*: gametophytic, 1 locus	Faberge, 1942; Lawrence, Afzal & Kendrick, 1978; Wright, 1979
Plantaginaceae	1	*P. lanceolata*: probably gamelophytic, 1 locus	Ross, 1973; possible dioecy in one sect. of *Plantago*, Yampolsky & Yampolsky, 1922

Polemonaiceae	8	Not known	Grant & Grant, 1965
Polygalaceae	1	Not known	Zapata & Arroyo, 1978
Resedaceae	1	Not known	—
Zingiberaceae	3	Not known	—
Less-well-established cases (*no genetic data*)			
Araceae	1		—
Asclepiadaceae	2		*A. verticillata*, Kephart, 1981; in *A. syriaca* inhibition is after fertilization. Sparrow & Pearson, 1948
Begoniaceae	?		—
Campanulaceae	2		—
Cistaceae	1		—
Cyanastraceae	1		Dulberger & Ornduff, 1980
Cyclanthaceae	1		Bateman, 1956
Lecythidaceae	1		Mori & Kallunki, 1976
Lentibulariaceae	1		—
Magnoliaceae	1		McDaniel, 1965; Thien, 1974
Musaceae	1		Kress, 1983
Nymphaeaceae	?		—
Orchidaceae	10		East, 1940; Singh & Thimappiah, 1980; possible dioecy in *Mormodes*, Croat, 1979
Tropaeolaceae	?		—

Table 5. *Claimed cases of homomorphic self-incompatibility (SI) in families with dioecy*

1. *Incompatibility well established*

Family	Number of genera with SI	Genetic control	Notes and references	Mixed genera	Notes and references
Agavaceae	3	Not known	Fryxell, 1957	—	
Annonaceae	1	Not known	Fryxell, 1957	—	
Bromeliaceae	2	*Ananas comosus*: probably gametophytic, 1 locus	Brewbaker & Gorrez, 1967	—	
Caryophyllaceae	2	*Cerastium arvense*: sporophytic, 1 locus	Pojar, 1974; Lundqvist, 1979	—	
Chenopodiaceae	1	*Beta vulgaris*: gametophytic, 4 loci	Larsen, 1977	—	
Compositae	43	4 species: sporophytic, 1 locus		*Vernonia*	Yampolsky & Yampolsky, 1922; East, 1940; Carlquist, 1974
				Senecio	
Convolvulaceae	2	*Ipomoea*: sporophytic, 1 locus	Kowyama, Shimomo & Kawasi, 1980	*Ipomoea*	Yampolsky & Yampolsky, 1922
Cornaceae	1	*Cornus sericea*: gametophytic, 1 locus	Hummel, Ascher & Pellett, 1982	—	
Cruciferae	80	4 species: sporophytic, 1 locus; *Erνca sativa*: sporophytic, ?3 loci	dioecy in *Lepidium*, Bateman, 1955*a,b*; see Lloyd, 1967; Lewis, 1977; de Nettancourt, 1977	–	
Euphorbiaceae	1	Not known	Muenscher, 1936	—	
Graminae	48	6 species, including *Festuca*: gametophytic, 2 loci	Cornish, Hayward & Lawrence, 1979; but see Osterbye, Larsen & Lundqvist, 1980; McGraw & Spoor, 1983	*Festuca* *Poa*	Connor, 1979; dioecy in Leucopoa, SI in *Poa colensoi*, H. E. Connor, personal communication

Leguminosae	20	3 species of *Trifolium*: gametophytic, 1 locus		*Erythrina*?	East, 1940; Hernandez & Toledo, 1979; Hernandez, 1982
Liliaceae	13	*Gasteria*: 1 locus, inh. in ovule	Brandham & Owens, 1978	—	
		Lilium longiflorum: gametophytic, 1 locus	Fett, Paxton & Dickenson, 1976		
Lobeliaceae	2	Not known	East, 1940	*Lobelia*	Dioecy in some Australian spp., Heywood, 1978
Malvaceae	3	Not known	East, 1940	—	
Myrtaceae	2	Not known	Hagman, 1975; Zapata & Arroyo, 1978	—	
Nyctaginaceae	2	Probably sporophytic	Zadoo & Khoshoo, 1975	—	
			East, 1940; Baker, 1964	—	
Onagraceae	10	*Oenothera organensis*: gametophytic, 1 locus	Raven, 1979	—	
Passifloraceae	1	Not known		—	
Ranunculaceae	4	Ranunculus acris: gametophytic, 4 loci	Osterbye, 1977		
Rosaceae	12	Several spp. including *Rubus arcticus*: gametophytic, 1 locus	Tammisola & Ryynanen, 1970	Fragaria	Staudt, 1952
				Potentilla	Clausen, Keck & Hiesey, 1940; Grewal & Ellis, 1972
				Rubus	Godley, 1979
Rutaceae	2	Not known	Li, 1980	—	
Sapotaceae	1	Not known	Bawa, 1974	—	
Scrophulariceae	9	4 species: gametophytic, 1 locus	Dioecy in *Hebe*, Godley, 1979	—	
Solanaceae	6	Several species: gametophytic, 1 locus but *Physalis ixocarpa* has 2 independent loci	Pandey, 1957	*Solanum*	Whalen & Anderson, 1981

Table 5 (*cont.*)

Table 5 (*cont.*)

2. *Incompatibility dubious (genetics not known)*

Family	Number of genera	Species	Notes and references	Mixed genera	Notes and references
Aceraceae	1	*Acer*		*Acer*	Fryxell, 1957; Gabriel, 1967
Anacardiaceae	2	*Spondias mombin*, *Mangifera indica* (inhibition after fertilization)	Sharma & Singh, 1970	*Spondias*	Bawa, 1974; Opler & Bawa, 1978
Capparaceae	1	*Capparis verrucosa* ?SI (incomplete incompatibility)	Zapata & Arroyo, 1978	—	—
Crassulaceae	1	*Sedum lanceolatum*	Pojar, 1974	*Sedum*	—
Dioscoriaceae	?	—	Carlquist, 1974	—	—
Epacridaceae	1	*Pentachondra pumila* (some plants)	Godley, 1966	—	—
Ericaceae	1	*Vaccinium myrtillus*	Meader & Darrow, 1944 (variable between plants, could be inbreeding depression)	—	—
Flacourtiaceae	1	*Prockia flava* (incomplete incompatibility)	Zapata & Arroyo, 1978	—	—
Hammamelidaceae	1	—	—	—	—
Labiatae	?	—	Dioecy in *Cuminia*, Carlquist, 1974	—	—
Moraceae	?	—	—	—	—
Portulacaceae	1	*Portulaca grandiflora* (some plants)	Tjebbes, 1930; possible dioecy in 2 genera	—	Yampolsky & Yampolsky, 1922
Rhamnaceae	1	*Zizyphus spina-christi*	Galil & Zeroni, 1967	—	—
Sapindaceae	1	*Urvillea ulmacea*	Zapata & Arroyo, 1978	—	—
Theaceae	1	*Thea sinensis*	Fryxell, 1957	—	—

Thymeleaceae	2	*Edgeworthia*	Fryxell, 1957	—	—
Tiliaceae	2	*Luehea speciosa*	Bawa, 1974	—	—
		Tilia	Anderson, 1976	—	—
Valerianaceae	1	—	East, 1940	*Valeriana*	Incompatibility not documented
Verbenaceae	Several	*Citherexylum tristachyum*? SI	East, 1940; Fryxell, 1957	*Citharexylum*	Tomlinson & Fawcett, 1972
Violaceae	?	—	—	—	—
Winteraceae	1	*Pseudowintera colorata* (inhibition after penetration of ovary)	Godley & Smith, 1981	—	—

Table 6. *Distribution of dioecy in angiosperm families*[a]

	Dioecy reasonably well established	Dioecy perhaps exists	Dioecy not known
No information about self-incompatibility	4 $T=157$	—	55 $T=225$
Only self-compatibility known	58 $T=2666$, $D=343$	—	55 $T=375$
Distyly known			
No homomorphic SI	7 $T=127$, $D=18$, $H=20$	1 $T=8$, $D=1?$, $H=5$	4 $T=203$, $H=19$
Homomorphic SI reported	7 $T=1440$, $D=94$, $I=28$, $H=101$	1 $T=70$, $D=1?$, $I=4$, $H=1$	2 $T=278$, $I=6$, $H=5$
All families with distyly	14	2	6
Homomorphic SI reported, but no distyly			
Well-established cases	24 $T=4058$, $D=337$, $I=251$	3 $T=270$, $D=3?$, $I=18$	18 $T=465$, $I=61$
Other cases	22 $T=1148$, $D=173$, $I\simeq 27$	—	14 $T=1224$, $I\simeq 24$

[a] Each cell gives the number of families in the appropriate category. Below these numbers are given *T*: the total numbers of genera in those families; *D*: the numbers with dioecious members; *I*: self-incompatible members; *H*: heterostyled members, where applicable.

Solanum, Senecio, Spondias, Lobelia), but dioecy is not certain in *Erythrina* and *Ipomoea*, self-incompatibility is not certain in *Valeriana, Citharexylum, Sedum, Ribes* (see Goldschmidt-Reischel, 1973) and *Acer* (see Piatnitsky, 1934), while neither dioecy nor self-incompatibility has been satisfactorily studied in *Olea* and *Vernonia*. This list includes the 11 genera claimed by Givnish (1982) except for *Passiflora* whose dioecious species are usually put in a separate genus (Masters, 1871). It should be noted that nearly all these genera are also known to have self-compatible members; the only exceptions are *Vernonia* and *Spondias*, both with only one or two non-dioecious species tested for self-incompatibility. Baker (1984) suggests that in *Rubus* and *Fragaria* dioecy has evolved only in polyploid members of the genera after loss of self-incompatibility.

It is hard to assess whether the facts on the mixed genera indicate any negative association of dioecy with homomorphic self-incompatibility. If the facts could be considered as established, and if we could assume that a reasonably random sample of genera in known mixed families had been studied, one could calculate the expected number of mixed genera on the assumption of no negative association by multiplying together the frequencies of genera with dioecy and with self-incompatibility. Preferably, this should be done within families, but if this cannot be done (because no family has been sufficiently well studied), it is not clear whether one should use only families with both systems known, or all families. The latter choice will of course lead to a low expected number of mixed genera, since there are only 90 families in which homomorphic self-incompatibility systems are claimed, and the number of families in which they have been well documented is even lower.

By confining ourselves to a closer examination of the mixed families we also avoid another problem, that families with dioecy but no self-incompatibility might formerly have been wholly dioecious, but have now re-evolved a cosexual state but not evolved self-incompatibility. This seems quite likely to happen, as the data suggest that the rate of evolution of new cases of dioecy is higher than that of homomorphic self-incompatibility. It has long been recognized that the homogeneity of the self-incompatibility system within families, compared with variety between them, suggests monophyletic origin of each families' incompatibility system (except perhaps for the co-occurrence of distyly and homomorphic incompatibility). The most recent data do not add anything to contradict this conclusion. We therefore cannot accept families with dioecy, but no self-incompatibility, as providing strong evidence for the evolution of dioecy from a self-compatible state, as these families may simply have lost self-incompatibility. It is probably best to avoid using these families in an analysis of this question, so as not to introduce this bias towards a negative association.

If we confine ourselves to the mixed families, with homomorphic self-incompatibility and dioecy, the expected number of genera with homomorphic incompatibility and dioecy, calculated as suggested above, is 20 or 26 (depending whether one uses data from only the 25 families in which homomorphic systems are reasonably well established, or includes those with doubtful claims). But it should be noted that of the 337 genera with dioecy in the 25 families that also have well-established homomorphic self-incompatibility, 187 are in the Euphorbiaceae, which has only one well-documented species with self-incompatibility. If we remove this family, the expected number of mixed genera is reduced to 16. The number of mixed genera known is 10–18 at most. For distyly and dioecy, we have 14 mixed families (omitting Iridaceae and Turneraceae, but including Leguminosae). For these families, the expected number of mixed genera is 9, compared with the known total of 10. If we omit the Rubiaceae (which contributes 500 genera, 91 with distyly and 36 with dioecy, including 5 mixed genera), the expected number of mixed genera is 2.1, and the number known is 5.

The correct interpretation of these figures is not obvious. The apparently good agreement between observed and expected numbers of mixed genera is probably fortuitous. If the true frequency of self-incompatibility in those families that have homomorphic incompatibility systems proves to be higher than the 5.0% obtained from the figures given above (as is almost certainly the case), then the expected number of mixed genera should be proportionately increased, and the apparent agreement between the observed and expected figures would disappear. Of course, it is likely that the true frequencies of dioecy and distyly, and of mixed genera of both kinds, have also been underestimated, but it seems extremely likely that the underestimation of self-incompatibility is much greater than that of the systems that can be detected morphologically. If this is correct, the true expected number of mixed genera would be higher than the figures given above for homomorphic incompatibility and dioecy, so that I suspect that there is probably a deficit of this class of mixed genera, but no, or a small, deficit of genera with both distyly and dioecy. A positive association of dioecy with distyly might be expected, as there is a clear evidence for a mechanism whereby distyly can evolve into dioecy (Baker, 1958; Opler *et al.*, 1975; Lloyd, 1979; Beach & Bawa, 1980; Casper & Charnov, 1982).

A further result that appears when one analyses the data at the genus level, is that there is a much lower frequency of dioecy in families with homomorphic self-incompatibility systems than in those where this system is not known. In the 24 families with dioecy and homomorphic incompatibility, but no distyly, the frequency of genera known to have dioecy is 0.071 (or 0.033 if we omit the Euphorbiaceae). In these families, 5.6% of the genera are known to have self-incompatibility (6% without Euphorbiaceae). The 22 families which have dioecy and may have homomorphic incompatibility have incompatibility claimed in only 2.4% of their genera, but dioecy in 15.1%. Where only self-compatibility is known, 12.9% of the genera have dioecious members. These differences are highly significant, by a χ^2 test; they remain so even if we include all families in which incompatibility has been claimed in the category 'self-incompatibility present', and there is still a significant difference if we compare the incidence of dioecy in all families with self-incompatibility (established or claimed), with that in all other families that are not predominantly dioecious. In families with dioecy and distyly (excluding Leguminosae), the frequency of genera with dioecy is also high (12.7%). These data suggest that there is a negative association between dioecy and homomorphic self-incompatibility.

Another way of posing the same question is to ask: when the close relatives of dioecious species are tested, are they self-incompatible or not? When within-genus studies like this are done, it might be possible to find the frequencies of self-incompatibility within the families studied, and so see whether the incidence is significantly different in species that are related to dioecious species. At present, however, the information about self-incompatibility is not good enough for this. Table 7 summarizes the available data, which I have subdivided into information from sub-dioecious species, and that from non-dioecious species in the same genera as dioecious species. The latter information should be combined with the data on mixed genera, containing both dioecious and self-incompatible species, given in Table 5. Even if we confine our attention to families in which self-incompatibility is known, there are many instances of self-compatibility among the relatives of dioecious species but, as already mentioned, it is not clear whether the number is higher or lower than might be found in a random sample from the same families. As Bawa (1980: p. 281) points out, the data from subdioecious species may indicate too high a frequency of self-compatibility, simply because dioecy has evolved and the selective value of the incompatibility system no longer exists, so that it has become weakened, or even disappeared. However, it seems surprising that no trace should remain in any of the species tested (which include eight species from families known to have self-incompatibility systems), particularly those such as *Euonymus europaeus* or the *Fuchsia* species, which have substantial frequencies of cosexual individuals in natural populations.

The data from members of genera with dioecious species in Table 7 must be combined with the 18 genera in Tables 3 and 5 that may contain both dioecious and self-incompatible species. The combined data indicate that, of 38 genera with dioecy where self-incompatibility has been tested, 16 are known to have self-incompatibility but also have self-compatible members, 20 have only self-compatibility known, while in two genera the other species tested proved to be self-incompatible. The total number of

Table 7. *Results of self-incompatibility tests in genera with dioecy (SC denotes self-compatibility, SI self-incompatibility)*

1. *Subdioecious species*

Family	Species	Self-incompatibility data	Reference
Caricaceae	*Carica papaya*	Occasional monoecious varieties, SC	East, 1940
Celastraceae	*Euonymus europeus*	Probably SC	J. Piper (personal communication)
Compositae	*Antennaria dioica*	SC	von Ubisch, 1930
	Cirsium arvense	SC	East, 1940
Ehretiaceae	*Cordia inermis*	SC	Opler, Baker & Frankie, 1975
Euphorbiaceae	*Mercurialis annua*	SC	Westergaard, 1958
	Aleurites cordata	SC	Simmonds, 1976
Liliaceae	*Asparagus officianlis*	SC	Westergaard, 1958
Nyssaceae	*Nyssa* species	SC	East, 1940
Onagraceae	*Fuchsia thymifolia*	SC	Arroyo & Raven, 1975
	Fuchsia microphylla	SC	Arroyo & Raven, 1975
Polygonaceae	*Spinacia oleracea*	SC	Westergaard, 1958
Ranunculaceae	*Thalictrum fendleri*	SC	Westergaard, 1958
	Thalictrum polygamum	SC	Westergaard, 1958
	Thalictrum dasycarpum	SC	Westergaard, 1958
Salicaceae	*Populus species*	SC	Westergaard, 1958
Simaroubaceae	*Simarouba glauca*	SC	East, 1940
Vitaceae	*Vitis*	SC	Westergaard, 1958

2. *Close relatives of dioecious species*

Family	Genus	Self-incompatibility data	References
Families with self-incompatibility known			
Caryophyllacaea	*Silene*	*S. otites, roemeri* can be selfed	Westergaard, 1958
Compositae	*Cotula*	3 spp. of section Leptinella SC	Lloyd, 1972
Cruciferae	*Lepidium*	*L. sisymbrioides* dioecious	Bateman, 1955*b*
		All 8 spp. tested SC	Bateman, 1955*a*
Graminae	*Buchloe*	*B. dactyloides* SC	East, 1940
Legunosae	*Bauhinia*	*B. pauletia* SC	Heithaus, Opler & Baker, 1974
Liliaceae	*Asparagus*	All tested species SC	East, 1940
Lobeliaceae	*Pratia*	*P. repens* SC	Moore, 1968
		Some Australian species dioecious	Heywood, 1978
Polygonaceae	*Coccoloba*	*C. caracasana* and *floribunda* dioecious	Opler & Bawa, 1978
		C. uvifera SC	East, 1940
	Triplaris	*T.* species dioecious	Yampolsky & Yampolsky, 1922
		T. americana SC	East, 1940
Rutaceae	*Zanthoxylum*	*Z. ciliata* and *setulosum* dioecious	Zapata & Arroyo, 1978
		Z. pterota SC	Zapata & Arroyo, 1978
Theaceae	*Symplocos*	17 species dioecious	Yampolsky & Yampolsky, 1922
		S. species SC	East, 1940
Verbenaceae	*Citharexylum*	30 species dioecious	Yampolsky & Yampolsky, 1922
		C. tristachyum perhaps SI, other species SC	East, 1940
Winteraceae	*Drimys*	*D. brasiliensis* SC	Godley & Smith, 1981
Families with no known self-incompatibility			
Cucurbitaceae	*Ecballium, Bryonia*	Monoecious spp. SC	Westergaard, 1958
Cyperaceae	*Carex*	SC	East, 1940

Table 7 (*cont.*)

Family	Genus	Self-incompatibility data	References
Dillenlaceae	*Actinidia*	SC	East, 1940
Meliaceae	*Trichilia*	*T. drageana*, *lanata* dioecious	Styles, 1972
		T. hirta SC	East, 1940
Moraceae	*Ficus*	Several spp. SC	East, 1940
Myricaceae	*Myrica*	*M. asplenifolia* SC	East, 1940
Myrsinaceae	*Ardisia*	*A. revoluta* SC	Bawa, 1974
Piperaceae	*Piper*	*P. auritum* SC	East, 1940

genera studied is, however, very low, and cannot yield any definite conclusion without a better knowledge of the frequency of self-incompatibility in related genera without dioecy.

Conclusions

In conclusion, the data, especially on self-incompatibility, do not permit an adequate test of the idea that there is a negative association between dioecy and self-incompatibility. It will probably be possible to obtain suitable data only by collecting data within families. One could then estimate the chance of self-incompatibility at the genus level, and compare it with the frequency found in genera with known dioecious members. Suitable families would probably include the composites, grasses, Liliaceae, Ranunculaceae and Rosaceae. Of course, one would still have to be very careful that an unbiassed taxonomic arrangement was used, and cases in which whole genera are defined on the basis of the existence of dioecy, would have to be carefully examined; as Givnish (1982) notes, this could generate a spurious negative association. A further advantage that would accrue from such a detailed within-family study is that one could take into account other breeding system information in addition to that on self-incompatibility. For example, dichogamy may sometimes strongly prevent self-fertilization (Cruden & Hermann-Parker, 1977); there are also other outbreeding mechanisms such as heterodichogamy in which there are two classes of individual, protandrous and protogynous, with sufficient synchrony to prevent selfing (reviewed by Gleeson, 1982; see also Galil & Zeroni, 1967).

Another question which the present study should illuminate is whether families with distyly also contain species with homomorphic incompatibility systems, and, if so, whether these are sporophytic in nature. This is important in relation to the idea that distyly evolved from sporophytic systems similar to those in the Composites and Crucifers, by loss of alleles. Table 3 shows that the existence of homomorphic incompatibility has been claimed in many of the families with distyly, but few cases have been studied in detail. The Leguminosae have a gametophytic system, and the same is probably true for the Acanthaceae, but these appear to be the only cases where genetic information is available, apart from the Sterculiaceae, in which the incompatibility system appears to be distinct from either the simple gametophytic or sporophytic type. As far as these limited data go, therefore, it seems that the families with distyly do not have homomorphic sporophytic incompatibility systems among their members, in agreement with the conclusions of Stevens & Murray (1982), based on the properties of the incompatibility reactions in distylic and homomorphic sporophytic systems (see also Bawa & Beach, 1983).

Summary

The available data on dioecy, homomorphic self-incompatibility and distyly are tabulated, in order to examine the question of a possible negative association between dioecy and self-incompatibility. The data on homomorphic self-incompatibility are so fragmentary that it seems premature to draw any conclusions from them, but the low frequency of dioecy in families known to have self-incompatibility compared with that in families with distyly or with no self-incompatibility, suggests that such an association may exist.

I should like to thank the following people for help in obtaining the information summarized in this paper: H. G. Baker, K. S. Bawa, H. E. Connor, R. Faden, F. R. Ganders, D. G. Lloyd, R. Ornduff, S. J. Owens, R. B. Primack. I am also very grateful to the staff of the Inter-Library loans section of the University of Sussex Library for their extremely valuable assistance throughout the course of this work.

References

Anderson, G. J. (1976). The pollination biology of *Tilia*. *American Journal of Botany*, **63**, 1203–12.

Anderson, G. J. (1979). Dioecious *Solanum* species of hermaphroditic origin as an example of a broad convergence. *Nature*, **282**, 836–8.

Anderson, R. C. & Beare, H. H. (1983). Breeding system and pollination ecology of *Trientalis borealis* (Primulaceae). *American Journal of Botany*, **70**, 408–15.

Arroyo, M. T. K. & Raven, P. H. (1975). The evolution of subdioecy in morphologically gynodioecious species of *Fuchsia* sect *Encliandra* (Onagraceae). *Evolution*, **29**, 500–11.

Baker, H. G. (1958). Studies in the reproductive biology of West African Rubiaceae. *Journal of the West African Science Association*, **4**, 9–24.

Baker, H. G. (1959). Reproductive methods as factors in speciation in flowering plants. *Cold Spring Harbor Symposia on Quantitative Biology*, **24**, 177–91.

Baker, H. G. (1962). Heterostyly in the Connaraceae with special reference to *Byrsocarpus coccineus*. *Botanical Gazette*, **123**, 206–11.

Baker, H. G. (1964). Variation in style length in relation to outbreeding in *Mirabilis* (Nyctaginaceae). *Evolution*, **18**, 507–12.

Baker, H. G. (1967). Support for Baker's law – as a rule. *Evolution*, **21**, 853–6.

Baker, H. G. (1984). The function of dioecy in seed plants. *American Naturalist*, **124**, 149–58.

Bateman, A. J. (1955*a*). Self-incompatibility in angiosperms. III. Cruciferae. *Heredity*, **9**, 53–68.

Bateman, A. J. (1955*b*). Note on dioecy in the Cruciferae. *Heredity*, **9**, 415.

Bateman, A. J. (1956). In *Handbook of Biological Data*, ed. W. S. Spector, p. 134. New York: W. B. Saunders.

Bawa, K. S. (1974). Breeding systems of tree species of a lowland tropical community. *Evolution*, **28**, 85–92.

Bawa, K. S. (1980). Evolution of dioecy in flowering plants. *Annual Reviews of Ecology and Systematics*, **11**, 15–39.

Bawa, K. S. & Beach, J. H. (1983). Self-incompatibility systems in the Rubiaceae of a tropical rain forest. *American Journal of Botany*, **70**, 1281–8.

Beach, J. H. & Bawa, K. S. (1980). Role of pollination in the evolution of dioecy from distyly. *Evolution*, **34**, 1138–42.

Blinkenberg, C., Brix, H., Schaffalitzky de Muckadell, M. & Vedel, H. (1958). Controlled pollinations in *Fagus*. *Silvae Genetica*, **7**, 116–22.

Brandham, P. E. & Owens, S. J. (1978). The genetic control of self-incompatibility in the genus *Gasteria* (Liliaceae). *Heredity*, **40**, 165–9.

Brewbaker, J. L. & Gorrez, D. D. (1967). Genetics of self-incompatibility in the monocot. genera, *Annanas* (pineapple) and *Gasteria*. *American Journal of Botany*, **54**, 611–16.

Carlquist, S. (1974). *Island Biology*. New York: Columbia University Press.

Casper, B. B. & Charnov, E. L. (1982). Sex allocation in heterostylous plants. *Journal of Theoretical Biology*, **96**, 143–9.

Chan, H. T. (1981). Reproduction of some Malaysian Dipterocarps. III. Breeding systems. *The Malaysian Forester*, **44**, 28–36.

Chapman, G. P. (1964). Some aspects of dioecism in Pimento (allspice). *Annals of Botany*, **18**, 451–8.

Charlesworth, D. (1984). Androdioecy and the evolution of dioecy. *Biological Journal of the Linnean Society*, **23**, 333–48.

Charlesworth, B. & Charlesworth, D. (1978). A model for the evolution of dioecy and gynodioecy. *American Naturalist*, **112**, 975–97.

Clausen, J., Keck, D. D. & Hiesey, W. M. (1940). Experimental studies on the nature of species. 1. Effect of varies environments on western North American plants. *Carnegie Institute of Washington Publication*, No. 520.

Clay, K. & Ellstrand, N. C. (1981). Stylar polymorphism in *Epigaea repens*, a dioecious species. *Bulletin of the Torrey Botanical Club*, **108**, 305–10.

Condon, M. A. & Gilbert, L. E. (1982). Reproductive biology and natural history of neotropical vines *Gurania* and *Psiguria* (= Anguria). In *Biology and Chemistry of the Cucurbitaceae*. Ithaca: Cornell University Press.

Connor, H. E. (1979). Breeding systems in the grasses: a survey. *New Zealand Journal of Botany*, **17**, 547–74.

Cope, F. W. (1962). The mechanisms of pollen incompatibility in *Theobroma cacao*. L. *Heredity*, **17**, 157–82.

Cornish, M. A., Hayward, M. D. & Lawrence, M. J. (1979). Self-incompatibility in rye-grass. 1. Genetic control in diploid *Lolium perenne* L. *Heredity*, **43**, 95–106.

Correns, C. (1928). Neue Untersuchungen an selbststerilen Pflanzen. *Biologisches Zentralblatt*, **48**, 759–68.

Cox, P. A. (1983). Search theory, random motion, and the convergent evolution of pollen and spore morphology in aquatic plants. *American Naturalist*, **121**, 9–31.

Croat, T. B. (1979). The sexuality of the Barro Colorado Island flora (Panama). *Phytologia*, **42**, 319–48.

Crowe, L. K. (1971). The polygenic control of outbreeding in *Borago officianalis*. *Heredity*, **27**, 111–18.

Cruden, R. W. & Hermann-Parker, S. M. (1977). Temporal dioecism: an alternative to dioecism. *Evolution*, **31**, 863–6.

Dahlgren, K. V. O. (1922). Selbststerilität innerhalb klonen von *Lysimachia nummularia. Hereditas*, **3**, 200–10.

Darwin, C. (1877). *The Different Forms of Flowers on Plants of the Same Species.* London: John Murray.

Darwin, C. (1878). *The Effects of Cross and Self Fertilisation in the Vegetable Kingdom.* London: John Murray.

Dulberger, R. & Ornduff, R. (1980). Floral morphology and reproductive biology of four species of *Cyanella* (Tecophileaceae). *New Phytologist*, **86**, 45–6.

East, E. M. (1929). Self-sterility. *Bibliographia genetica*, **5**, 331–70.

East, E. M. (1940). The distribution of self-sterility in the flowering plants. *Proceedings of the American Philosophical Society*, **82**, 449–518.

Faberge, A. C. (1942). Genetics of the Scapiflora section of *Papaver*. I. The garden Iceland poppy. *Journal of Genetics*, **44**, 169–93.

Fett, W. F., Paxton, J. P. & Dickinson, D. B. (1976). Studies on the self-incompatibility response of *Lilium longiflorum. American Journal of Botany*, **63**, 1104–8.

Forman, L. L. (1962). *Aethiolirion*, a new genus of Commelinaceae from Thailand, with notes on allied genera. *Kew Bulletin*, **16**, 209–21.

Free, J. B. (1970). *Insect Pollination of Crops.* London: Academic Press.

Fryxell, P. A. (1957). Mode of reproduction of higher plants. *Botanical Review*, **23**, 135–233.

Fyfe, J. L. (1957). Relational incompatibility in diploid and tetraploid lucerne. *Nature*, **179**, 591–2.

Gabriel, W. J. (1967). Reproductive behaviour in sugar maple: self-compatibility, cross compatibility, agamospermy, and agamocarpy. *Silvae Genetica*, **16**, 165–8.

Galil, J. & Zeroni, M. (1967). On the pollination of *Zizyphus spina-Christi* (L.) Willd. *Israel Journal of Botany*, **16**, 71–7.

Ganders, F. R. (1976). Self-incompatibility in the Cactaceae. *Cactus and Succulent Journal Great Britain*, **38**, 39–40.

Ganders, F. R. (1979). The biology of heterostyly. *New Zealand Journal of Botany*, **17**, 607–35.

Germain, E., Leglise, P. & Delorty, F. (1981). Analyse du système d'incompatibilité pollinique observé chez le noisetier *Corylus avellana* L. 1er colloque sur les récherches fruitières. Bordeaux, pp. 197–216.

Givnish, T. J. (1980). Ecological constraints on the evolution of breeding systems in seed plants: dioecy and dispersal in gymnosperms. *Evolution*, **34**, 959–72.

Givnish, T. J. (1982). Outcrossing versus ecological constraints in the evolution of dioecy. *American Naturalist*, **119**, 849–65.

Gleeson, S. K. (1982). Heterodichogamy in walnuts: inheritance and stable ratios. *Evolution*, **36**, 892–902.

Godley, E. J. (1964). Breeding systems in New Zealand plants. 3. Sex ratios in some natural populations. *New Zealand Journal of Botany*, **2**, 205–12.

Godley, E. J. (1966). Breeding systems in New Zealand plants. 4. Self-sterility in *Pentachondra pumila. New Zealand Journal of Botany*, **4**, 249–54.

Godley, E. J. (1979). Flower biology in New Zealand. *New Zealand Journal of Botany*, **17**, 441–66.

Godley, E. J. & Smith, D. H. (1981). Breeding systems in New Zealand plants. 5. *Pseudowintera colorata* (Winteraceae). *New Zealand Journal of Botany*, **19**, 151–6.

Goldschmidt-Reischel, E. (1973). Selbststerilitat bei *Ribes aureum*. I. Pollinierung, Fruchtansatz und Samenanlage. *Zeitschrift für PflanzenZüchtung*, **68**, 225–52.

Gornall, R. J. & Bohn, B. A. (1984). Breeding systems and relationships among species of *Boykinia* and related genera (Saxifragaceae). *Canadian Journal of Botany*, **62**, 33–7.

Grant, V. & Grant, K. A. (1965). *Flower Pollination in the Phlox Family*. New York: Columbia University Press.

Gregg, K. B. (1975). The effect of light intensity on *Cycnoches* and *Catasetum* (Orchidaceae). *Selbyana*, **1**, 101–13.

Grewal, M. S. & Ellis, J. R. (1972). Sex determination in *Potentilla fruticosa*. *Heredity*, **29**, 359–62.

Hagman, M. (1975). Incompatibility in forest trees. *Proceedings of the Royal Society B*, **188**, 313–26.

Heithaus, E. R., Opler, P. A. & Baker, H. G. (1974). Bat activity and pollination of *Bauhinia pauletia*: bat-pollinator coevolution. *Ecology*, **55**, 412–19.

Hernandez, H. M. (1982). Female sterility in *Erythrina montana*. *Allertonia*, **3**, 71–6.

Hernandez, H. M. & Toledo, V. M. (1979). The role of nectar robbers and pollination in the reproduction of *Erythrina leptorhiza*. *Annals of the Missouri Botanical Gardens*, **66**, 512–20.

Heywood, V. M. (ed.) (1978). *Flowering Plants of the World*. Oxford: Oxford University Press.

Hummel, R. L., Ascher, P. D. & Pellett, H. M. (1982). Genetic control of self-incompatibility in red-osier dogwood. *Journal of Heredity*, **73**, 308–9.

Jacob, V. J. (1980). Pollination, fruit setting and incompatibility in *Cola nitida*. *Incompatibility Newsletter*, **12**, 50–6.

Keng, H. (1959). Androdioecism in the flowers of *Trochodendron aralioides*. *Journal of the Arnold Arboretum, Harvard University*, **40**, 158–60.

Kephart, S. R. (1981). Breeding systems in *Asclepias incarnata* L., *A. syriaca* L. and *A. verticillata* L. *American Journal of Botany*, **68**, 226–32.

Knight, R. & Rogers, M. M. (1955). Incompatibility in *Theobroma cacao*. *Heredity*, **9**, 69–77.

Kowyama, Y., Shimano, N. & Kawasi, T. (1980). Genetic analysis of incompatibility in the diploid *Ipomoea* species closely related to the sweet potato. *Theoretical and Applied Genetics*, **58**, 149–55.

Kress, W. J. (1983). Self-incompatibility in Central American *Heliconia*. *Evolution*, **37**, 735–44.

Krug, C. A. & Carvalho, A. (1951). The genetics of *Coffea*. *Advances in Genetics*, **4**, 128–58.

Kubitzki, K. & Baretta-Kuipers, T. (1969). Pollendimorphismus und Androdiöezie bei *Tetracera* (Dilleniaceae). *Naturwissenschaften*, **56**, 219–20.

Larsen, K. (1977). Self-incompatibility in *Beta vulgaris* L.1. Four gametophytic, complementary S-loci in sugar beet. *Hereditas*, **85**, 227–48.

Lawrence, M. J., Afzal, M. & Kenrick, J. (1978). The genetical control of self-incompatibility in *Papaver rhoeas*. *Heredity*, **40**, 239–53.

Lewis, D. (1977). Sporophytic incompatibility with 2 and 3 genes. *Proceedings of the Royal Society, B*, **196**, 161–70.

Li, S.-J. (1980). Self-incompatibility in 'Matou' Wentan (*Citrus grandis* L. Osb.). *HortScience*, **15**, 298–300.

Lloyd, D. G. (1967). The genetics of self-incompatibility in *Leavenworthia crassa*

Rollins (Cruciferae). *Genetica*, **438**, 227–42.

Lloyd, D. G. (1972). Breeding systems in *Cotula* L. (Compositae, Anthemideae). I. The array of monoecious and diclinous systems. *New Phytologist*, **71**, 1181–94.

Lloyd, D. G. (1975). The maintenance of gynodioecy and androdioecy in angiosperms. *Genetica*, **45**, 325–39.

Lloyd, D. G. (1979). Evolution towards dioecy in heterostylous plants. *Plant Systematics and Evolution*, **131**, 71–80.

Lueck, E. E. & Miller, J. M. (1982). Pollination biology and chemotaxonomy of the *Echinocerus viridiflorus* complex (Cactaceae). *American Journal of Botany*, **69**, 1669–72.

Lundqvist, A. (1979). One-locus sporophytic self-incompatibility in the carnation family, Caryophyllaceae. *Heredity*, **91**, 307 (Abstract).

McComb, J. A. (1966). The sex forms of species in the flora of the Southwest of Western Australia. *Australian Journal of Botany*, **14**, 303–16.

McDaniel, J. M. (1963). Securing seed production in *Magnolia acuminata* and *M. cordata*. *Proceedings of the Annual Meeting of the International Plant Propagation Society*, **13**, 120–3.

McGraw, J. M. & Spoor, W. (1983). Self-incompatibility in *Lolium* species. 2. *Lolium perenne* L. *Heredity*, **50**, 21–7.

Masters, M. T. (1871). Contributions to the natural history of the Passifloraceae. *Transactions of the Linnean Society of London*, **27**, 593–645.

Meader, E. M. & Darrow, G. M. (1944). Pollination of the rabbit eye blueberry and related species. *Proceedings of the American Society for Horticultural Science*, **45**, 267–74.

Moore, D. M. (1968). The vascular flora of the Falkland Islands. *British Antarctic Survey Science Reports*, **60**, 1–202.

Mori, S. A. & Kallunki, J. A. (1976). Phenology and floral biology of *Gustavia superba* (Lecythidaceae) in central Panama. *Biotropica*, **8**, 184–92.

Muenscher, W. C. (1936). The production of seed by *Euphorbia cyparissias*. *Rhodora*, **38**, 161–3.

Muzik, T. J. (1948). What is the pollinating agent for *Heavea brasiliensis*? *Science*, **108**, 540.

de Nettancourt, D. (1977). *Incompatibility in Angiosperms*. Berlin: Springer-Verlag.

Opler, P. A., Baker, H. G. & Frankie, G. W. (1975). Reproductive biology of some Costa Rican *Cordia* species (Boraginaceae). *Biotropica*, **7**, 234–47.

Opler, P. A. & Bawa, K. S. (1978). Sex ratios in tropical forest trees. *Evolution*, **32**, 812–21.

Ornduff, R. (1966). The origin of dioecism from heterostyly in *Nymphoides* (Menyanthaceae). *Evolution*, **20**, 309–14.

Ornduff, R. (1983). Studies on the reproductive system of *Nivenia corymbosa*, an apparently androdioecious species. *Annals of the Missouri Botanical Gardens* (in press).

Osterbye, U. (1977). Self-incompatibility in *Ranunculus acris* L. II. Four S-loci in a German population. *Hereditas*, **87**, 173–8.

Osterbye, U., Larsen, K. & Lundqvist, A. (1980). Comments on self-incompatibility in the *Gramineae*. *Incompatibility Newsletter*, **12**, 44–9.

Owens, S. J. (1981). Self-incompatibility in the Commelinaceae. *Annals of Botany*, **47**, 567–81.

Pandey, K. K. (1957). Genetics of self-incompatibility in *Physalis ixocarpa* Brot. a new system. *American Journal of Botany*, **44**, 879–87.

Philipp, M. & Schou, O. (1981). An unusual heteromorphic incompatibility system. Distyly, self-incompatibility, pollen load and fecundity in *Anchusa officinalis*. *New Phytologist*, **89**, 693–703.

Piatnitsky, S. S. (1934). Experiments on self-pollination of *Larix*, *Acer* and *Quercus* (in Russian). *Acta Instituti Botanic Akademiae Scientiarum URSS*, Series **4**, 297–318.

Pojar, J. (1974). Reproductive dynamics of four plant communities of southwestern British Columbia. *Canadian Journal of Botany*, **52**, 1819–34.

Policansky, D. (1981). Sex choice and the size advantage model in jack-in-the-pulpit. *Proceedings of the National Academy of Sciences (USA)*, **78**, 1306–8.

Raghuvanshi, S. S. & Pathak, C. S. (1980). Breakdown of self-incompatibility in *Arrhostoxylum elegans*. *Journal of Heredity*, **7**, 143–5.

Raven, P. H. (1979). A survey of reproductive biology in Onagraceae. *New Zealand Journal of Botany*, **17**, 575–93.

Robertson, K. R. (1975). The Oxalidaceae in the southeastern United States. *Journal of the Arnold Arboretum, Harvard University*, **56**, 223–39.

Rosenthal, C. (1938). Uber Selbststerilität des medizinischen Rhabarbers. *Zeitschrift für Pflanzenzuchtung*, **22**, 317–22.

Ross, M. D. (1973). Inheritance of self-incompatibility in *Plantago lanceolata*. *Heredity*, **30**, 169–76.

Schroeder, S. A. (1951). Heterostyly and sterility in *Carissa grandiflora*. *Proceedings of the American Society for Horticultural Science*, **57**, 419–22.

Schulz, O. E. (1970). Erythroxylaceae. *Das Pflanzenreich*, **14(134)**, ed. A. Engler, pp. 1–164. Leipzig: W. Engelmann.

Sears, E. R. (1937). Cytological phenomena associated with self-sterility in the flowering plants. *Genetics*, **22**, 130–81.

Sharma, D. K. & Singh, R. N. (1970). Self-incompatibility in Mango (*Mangifera indica* L.). *Horticultural Research*, **10**, 108–18.

Simmonds, N. W. (1976). *Evolution of Crop Plants*. London: Longmans.

Singh, F. & Thimmappiah. (1980). Nature and breakdown of self-incompatibility in *Dendrobium aggregatum* (Orchidaceae) with pollen irradiation. *Incompatibility Newsletter*, **12**, 30–5.

Sohmer, S. M. (1972). Revision of the genus *Charpenteria* (Amarantaceae). *Brittonia*, **24**, 283–312.

Sparrow, F. K. & Pearson, N. L. (1948). Pollen compatibility in *Asclepias syriaca*. *Journal of Agricultural Research*, **77**, 187–99.

Staudt, G. (1952). Zur selbsterilität von *Fragaria viridis* DUCH. *Naturwissenschaften*, **39**, 572–3.

Stephenson, A. G. (1981). Flower and fruit abortion: proximate causes and ultimate functions. *Annual Reviews of Ecology and Systematics*, **12**, 253–79.

Stevens, V. A. M. & Murray, B. G. (1982). Studies on heteromorphic self-incompatibility systems: physiological aspects of the incompatibility system of *Primula obconica*. *Theoretical and Applied Genetics*, **61**, 245–56.

Styles, B. T. (1972). The flower biology of the Meliaceae and its bearing on tree breeding. *Silvae Genetica*, **21**, 175–82.

Tammisola, J. & Ryynanen, A. (1970). Incompatibility in *Rubus arcticus* L. *Hereditas*, **66**, 269–78.

Taroda, T. & Gibbs, P. E. (1982). Floral biology and breeding system of *Sterculia chicha* St. Hil. *New Phytologist*, **90**, 735–43.

Thien, L. B. (1974). Floral biology of *Magnolia. American Journal of Botany*, **61**, 1037–45.

Thien, L. B., White, D. A. & Yatsu, Y. (1983). The reproductive biology of a relict – *Illicium floridanum* Ellis. *American Journal of Botany*, **70**, 719–27.

Thompson, M. M. (1979). Genetics of incompatibility in *Corylus avellana* L. *Theoretical and Applied Genetics*, **54**, 113–16.

Thomson, J. D. & Barrett, S. C. H. (1981). Selection for outcrossing, sexual selection, and the evolution of dioecy in plants. *American Naturalist*, **118**, 443–9.

Tjebbes, K. (1930). Interfertile gruppen innerhalb einer selststerilen Form von *Portulaca grandiflora* Lindl. *Botaniska Notiser*, 48–52.

Tomlinson, P. B. & Fawcett, P. (1972). Dioecism in *Citharexylum* (Verbenaceae). *Journal of the Arnold Arboretum, Harvard University*, **53**, 386–89.

Ubisch, G. von (1936). Genetic studies on the nature of hermaphroditic plants in *Antennaria dioica* (L.) Gaertn. *Genetics*, **21**, 282–94.

Urban, I. (1885). Morphologie der Gattung *Bauhinia. Berichte der Deutsch Botanische Gesellschaft*, **3**, 81–101.

van Wyk A. E. (1982). A new species of *Eugenia* (Myrtaceae) from Southern Natal. *South African Journal of Botany*, **1**, 158–62.

Varopoulos, A. (1979). Breeding systems in *Myosotis scorpioides* (Boraginaceae). I. Self-incompatibility. *Heredity*, **42**, 149–57.

Vogel, S. (1955). Uber den Blutendimorphismus einiger Sudafricanischen Pflanzen. *Oesterreichiche Botanische Zeitung*, **102**, 486–500.

Westergaard, M. (1958). The mechanism of sex determination in dioecious flowering plants. *Advances in Genetics*, **9**, 217–81.

Whalen, M. D. & Anderson, G. J. (1981). Distribution of gametophytic self-incompatibility and infrageneric classification in *Solanum. Taxon*, **30**, 761–7.

Wright, G. M. (1979). Self-incompatibility in *Eschscholzia californica. Heredity*, **43**, 429–31.

Yampolsky, C. & Yampolsky, H. (1922). Distribution of sex forms in the phanerogamic flora. *Bibliotheca genetica*, **3**, 1–62.

Young, D. A. (1972). The reproductive biology of *Rhus integrifolia* and *Rhus ovata*. (Anacardiaceae). *Evolution*, **26**, 406–14.

Zadoo, S. N. & Khoshoo, T. N. (1975). Nature of self-incompatibility in the cultivated *Bougainvilleas. Incompatibility Newsletter*, **5**, 73–5.

Zapata, T. R. & Arroyo, M. T. K. (1978). Plant reproductive ecology of a secondary deciduous tropical forest in Venezuela. *Biotropica*, **10**, 221–30.

18

Natural selection and the evolutionary ecology of sex allocation

E. L. CHARNOV

Introduction

The subtitle of this paper might well be: 'What behavioural or evolutionary ecologists, or sociobiologists really do'. To me, all three labels refer to essentially the same enterprise: the use of natural selection to understand the distribution of sex ratios, social and breeding systems, reproductive rates and so forth in nature.

I will first discuss in a general way the use of natural selection as an explanatory principle. In attempting to understand, for example, the distribution of separate versus combined sexes (dioecy versus hermaphroditism), we seek theoretical predictions (e.g. when should an organism be a hermaphrodite?) in terms of the reproductive consequences of possible alternative states or the transition between states. In short, we ask when or under what environmental, social or life-history conditions natural selection favours one or the other form of sexuality. We test our theory by arranging experiments, or geographic or taxonomic comparisons, to see if selection acts as we think it does or if a particular form of sexuality is matched to the predicted environmental condition. Thus we seek to understand sexuality in terms of the ultimate causes (the *why* questions) rather than the proximate mechanisms (the physiologic *how* questions). Seeking answers to the way in which nature is structured, in terms of *why* questions, I shall term *selection thinking.*

This paper really has three simple messages, simple yet important if we are to understand what evolutionary ecology does, and does not do. First, I will demonstrate and explain the value of ultimate as opposed to proximate questions. Second, I will show that selection thinking very often involves not *genetic determinism* in the sense of (say) genes controlling toe number, but rather the evolution of facultative responses. A facultative response is a short-term (within-generation) response to altered environmental conditions. Restated, this second message is that genes often 'say things like': look around this breeding season and produce the *best* number of babies for the prevailing food conditions (where best relates to the reproductive interests of the parents). In particular, selection theory often allows us to predict what the response will be, that is to determine just what an organism will do in response to altered conditions (of course, within certain limits). Third, I want to emphasize that the trait selected for in social situations often differs, depending upon whose interests selection is acting on; that is, *conflict of interest* often characterizes problems in the evolution of social behavior.

To illustrate these three notions, I will discuss one, classic historical example (Lack's theory of bird-clutch size) and then go into some detail in the evolution of sex ratio, and related problems – an area known as the *theory of sex allocation* (Charnov, 1982).

In the 1940s, the great British ecologist, David Lack, became interested in the factors affecting clutch size in birds. Why did some species attempt to rear but a single offspring, while some titmice produced a dozen? And why did clutch size within a species alter from year to year or with latitude? His approach to the determinants of clutch size consisted of asking the *ultimate question* of what were the consequences (to the parents' fitness) of rearing a clutch of a particular size. The fundamental idea was quite simple. Suppose that the survivorship to adulthood of each offspring declined with increasing clutch size (perhaps due to less food being available for each child), as shown in Fig. 1. Disregarding clutch-size influences on parental survival, this survival decline would mean that some intermediate clutch

Fig. 1. Lack's hypothesis for evolution of clutch size. If the survival of an individual offspring to adulthood declines with increasing clutch size, an intermediate clutch size (shown on the figure) results in the greatest number of surviving offspring; this is a measure of parental fitness. Natural selection is expected to favour parents who produce this clutch size.

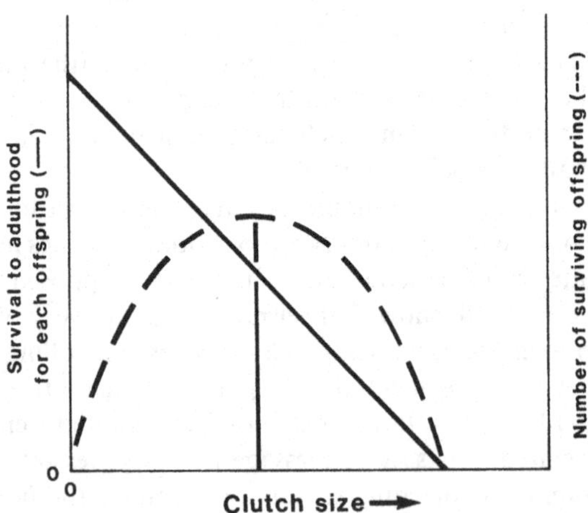

size would be the value which resulted in the largest number of surviving offspring, a fairly good measure of parental fitness. Reviews of this idea are to be found in most ecology texts, and of course, Lack's own writings (especially 1954, 1966, 1968).

It is not my purpose to discuss the many additions and alterations which have been made to this basic framework; needless to say the idea has generated a tremendous amount of very fruitful research. I think the importance of the idea lies in its formulation as an ultimate question. Between-species variation and within-species variation from year to year (or among habitats) are to be understood in terms of alterations in shape of the survival curve of Fig. 1. The shape of the curve itself is to be understood as a function of environmental variables such as food availability or nest-predation patterns. Nowhere in the conception do the underlying physiological mechanisms play much of a role; that is, to understand the clutch size is to first understand the tradeoff curve. For some species (e.g. *Parus major*, the great tit) the year-to-year variation in clutch size may be two-fold; in years of plentiful food the parents attempt to rear twice as many children (Perrins, 1965). Thus, selection on clutch size often favours temporal plasticity and, in theory, this is to be understood as year-to-year shifts in the tradeoff relations.

In 1974, Robert Trivers noted that virtually all models for the evolution of life-history attributes (such as clutch size) assigned control to the parent. That is, the distribution of resources among offspring had classically been assumed to evolve to increase or maximize the parents' fitness. He pointed out that if we allow the offspring to control the resources it gets from the mother, selection operates on it to take more than the amount which the parent is selected to give. Thus, parent and offspring disagree over the quantity of resource to be given to the offspring. This perspective on life-history evolution was unknown prior to 1974, and raises many new questions, particularly for organisms such as birds which display much parental care. Conflict of interest has proven important in understanding many other processes. Examples are male–female relations, animal fighting, and even interactions within an individual, i.e. intragenomic conflict, as discussed later. I now turn to the problem of sex allocation.

Sex-allocation theory

Let me introduce the problem of *sex allocation* (Charnov, 1982) through a simple classification of the forms of sexual reproduction and five general questions. While certainly not exhaustive, the scheme used here is sufficient for our purposes. Most animal or plant species produce only two types of gametes: large and small. In *dioecious* (=gonochoric) organisms, males and females are separate throughout their lives (an individual produces only one size of gamete in its lifetime). In *hermaphroditic* organisms, a single individual produces both large and small gametes in its lifetime. Hermaphroditism takes two forms: *Sequential*: an individual functions early in life as one sex and then switches to the other sex for the rest of its life. *Simultaneous*: an individual produces both kinds of gametes in each breeding season (more or less at the same time). Within sequential hermaphroditism we distinguish protandry (male first) from protogyny (female first).

In relation to the above scheme, the problem of *sex allocation* may be stated as follows:

(i) For a dioecious species, what is the equilibrium *sex ratio* (proportion of males among the offspring) maintained by natural selection?
(ii) For a sequential hermaphrodite, what is the equilibrium *sex order* (protandry or protogyny) and time of *sex change*?
(iii) For a simultaneous hermaphrodite, what is the equilibrium allocation of resources to male versus female function in each breeding season?
(iv) Under what conditions are the various states of hermaphroditism or dioecy evolutionarily stable? For example, when does selection favour genes for protandry over dioecy or when is a *mixture* of sexual types stable?
(v) When does selection favour the ability of an individual to alter its allocation to male versus female function, in response to particular environmental or life-history situations?

These problems are very similar to one another in that each involves working out an equilibrium under natural selection where the possible genotypes have different genetic contributions through male versus female function. Answers to these questions must, of course, consider the biology of the organisms – growth, morphology, mortality, competition (inter- and intraspecific), predation, patchiness in the environment, etc. – as well as possible genetic factors (e.g. inbreeding, autosomal versus cytoplasmic inheritance, etc.). Nevertheless, my phrase, 'the problems are very similar to one another' is meant to indicate that all five questions are really one question, phrased in different forms.

Consider a typical diploid organism. R. A. Fisher (1930) noted the seemingly trivial fact that, with respect to autosomal genes, each zygote gets half of its genome from its father and half from its mother. To put it simply: *everyone has exactly one father and one mother*. However, far from being trivial, this fact holds the key to understanding sex allocation in diploids (and a similar principle holds for haplodiploids). Regardless of whether we are concerned with dioecy, simultaneous hermaphroditism or sequential hermaphroditism, half the autosomal genes come via the male function and half via the female. This fact has two important implications. First, an individual's reproductive success through male function (sperm) is to be measured relative to the male function of other individuals and vice versa for the female function. Second, since half the zygote genes come via each pathway, male and female function are in a real sense equivalent means to reproductive success. Consider, for example, dioecy and the sex ratio. If many daughters are being produced, then large reproductive gains accrue to the producers of sons which are relatively scarce. Selection then favours more sons and an equilibrium will be established where reproductive gains through male and female offspring are equalized. Note that the process generates its own natural selection: it is the scarcity of one sex which itself makes increased production of that sex worthwhile. This is a form of frequency-dependent natural selection and is based on the inevitable fact of reproductive biology that everyone has one mother and one father. This frequency dependence makes sex allocation a prime candidate for ESS techniques (first applied here by Shaw & Mohler, 1953). And John Maynard Smith has been a prime mover in the development of the theory (summary in 1978, 1982).

Theoretical analyses of sex allocation

The problem of sex allocation may be visualized as in Fig. 2. The axes are fitness through male function and through female function. The curve represents the possible tradeoffs, with the endpoints representing *all* male function and *all* female function. Consider three cases: (i) In a species that is dioecious, the curve represents sons and daughters, and we are concerned about the evolution of the sex ratio; (ii) in a species that is a sex reverser, the axes represent reproductive gain as a male versus reproductive gain as a female (the tradeoff curve then represents various proportions of the lifetime spent as a male versus a female); (iii) in a species that is a simultaneous hermaphrodite (consider it a plant), the axes represent reproductive gain via seeds versus pollen. Using population-genetic arguments (Charnov, 1982), one can show that the equilibrium favoured by natural selection is *often* that value on the tradeoff curve which maximized the *product* of the fitness through male function multiplied by the fitness through female function. By casting the problem of sex allocation in the broad form shown in Fig. 2, we find a very general answer to the ultimate questions and an answer that is quite independent of any detailed considerations of proximate mechanisms. Indeed, only at this level of analysis do the five questions posed earlier turn into different forms of the same question. While there are several good examples for testing this theory (Charnov, 1982), I will restrict discussion here to one particular situation, that of sex-ratio control in spatially highly structured populations of wasps and mites.

Sex ratio control in wasps and mites: local mate competition

In 1967, W. D. Hamilton constructed the first sex-ratio theory for spatially structured populations. Here I shall calculate the sex ratio favoured by selection for Hamilton's situation. He proposed the following population structure (labelled 'Local Mate Competition' and abbreviated LMC): The world consists of a large number of islands which, in one generation, are colonized by n fertilized females. These females produce sons and daughters and mating takes place only within each island. The males die and the newly fertilized females (the next generation) then disperse to again colonize islands. Since germinations are discrete, the islands are vacant at the end of each generation. Suppose that we introduce into a large population a mutant female who alters her sex ratio from a proportion r of sons to $\hat{r}$. While rare, this mutant will occur in a group with $n-1$ normal females. Her fitness will be the number of migrants to which she contributes genes through her production of daughters and through females inseminated by her sons. If each mother produces b offspring, the mutant's fitness (W_t) will be

$$W_t = (\text{her daughters}) + (\text{females fertilized by her sons}),$$

Fig. 2. Sex allocation. The figure graphs a hypothetical tradeoff between male fitness and female fitness. The axes may be labelled in any of a number of ways. For example, they may be son production versus daughter production, the problem of interest being selection on the sex ratio. Other interpretations (see text) include reproductive gains for time as a male or female in a sex changer; or reproductive gains for pollen versus seed in a hermaphroditic plant. For all of these *seemingly different* cases, the equilibrium favoured by selection takes the same form shown by the dot on the tradeoff curve. This dot marks the value which maximizes the product of the gain through male fitness multiplied by the gain through female fitness.

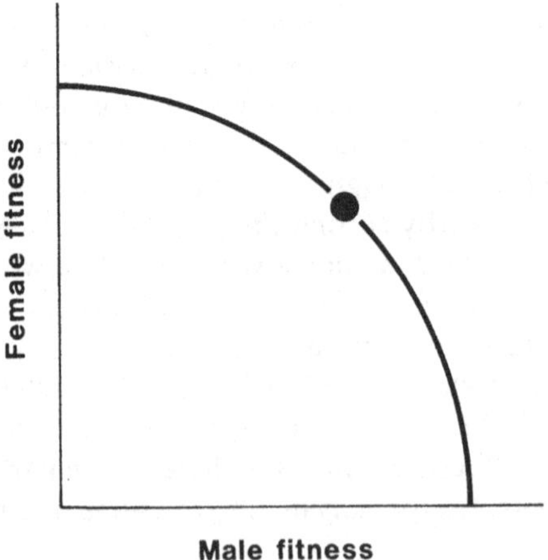

or

$$W_t = b(1-\hat{r}) + \left(\frac{\hat{r}b}{b\hat{r}+(n-1)rb}\right)[b(1-\hat{r})+(n-1)(1-r)b].$$

If $\hat{r}=r$, the mutant's fitness equals that of a normal female. The equilibrium, or favoured sex ratio, is r^*, for which the mutant cannot do better by setting $\hat{r}\neq r$. We require that $\partial W_t/\partial\hat{r}=0$ when $\hat{r}=r$; if this is a maximum, then W_t will decrease for $\hat{r}\neq r$ (at least near r). Applying this rule to W_t, we find the value to be:

$$r^* = \frac{n-1}{2n}. \tag{1}$$

If $n=1$, we have strict sib mating and $r^*=0$, which in practice means that a female should only produce enough sons to ensure the insemination of her daughters. We get the result here that $r^*=0$ since the model implicitly assumed a very large family size. As n gets large, $r^*\rightarrow 1/2$ as shown in Fig. 3. This is the result Hamilton (1967) first derived and applied to a diploid population. It is possible to show that this value maximizes the following product relation where m = number of sons produced and f = number of daughters produced: $f\cdot m^{(n-1)/(n+1)}$. The exponent $(n-1)/(n+1)$ shows how sons are devalued as a function of mating group size, relative to daughters.

Many parasitoid wasps (and some mites) meet or at least approximate the breeding structure assumed in this model. It is also interesting that many variations in the detailed assumptions about population structure do not alter the basic prediction of a steady rise in the sex ratio with increasing group size from female biassed to an asymptote near 1/2. Does nature obey

Fig. 3. Local mate competition and the sex ratio. As given in Eqn (1), increasing numbers of associated ovipositing females cause the equilibrium sex ratio to go from very female biassed to near equality.

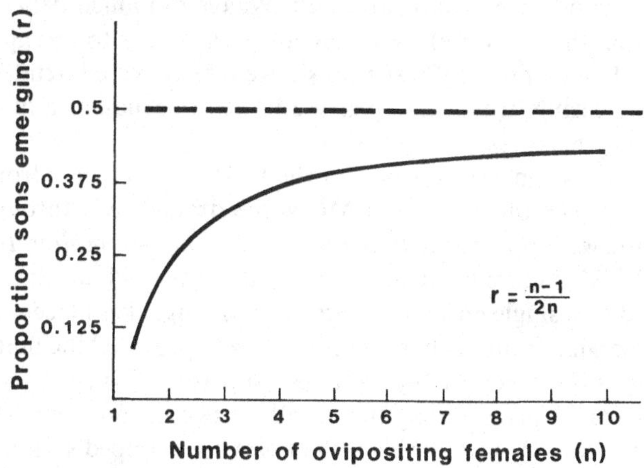

these rules? In what follows I will briefly discuss some of the key LMC data (a more complete discussion is in Charnov, 1982).

In his original formulation of the LMC hypothesis, Hamilton (1967) listed 26 insect and mite species which were haplodiploid and typically had sib mating. All had a strong female bias in the sex ratio. Scolytid bark beetles were of particular interest since they could be divided into two general life histories. In the first, a female and her brood occupied a gallery under bark and mating took place before dispersal from the larval host. In the second type, mating took place after dispersal, upon arrival at the new host. In the first, there is typically sib mating, the sex ratio is strongly female biassed, and males are often flightless (reviewed in Beaver, 1977). In the second, sex ratio at emergence from the larval host is near equality (data in Bakke, 1968; Bartels & Lanier, 1974).

Parasitoid wasps

In a later paper on fig wasps, Hamilton (1979) showed a general positive relationship between the presence of winged males (an indicator of how much mating takes place after dispersal from the host fig) and the proportion of males. Waage (1982) studied LMC in the hymenopteran family Scelionidae. The large majority of species parasitize the eggs of Lepidoptera and Hemiptera. A single wasp egg is laid per host egg, but Waage argued that the degree of LMC was probably related to the dispersion of host eggs. The hosts ranged from those depositing eggs singly to those with masses of up to ~1000 eggs. Wasp species attacking eggs deposited singly should show little LMC, as should those attacking large masses. Here, the mass exceeds the egg-laying capacity of a single female, and the mass would most likely attract several wasp females. Excluding species attacking singly deposited eggs, LMC should generally decline with increasing size of the egg mass attacked. Waage examined field data (his own and from the literature), and assigned each species to an egg mass of typical size. Data for nearly 30 species showed the expected trend, with the sex ratio increasing from female biassed to nearly equal, as a function of increasing egg-mass size.

Probably the strongest evidence for the LMC model comes from wasps and mites in which the degree of LMC varies dramatically through space and time, and *where theory predicts a facultative, short-term alteration in the sex ratio.* LMC is more likely in a gregarious parasitoid, in which several eggs are laid on a single host and where most mating takes place among the children emerging from the host. This effectively provided the material for Hamilton's 1967 list. Here, LMC declines if superparasitism occurs because mating then takes place among the broods of two or more females. LMC may occur in solitary parasitoids if the hosts are clumped so that mating mostly takes place among the offspring emerging from a single clump, as just discussed for the Scelionidae (Waage, 1982). LMC declines for both gregarious and solitary species if the hosts are clumped with several mothers ovipositing on a clump. One wasp species (*Nasonia*) has been well studied with respect to these effects.

Nasonia vitripennis

This is a small (1–3 mm) gregarious, parasitoid wasp (family Pteromalidae) which attacks the pupae of cyclorrhaphous flies (blowflies). It has been much studied (particularly its genetics) and the general biology is well known (reviews in Whiting, 1967; Cassidy, 1975). Edwards (1954) and Wylie (1958) describe the host-finding and oviposition behaviour. *Nasonia* females parasitize pupae from one to several days old (Wylie, 1963; Chabora & Pimentel, 1966). Upon locating a host puparium, a female climbs on and searches its surface, tapping rapidly with her antennae. She drills through the puparium wall with her ovipositor which then is plunged deep within the pupa. At this time, she presumably assessed host suitability for oviposition and also injects a venom which kills and preserves the pupa (Wylie, 1958; Beard, 1964). Eggs are laid in a circle around the sting site in the space between the pupa and the puparium. Following oviposition, females feed on host fluids, which are necessary to mature additional eggs (Edwards, 1954; King & Hopkins, 1963). Multiple attacks are made upon a single host before oviposition is complete (Wylie, 1965). During the interval between attacks, while new eggs are maturing, a female may rest quietly or move about the immediate vicinity of the host, possibly assessing the presence of nearby hosts. After several batches of eggs are laid, a female leaves in search of new hosts. The final clutch size ranges from 10 to 50 eggs. The sex ratio of the primary parasite is 10–15% males. Superparasitism occurs when a female oviposits on a previously parasitized host.

Nasonia was of particular interest to Hamilton (1967) since the superparasite increased the proportion of males among her brood, as is predicted by LMC theory (Wylie, 1966; Holmes, 1970, 1972; Werren, 1980*a*, 1983). Males have short wings and cannot fly, so that mating takes place on the host or, if hosts are clumped, in the immediate vicinity. There are several laboratory studies of sex-ratio alteration related to the degree of crowding of females. Wylie (1965, 1966) studied *Nasonia* on a house-fly host (*Musca*). (Note, however, that *Musca* is not a natural host for *Nasonia* and mortality was rather high compared with studies on blowfly hosts.) When he confined increasing numbers of wasps with 10 hosts, the sex ratios clearly increase from female biassed to nearly equal. Walker (1967) and Velthuis, Velthuis-Kluppell & Bossink (1965) carried out similar experiments (using blowfly hosts) with similar results. Walker's (1967) data were of particular interest since they showed an asymptote at about 60% males. In addition, data from another of her experiments, which varied both host and parasite number, showed that the proportion of males increases (for a fixed number of wasps) as the number of hosts increased, or (for a fixed number of hosts) as the number of wasps increased. This shift is predicted by theory since LMC declines with increased spacing of the broods, brought about by either decreasing the number of wasps or increasing the number of hosts.

Werren (1980*b*, 1983) carried out similar experiments, also with blowfly hosts. Again (Fig. 4), we see a clear rise in sex ratio to near equality. His work also eliminated differential mortality as a factor (generally unknown in previous work). In all these situations, LMC was altered both because of superparasitism and because the hosts were clumped. In laboratory cultures of *Nasonia*, where several females' broods were reared together (details in Werren, 1980*b*, 1983), the emergent sex ratio was approximately 50% males.

In nature, *Nasonia* attacks a wide range of blowfly species (Whiting, 1967). Werren (1980*b*, 1983) has shown that the host situations which are utilized range from single pupae to thousands of pupae in a patch (i.e. under a large carcass). In general, more females are attracted to larger carcasses. While it is not possible to know the number of females (foundresses) at a

Fig. 4. Sex ratio from patches containing 1–12 associated ovipositing females for the wasp *Nasonia* attacking blowfly pupae. Mean values ± 2 standard errors are represented (at least eight replicates per density). The mean values are rather close to those predicted by LMC theory. These experiments eliminated sex specific mortality as a causal factor. (Redrawn from Werren, 1980*b*, 1983.)

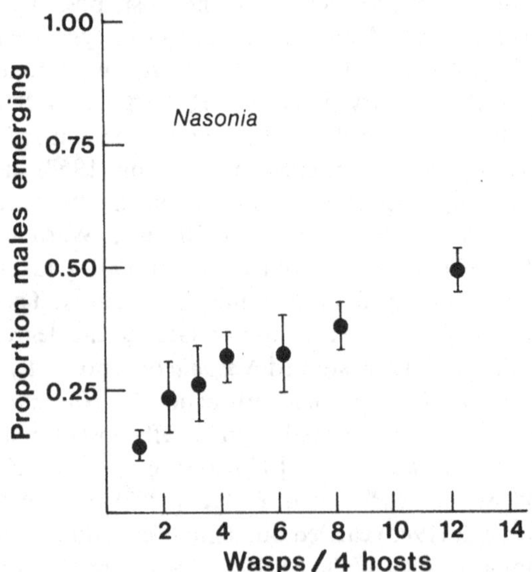

given patch, it is possible to estimate the number of wasps emerging from a patch. This should generally be correlated with number of foundresses. Field data (Werren, 1980*b*, 1983) show a clear rise to a sex ratio near equality. These data also eliminate superparasitism and sex-differential mortality as factors in the sex-ratio shift. Data on many other parasitoids also support the prediction that females kept in groups may shift the sex ratio from female biassed towards equality. Some of these are reviewed by Kochetova (1977) and Waage (1982).

For example, *Trissolcus grandis* is a solitary-egg parasite. Viktorov & Kochetova (1973*a*) showed that females kept alone produced about 13% sons. Females simply kept in tubes which contained traces of other females (previous occupancy) altered their sex ratio to 37% sons. In this same species, females kept in groups also produced more sons (Viktorov, 1968). Viktorov & Kochetova (1973*b*) showed a similar result for *Dahlbominus fuscipennis*, a gregarious ectoparasite of sawfly pupae. They also eliminated differential mortality as a cause of the sex ratio shift. I now discuss sex-ratio data for mites.

Mites

Spider mites are small, haplodiploid creatures (~0.5 mm length) that suck out the contents of leaf cells. They are of immense economic importance (Welch, 1979; Wrensch, 1979). The general life history (with reference to the genus *Tetranychus*) has been reviewed by Mitchell (1973), from which most of this discussion is drawn. A post-dispersal, fertilized female mite feeds in a restricted area, which she marks out with silk. All of her eggs are laid in this area, and the young feed and develop within the territory. At the end of immature development, the young enter a quiescent deutonymph stage, and most are still in the original territory or very near (McEnroe, 1969, 1970). Adult males actively guard female deutonymphs, and mate with them upon emergence as adults (Potter, Wrench & Johnston, 1976*a*,*b*). Since sperm competition favours the first male to mate (Helle, 1967), such guarding is of much importance to male reproductive success.

Only at this one stage are females available for mating, and males often much outnumber available females and actively fight for possession of the quiescent deutonymphs, particularly at high population densities (Potter, 1978, 1979). At low population densities, sib mating is probably the rule (McEnroe, 1969; Mitchell, 1973). However, at high densities, the territories are more closely packed and much of the mating is among non-relatives. After mating, on the first day of adult life, females become restless and disperse. This response is strong in crowded conditions but weak when resources are abundant. After dispersal, females settle down to begin a new cycle. Males do not disperse. They are much smaller than females (15–30%

of a female's weight), and stop growing at maturity. Females continue growing as adults. The sex ratio at maturity is often very female biassed (Mitchell, 1972), and McEnroe (1969) suggested that this was due to LMC. If we consider the territory a feeding resource which a mother allocates to sons and daughters (under low population density), a sex ratio of 20% males translates into a resource allocation of about 5% into males.

Two published studies have considered the impact of crowding on the sex ratio at maturity (Wrensch & Young, 1978; Zaher, Shehata & El-Khatib, 1979). Both confined various numbers of fertilized females to leaf discs and looked at the sex ratio produced among the progeny. Since leaf areas were very similar in the two studies, I have presented the data in Fig. 5 as sex ratio versus female-mite density. As theory predicts, the sex ratio increases with density almost to equality.

I think that these wasp and mite data very strongly suggest that the sex ratio is adjusted (*often facultatively*) to the conditions of the breeding scheme. With mostly sib mating, we get mostly daughters; with more random mating, the ratio rises to near equality.

An aside to higher plants

Most higher plants are hermaphroditic, producing pollen and seeds more or less at the same time. It has been realized for some years (reviewed in

Fig. 5. Crowding of females versus the sex ratio in spider mites (*Tetranychus*). The experiments confined females to leaf discs and looked at the sex ratio among the progeny produced. Leaf discs were ~3 cm² in both studies, so data are represented as female (associated ovipositing females) mite density. Data show a clear rise in the sex ratio from few males to near equality. (Data from Wrensch & Young, 1978; Zaher *et al.*, 1979.)

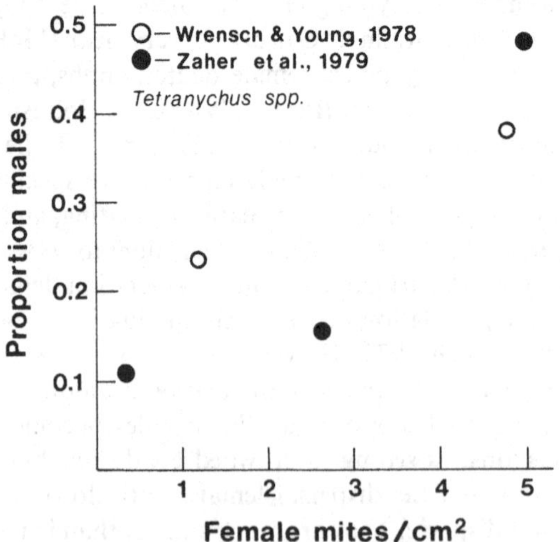

Charnov, 1982) that self-fertilization was rather like sib mating and we predicted that highly selfed plants should allocate little resource to pollen. Consider the work of Lloyd (1972*a*,*b*) on the genus *Cotula* (Compositae). There are about 80 species with gender expression as follows: dioecy ~10; monoecy ~24; gynomonoecy ~35; perfect flowers ~10. In the species tested there were no physiological barriers to selfing. In the monoecious species, a floral head has both male and female florets. Floret length was an approximate indicator of the level of selfing (shorter florets meaning more selfing). As first noted by Lloyd (1972*b*), there is a significant linear average regression (correlation coefficient = 0.61, sample size = 17) of average proportion male florets versus floret length, although visual inspection of the data suggests (with the exception of one point) a relation with an asymptote at about 60% male flowers (Charnov, 1982). Perhaps some of the scatter is related to variation among the species in seed size. This positive relation is predicted by LMC theory.

In sum, these LMC examples drawn from wasps, beetles, mites and higher plants illustrate well the power of an ultimate-question perspective. Some of the examples are a good demonstration of selection favouring facultative shifting of a major life-history parameter (sex ratio), in response to shifts in a social environment.

Sex allocation and intragenomic conflict of interest

The calculation for Fig. 3 (Eqn 1) assumed that the genes controlling the sex ratio are autosomal, nuclear genes. The wasps and mites are haplodiploid, but the autosomal control assumption gives almost the same answer as mother control under haplodiploidy. However, other forms of inheritance are possible; for example, genes located on sex chromosomes or non-nuclear, cytoplasmic DNA. There are a wide array of possibilities. Figure 6 shows the equilibria which would result from various forms of control. XX♀ is haplodiploidy with mother control; note how close this is to the result predicted under autosomal control.

Figure 6 illustrates well the degree of conflict over sex ratio shown by the various 'genes' which may be present in a single individual. Many of these control possibilities are indeed known in nature (review in Charnov, 1982; Bull, 1983). Cytoplasmic control of pollen production (allocation to male function) is well known in hermaphroditic plants. In many cases, interaction between the various genetical elements has also been studied. For example, cytoplasmic inheritance of being a female (turning off pollen production) is often accompanied by nuclear genes (called 'restorer genes') which 'combat the cytoplasmic DNA to restore pollen production'. Indeed, at least two extrachromosomal factors affecting sex ratio are now known from *Nasonia* (Werren, Skinner & Charnov, 1981; Skinner, 1982) and their affects are now being studied in my laboratory. I used the autosomal perspective for much of the previous discussion of data because, for most sex allocation problems, it appears at present to be the most correct one. However, *conflict of interest* is always potentially part of a sex-allocation pattern in nature.

Conclusion

This chapter discussed three major topics. First was the value of *ultimate* questions, contrasted to *proximate* questions. While I stress what one gains by casting problems such as *sex allocation* in the ultimate framework, we must also be aware that the physiological mechanisms are themselves of much interest. It is that the ultimate questions are often just different questions. However, there is no reason we cannot ask natural selection questions about the proximate mechanisms themselves. Indeed, Bull (1983) has done just this for sex-determining mechanisms under dioecy. The second topic is that natural selection often endows the organism with plastic responses to the environmental change which an individual may see during its lifetime. What we can do here is to calculate what the response

Fig. 6. Evolutionary stable strategy (ESS) sex ratios under local mate competition (LMC). Assuming Hamilton's (1967) original population structure, the graph shows the sex-ratio equilibrium as a function of the number of associated ovipositing (founding) females (*n*), and the particular part of the genome assumed to control the sex ratio. Note how close the answers are for autosomal and X, in XX♀ (=haploidiploidy), control. Also note the conflict of interest shown by various genomic components. Graph redrawn from Hamilton (1979). See this reference for detailed discussion, and method of derivation of results.

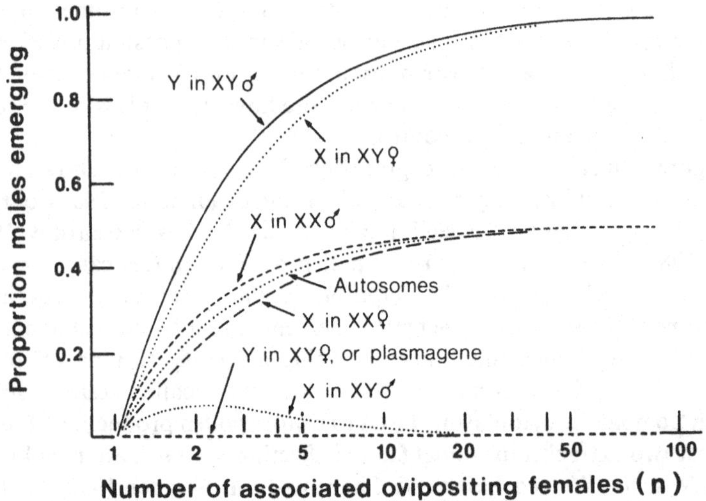

ought to be, using fitness enhancement as a guide to the options 'chosen'. And, finally, conflict of fitness interest is an important part of all social interactions. Sex allocation provides some of the best examples of intragenomic conflict, where an 'individual' is not even an 'individual' with respect to fitness interests (Charnov, 1982).

References

Bakke, A. (1968). Field and laboratory studies on sex ratio in *Ips acuminatus* (Coleoptera: Scolytidae) in Norway. *Canadian Entomologist*, **100**, 640–8.

Bartels, J. M. & Lanier, G. N. (1974). Emergence and mating in *Scolytus multistriatus* (Coleoptera: Scolytidae). *Annals of the Entomological Society of America*, **67**, 365–9.

Beard, R. L. (1964). Pathogenic stinging of housefly pupae by *Nasonia vitripennis* (Walker). *Journal of Insect Pathology*, **6**, 107.

Beaver, R. A. (1977). Bark and Ambrosia beetles in tropical forests. *Proceedings of the Symposium on Forests Pests and Diseases in Southeast Asia. Biotropica Special Publications*, **2**, 133–47.

Bull, J. J. (1983). *The Evolution of Sex Determining Mechanisms and Sex Chromosomes.* Menlo Park, California: Benjamin Cummings.

Cassidy, J. D. (1975). The parasitoid wasps, *Habrobracon* and *Mormoniella.* In *Handbook of Genetics*, ed. R. C. King, vol. 3, pp. 173–203. New York: Plenum Press.

Chabora, P. C. & Pimental, D. (1966). Effect of host (*Musca domestica* L.) age on the Pteromalid parasite (*Nasonia vitripennis* Walker). *Canadian Entomologist*, **98**, 1226–31.

Charnov, E. L. (1982). *The Theory of Sex Allocation.* Princeton Monograph in Population Biology No. 18. Princeton: Princeton University Press.

Edwards, R. L. (1954). The host finding and oviposition behavior of *Mormoniella vitripennis* (Walker) (Hym.: Pteromalidae), a parasite of muscoid flies. *Behavior*, 7, 88–112.

Fisher, R. A. (1930). *The Genetical Theory of Natural Selection.* Oxford: Clarendon Press.

Hamilton, W. D. (1967). Extraordinary sex ratios. *Science*, **156**, 477–88.

Hamilton, W. D. (1979). Wingless and fighting males in fig wasps and other insects. In *Reproductive Competition and Sexual Selection in Insects*, ed. M. S. Blum & N. A. Blum, pp. 167–208. New York: Academic Press.

Helle, W. (1967). Fertilization in the two-spotted spider mite (*Tetranychus urticae*: Acari). *Entomological experimentalis et applicata*, **10**, 103–10.

Holmes, H. B. (1970). *Alteration of sex ratio in the parasitic wasp, Nasonia vitripennis.* PhD thesis, University of Massachusetts, Amherst.

Holmes, H. B. (1972). Genetic evidence for fewer progeny and a higher percent males when *Nasonia vitripennis* oviposits in previously parasitized hosts. *Entomophaga*, **17**, 79–88.

King, P. E. & Hopkins, C. R. (1963). Length of life of the sexes in *Nasonia vitripennis* (Walker) (Hymenoptera: Pteromalidae) under conditions of starvation. *Journal of Experimental Biology*, **40**, 751–61.

Kochetova, N. I. (1977). Factors determining the sex ratio in some entomophagous hymenoptera. *Entomological Review*, **56**, 1–5.

Lack, D. (1954). *The Natural Regulation of Animal Numbers.* Oxford: Clarendon Press.

Lack, D. (1966). *Population Studies of Birds.* Oxford: Clarendon Press.

Lack, D. (1968). *Ecological Adaptations for Breeding in Birds.* London: Methuen.

Lloyd, D. G. (1972*a*). Breeding systems in *Cotula* L. (Compositae, Anthemideae). I. The array of monoclinous and diclinous systems. *New Phytologist*, **71**, 1181–94.

Lloyd, D. G. (1972*b*). Breeding systems in *Cotula* L. (Compositae, Anthemideae). II. Monoecious populations. *New Physiologist*, **71**, 1195–202.

McEnroe, W. D. (1969). Spreading and inbreeding in the spider mite. *Journal of Heredity*, **60**, 343–5.

McEnroe, W. D. (1970). A natural outcrossing swarm of *Tetranychus urticae. Journal of Economic Entomology*, **63**, 343–5.

Maynard Smith, J. (1978). *The Evolution of Sex.* Cambridge University Press.

Maynard Smith, J. (1982). *Evolution and the Theory of Games.* Cambridge University Press.

Mitchell, R. (1972). The sex ratio of the spider mite, *Tetranychus urticae. Entomologica experimentalis et applicata*, **15**, 299–304.

Mitchell, R. (1973). Growth and population dynamics of a spider mite (*Tetranychus urticae*; Acarina: Tetranychidae). *Ecology*, **54**, 1349–55.

Perrins, C. N. (1965). Population fluctuations and clutch size in the great tit, *Parus major. Journal of Animal Ecology*, **34**, 601–47.

Potter, D. A. (1978). Functional sex ratio in the carmine spider mite. *Annals of the Entomological Society of America*, **71**, 218–22.

Potter, D. A. (1979). Reproductive behavior and sexual selection in Tetranychine mites. *Recent Advances in Acarology*, **1**, 137–45.

Potter, D. A., Wrensch, D. L. & Johnston, D. E. (1976*a*). Guarding, aggressive behavior and mating success in male two-spotted spider mites. *Annals of the Entomological Society of America*, **69**, 707–11.

Potter, D. A., Wrensch, D. L. & Johnson, D. E. (1976*b*). Aggression and mating success in male spider mites. *Science*, **193**, 160–1.

Shaw, R. F. & Mohler, J. D. (1953). The selective advantage of the sex ratio. *American Naturalist*, **87**, 337–42.

Skinner, S. W. (1982). Maternally inherited sex ratio in the parasitoid wasp, *Nasonia vitripennis. Science*, **215**, 1133–4.

Trivers, R. L. (1974). Parent–offspring conflict. *American Zoologist*, **14**, 249–65.

Velthuis, H. H. W., Velthuis-Kluppell, F. M. & Bossink, G. A. H. (1978). Some aspects of the biology and population dynamics of *Nasonia vitripennis* (Walker) (Hymenoptera: Pteromalidae). *Entomologica experimentalis et applicata*, **8**, 205–27.

Viktorov, G. A. (1968). Effect of population density on sex ratio in *Trissolcus grandis* Thoms. (Hymenoptera, Scelionidae). *Zoologlscheskii Zhurnal*, **47**, 1035–9.

Viktorov, G. A. & Kochetova, N. I. (1973*a*). The role of trace pheromones in regulating the sex ratio in *Trissolcus grandis* (Hymenoptera, Scelionidae). *Zhurnal Obschchel Biologii*, **34**, 559–62.

Viktorov, G. A. & Kochetova, N. I. (1973*b*). On the regulation of the sex ratio in

Dahlbominus fuscipennis Zett. (Hymenoptera, Eulophidae). *Entom. obozr.*, **52**, 651–7.

Waage, J. K. (1982). Sib-mating and sex ratio strategies in Scelionid wasps. *Ecological Entomology*, 7, 103–12.

Walker, I. (1967). Effect of population density of the viability and fecundity in *Nasonia vitripennis* Walker (Hymenoptera, Pteromalidae). *Ecology*, **48**, 294–301.

Welch, S. M. (1979). The application of simulation models to mite pest management. *Recent Advances in Acarology*, **1**, 31–40.

Werren, J. (1980*a*). Sex ratio adaptations to local mate competition in a parasitic wasp. *Science*, **208**, 1157–9.

Werren, J. (1980*b*). *Studies in the evolution of sex ratios.* PhD thesis, University of Utah, Salt Lake City, Utah.

Werren, J. (1983). Sex ratio evolution under local mate competition in a parasitic wasp. *Evolution*, **37**, 116–24.

Werren, J. H., Skinner, S. W. & Charnov, E. L. (1981). Paternal inheritance of a daughterless sex ratio factor. *Nature*, **293**, 467–8.

Whiting, A. R. (1967). The biology of the parasitic wasp, *Mormoniella vitripennis. Quarterly Review of Biology*, **42**, 333–406.

Wrensch, D. L. (1979). Components of reproductive success in spider mites. *Recent Advances in Acarology*, **1**, 155–64.

Wrensch, D. L. & Young, S. S. Y. (1978). Effects of density and host quality on rate of development, survivorship and sex ratio in the carmine spider mite. *Environmental Entomology*, 7, 499–501.

Wylie, H. G. (1958). Factors that affect host finding by *Nasonia vitripennis* (Walker) (Hymanoptera: Pteromalidae). *Canadian Entomologist*, **90**, 597–608.

Wylie, H. G. (1963). Some effects of host age on parasitism by *Nasonia vitripennis* (Walker) (Hymenoptera: Pteromalidae). *Canadian Entomologist*, **95**, 881–6.

Wylie, H. G. (1965). Some factors that reduce the reproductive rate of *Nasonia vitripennis* (Walk.) at high adult population densities. *Canadian Entomologist*, **97**, 970–7.

Wylie, H. G. (1966). Some mechanisms that affect the sex ratio of *Nasonia vitripennis* (Walk.) (Hymenoptera: Pteromalidae) reared from superparasitized housefly pupae. *Canadian Entomologist*, **98**, 645–53.

Zaher, M. A., Shehata, K. K. & El-Khatib, H. (1979). Population density effects on biology of *Tetranychus arabicus*, the common spider mite in Egypt. *Recent Advances in Acarology*, **1**, 507–9.

19

The evolution of sexual size dimorphism in birds and mammals: a 'hot blooded' hypothesis

P. J. GREENWOOD AND P. WHEELER

The degree of difference in the sizes of the sexes varies enormously across the animal kingdom. At one extreme there are those species of numerous taxa where females are relatively large and males are dwarfs. A few species of deep sea angler fish are bizarre and familiar examples with the male a tiny, fused appendage on the body of the female. At the other extreme, though somewhat less extravagant, male elephant seals can be eight times the weight of reproductive females.

It is clear that throughout most of the animal kingdom there is an overwhelming preponderance of species in which females are the larger sex (Darwin, 1971; Ghiselin, 1974). This size relation is characteristic of most gonochoristic invertebrates and lower vertebrates and despite obvious exceptions in these taxa the overall sex bias is unquestionable. For example, females are larger than males in the nematodes, one of the most abundant groups of animals, and in most insects, the group with the largest number of species.

The fact that females have evolved as the larger sex in the vast majority of species is not surprising given the basic asymmetry in gamete size between the sexes. Females produce large, immobile eggs with attendant food reserves; the energetic costs of reproduction are high. Males produce small, motile sperm. Reproductive output solely in terms of the production of gametes will be more strongly related to body size in females than males. Unless there is strong selection on males for an enormous increase in sperm production, females will evolve a larger body size to accommodate their greater relative investment in gametes. Darwin (1871) was well aware of the importance of this factor in his discussions of the evolution of sexual size dimorphism. Two other factors may also operate to enhance the degree of dimorphism in favour of females. First, any prolongation of female investment beyond that of the egg stage, such as gestation or brooding of young, may be an additional force acting to increase female size. Second, in environments (particularly aquatic ones) where motility is the primary means whereby males achieve access to and fertilization of females, there may be selection for a decrease in male size (Ghiselin, 1974).

One consequence of the difference in investment of the sexes is that the sex investing the least will have the potential to compete for access to members of the other sex. Whereas females are limited to their own egg production or the number of offspring they can rear, a male's reproductive output is limited to the number of females he can fertilize. Males should therefore be expected to compete for access to females. This difference between the sexes provides the cornerstone of Darwin's theory of sexual selection and subsequent refinements of it (Fisher, 1930; Trivers, 1972).

In species where males compete intensively for mates there may be selection for an increase in male size, weaponry and aggression, if the outcome is higher reproductive success. Larger and more combative individuals may outcompete other less-well-endowed males. In addition, females may select such males to pass on those qualities to their sons. If reproductive success becomes more strongly and steeply related to male size than female size, then males may evolve as the larger sex. This will occur in those species where inter-male competition is most intensive, particularly those with polygynous mating systems. However, two important caveats must be made. When males do compete intensively for females (e.g. common toad, Davies & Halliday, 1979) it does not necessarily follow that males will evolve as the larger sex. Nor when males are the larger sex does it necessarily mean that it is the result of male–male competition (e.g. *Gammarus pulex*, Adams & Greenwood, 1983).

Nevertheless, amongst some groups of lower vertebrates and invertebrates there may be an association between male–male competition, the evolution of larger male size and the development of elaborate male weaponry. Examples include the horned beetles, some crabs, amphibians and reptiles (Darwin, 1871; Shine, 1978, 1979; Berry & Shine, 1980). But this apparent association in a few groups does not override the prevalent pattern of greater female size in the vast majority of species.

Many authors who have recently considered the evolution of larger male size in the higher vertebrates have relied almost exclusively on Darwin's theory of sexual selection. One correlation, frequently pointed out, is that between the degree of dimorphism and polygamy. Males are considerably larger than females in highly polygynous species of primates (Clutton-Brock, Harvey & Rudder, 1977), pinnipeds and ungulates (Alexander *et al.*, 1979) and grouse (Wiley, 1974), although this association is not so clear cut in other groups of higher vertebrates (Ralls, 1977). The increase in relative male size coincides with the development of enlarged canines in primates (Harvey, Kavanagh & Clutton-Brock, 1978), antlers for fighting in ungulates (Clutton-Brock, Albon & Harvey, 1980) and elaborate bright plumage in grouse. In contrast, males tend to be only slightly larger in

Table 1. *Sexual size dimorphism in amphibians and reptiles*[a]

	Female larger sex	Male = female	Male larger sex
Amphibians			
Order: Salientia – frogs and toads	514	41	21
Order: Urodela – salamanders	47	18	13
Reptiles			
Order: Squamata – snakes	149	14	61
Order: Chelonia – turtles	36	10	25

Data adapted from Shine (1978, 1979) and Berry & Shine (1980).
[a] All figures represent numbers of species.

monogamous species, male weaponry is absent or less well developed and plumage and pelage are often similar in the two sexes. In a few groups of both taxa, females are larger than males.

By concentrating on those highly polygynous species where sexual selection is clearly important and male size a major influence on reproductive success, zoologists have tended to ignore the fundamental distinction between higher vertebrates and other groups of animals. Females are usually the larger sex in most invertebrates and lower vertebrates (see Table 1), males are usually the larger sex in the vast majority of birds and mammals (see Table 2).

Why should the normal pattern be reversed in birds and mammals? Four possible reasons have been suggested. First, in both birds and mammals there is greater parental investment in the care and rearing of eggs and young. Most species of birds and mammals have clutch or litter sizes of less than ten and many larger species raise only one or two young each breeding season. Large size may be relatively unimportant to females if reproductive success is determined more by the number of progeny they can rear to independence than by the number of eggs which the female can produce (Trivers, 1972). However, whilst there may be reduced selection for larger females in higher vertebrates, compared with other groups where egg number increases with body size and growth may be indeterminate, the hypothesis cannot account for females becoming smaller than males nor for the finding that large female size may be an important influence on egg size in birds (Perrins, 1979) and neonatal size and lactation in mammals (Ralls, 1976*a*).

Second, factors which favour the evolution of parental care may also predispose species towards greater degrees of male–male competition. Ghiselin (1974) suggests that males compete for access to females in species

Table 2. *Sexual size dimorphism in avian and mammalian orders*

Birds		Mammals	
Apterygiformes	F	Monotremata	M[a]
Struthioniformes	M	Marsupialia	M
Rheiformes	M	Insectivora	M
Casuariiformes	M	Dermoptera	M
Tinamiformes	F[b]	Chiroptera	F
Podicepediformes	M	Primates	M
Gaviiformes	M	Edentata	M
Spheneisciformes	M	Pholidota	M
Procellariiformes	M	Lagomorpha	F
Pelecaniformes	V[c]	Rodentia	M
Ciconiiformes	M	Mysticeti	F
Anseriformes	M	Odontoceti	M
Falconiformes	F	Carnivora	M
Galliformes	M	Pinnipedia	M
Gruiformes	M	Tubulidentata	M
Charadriiformes	V	Proboscidea	M
Columbiformes	M	Hyracoidea	M
Psittaciformes	M	Sirenia	F
Cuculiformes	M	Perissodactyla	M
Strigiformes	F	Artiodactyla	M
Caprimulgiformes	M		
Apodiformes	M		
Trogoniformes	M		
Coliiformes	M		
Coraciiformes	M		
Piciformes	M		
Passeriformes	M		

Data on mammals modified from Ralls (1977).
[a] M = males larger than females in majority of species in order.
[b] F = females larger than males in majority of species in order.
[c] V = trend unclear, large numbers of species in M and F categories.

with maternal care with a resultant increase in relative male size whereas, in the absence of parental care, male reproductive success is determined more by mobility than direct competitive ability. However, it is difficult to reconcile the hypothesis with the varied patterns of parental care in higher vertebrates and larger male size, particularly in those monomorphic monogamous species with bi-parental care.

Third, selection for larger male body size and increased aggressiveness may be more prevalent in diurnal, active and visually oriented animals (Trivers, 1972). Again, whilst there may appear to be a superficial link between the hypothesis and patterns of sexual size dimorphism, it cannot account for the overall distinction between birds and mammals and other groups such as reptiles and insects with comparable sensory adaptations.

Finally, Gould (1983) has suggested that the relatively large brain size of birds and mammals may result in the evolution of complex behavioural interactions which facilitate increases in male size through male–male competition. But this hypothesis could equally well apply to many invertebrates and lower vertebrates which have complex, if stereotyped, courtship and competitive repertoires which should, but do not, result in similar patterns of sexual size dimorphism.

In summary, no one hypothesis can yet explain the evolution of smaller females in birds and mammals. Sexual selection is clearly implicated in extreme cases of dimorphism but not necessarily for the bulk of species where males are only slightly bigger than females. The similar pattern in the two groups transcends major life history differences between them, particularly the fact that most birds are monogamous, most mammals polygamous. The similarity is made the more intriguing since each group has evolved independently from reptiles in which, at least in extant forms, females are usually larger than males. Two other trends are also apparent amongst the higher vertebrates. First, polygynous mammals tend to exhibit more extreme forms of sexual size dimorphism than polygynous birds (Ralls, 1976*b*). Second, these extremes occur in those orders of mammals which have a large modal size (e.g. primates, pinnipeds, artiodactyls; Ralls, 1977).

Sexual size dimorphism and hyperthermia

The major common difference between higher vertebrates and their reptilian ancestors is their mode of temperature regulation. Both birds and mammals are tachymetabolic ('homeothermic endotherms'), possessing high levels of endogenous heat production and low thermal conductances due to an insulatory layer of feathers or fur. Although under most conditions this confers the advantage of exceptional stability in body temperature, it does make them susceptible to hyperthermia, particularly during periods of muscular activity when body temperature may rise several degrees above its normal level (Taylor *et al.*, 1971; Taylor & Rowntree, 1973; Hudson & Bernstein, 1981). While most body tissues will function normally during such elevations, others, such as the testes and developing embryo, are more sensitive to any temperature increase. The crucial factor in this is that the reproductive processes of the two sexes are differentially vulnerable to hyperthermia.

In the male, an increase in body temperature of only a few degrees centigrade above normal will reduce the viability of stored spermatozoa and can damage the spermatogenic epithelium itself (Moore, 1926). Most mammals have alleviated the problem by externalizing the testes, either permanently or periodically during the reproductive season, within a scrotal sac protruding from the body wall. The scrotum (which is thermally insulated from the rest of the body by vascular counter-current heat exchange and is additionally cooled by cutaneous sweat glands) is normally maintained 2–10 deg C below the temperature of the abdomen. In cetaceans and pinnipeds the testes are not externalized but they may be cooled by venous blood returning from the surface of the flippers. Although, for aerodynamic streamlining, birds possess intraabdominal testes, their location between the posterior air sacs (into which inspired air is first drawn) does suggest some localized cooling might be achieved (Cowles & Nordstrom, 1946; Cowles, 1965). In some species of birds the caudal sperm-storage region of the vas deferens is expanded into a cloacal protuberance, the temperature of which is several degrees lower than that of the abdomen (Wolfson, 1954). There is also evidence that avian spermatogenesis proceeds mainly at night, when body temperature is lowest and periods of hyperthermia are not likely to be experienced (Riley, 1937, 1940; Morton, Lofts & Orr, 1970*a*; Murton, Lofts & Westwood, 1970*b*). If, despite the adaptations, the testes do experience hyperthermia, a period of infertility may result, although this is usually only temporary as lost spermatozoa are soon replaced and damage to the spermatogenic epithelium can normally be repaired (Bowler, 1967).

The temperature sensitivity of the embryo means that the problems posed by hyperthermia to the reproductive processes of the female are considerably greater. First, the greater mass and abdominal location of the tissues at risk means they cannot be thermally isolated and selectively cooled, and are therefore directly exposed to any elevations of maternal body temperature. Although in birds most embryonic development takes place outside the body cavity of the female, the early stages (up to or including gastrulation) occur while the egg is still in the reproductive tract. Also, after insemination of the female, the viability of the spermatozoa must be preserved while they are stored in a pouch in the lower oviduct for varying periods, of up to several months in some species, before fertilization of the eggs (Sturkie, 1965). It should be noted that since most birds produce clutches of more than one egg, laid at intervals over many days, the total time the reproductive process of the female is sensitive to hyperthermia is considerably longer than the period of vulnerability of an individual embryo. Second, the effects of hyperthermia are irreversible, resulting in permanent damage or loss of the progeny (Alsop, 1919; MacFarlane, Pennycuik & Thrift, 1957; Skreb & Frank, 1963; Edwards, 1967, 1968). The precise effects vary considerably depending on the species, stage of development at which hyperthermia is experienced and the extent and duration of the elevation in temperature. It is the earlier stages of avian and mammalian ontogeny which appear to be especially vulnerable (Moreng & Shaftner, 1951; Edwards, 1969*a*).

Among mammals, maternal thermal stress results in an increased incidence of resorption, abortion and still-birth of the young. Hyperthermia-induced congenital defects include abnormal limb growth and, in particular, damage to the central nervous system (Edwards, 1969*b*; Edwards, Penny & Zevnik, 1971), which in some species may be caused by brief elevations in maternal temperature of only a few degrees centigrade. For example, in the guinea pig it has been demonstrated that maternal temperatures of 42 °C and above result in a wide range of serious or fatal thermal injuries to the embryos (Edwards, 1969*a*,*b*; Jonson *et al*., 1976; Wanner & Edwards, 1983). Experiments with this species have shown that, across a wide range of environmental temperatures, running causes a rapid rise in body temperature which continues as long as activity is maintained. Body temperatures in excess of 42 °C can be attained within 15 minutes of the onset of exercise (Caputa, Kadziela & Narebski, 1983).

The problem of hyperthermia for the female is made even greater by the additional heat production of the enlarged reproductive organs and unborn young. These metabolically active tissues will generate a considerable amount of heat requiring dissipation through the body surface of the female. For example, the basal metabolism of the human female increases throughout pregnancy to 20% above its normal level. The thermoregulatory problem imposed by this extra endogenous heat production will increase the risk of thermal stress to both embryo and the female herself.

There is consequently a difference in the vulnerability of the reproductive processes of the sexes to hyperthermia; the female is more likely to experience heat stress than the male and the detrimental effects of this on successful reproduction are potentially more serious. In seasonally breeding species it could entail a total loss of reproductive output for the year. We therefore propose that there is greater selection pressure on the female of a species to avoid damaging elevations of core temperature during the reproductive period. This will favour individuals with higher conductances, which are most able to dissipate excess endogenous heat production. Within a taxonomic group the thermal conductance of an animal is related to its surface area to volume ratio; smaller animals have higher conductances (Herreid & Kessel, 1967; Bradley & Deavers, 1980) and are consequently better able to cope with endogenous heat stress. A fractional decrease in linear body size will produce a reciprocal increase in the surface area to volume ratio. For example, a female 10% shorter than a male will have an approximately 12% greater ratio. Therefore, if the described problems of hyperthermia are significant influences on reproductive success this thermoregulatory advantage could be the physiological basis for smaller female body size in birds and mammals.

This hypothesis satisfactorily explains the three observed trends in sexual dimorphism. First, smaller female size would only be expected to predominate in higher vertebrates since these are the groups possessing the low thermal conductances and high levels of metabolic heat production which make hyperthermia a major problem. This does not imply that sensitivity to high temperatures is confined to avian and mammalian embryos. For example, in the viviparous lizard *Lacerta vivipara* studies *in vitro* indicate the optimum temperature for embryonic development is 27 °C (Maderson & Bellairs, 1962), below the species' normal preferred body temperature of around 30 °C. However, the problem is solved by the female behaviourally thermoregulating at a lower temperature during the reproductive period (Patterson & Davies, 1978). Second, since in mammals the female retains the embryo for more of its development and the testes of the male are better protected against hyperthermia, there is a greater difference in sensitivity to heat stress between the sexes than in birds. This would explain why sexual size dimorphism is generally more marked in mammals. Third, as hyperthermia is a greater problem for larger species, due to their lower surface area to volume ratio, these will experience stronger selection for reduced female size. This is consistent with the observation that in many groups of birds and mammals the degree of sexual dimorphism increases with body size.

Although it might appear that the problem of maternal hyperthermia would be less for species inhabiting cooler regions, the degree of sexual dimorphism will not in fact be strongly influenced by climate. The reason is that animals have adapted to low-temperature environments with increased pelage and subcutaneous fat thickness to reduce the energetic cost of thermoregulation (Scholander *et al.*, 1950*a*,*b*,*c*), which means that they are still vulnerable to hyperthermia during activity. A possible exception could be some aquatic mammals for which the low temperature and high thermal conductivity of their environment will make damaging elevations of body temperature largely avoidable, thereby reducing the constraint on female size. Since these physical conditions also make the production of large young at birth extremely advantageous for thermoregulatory reasons, there may actually be selection for increased female size. This could explain why among non-polygynous pinnipeds, cetaceans and sirenians females are often the larger sex (Ralls, 1976*a*).

Sexual size dimorphism and loading

So far we have considered two factors, sexual selection and hyperthermia, which have influenced the major trends in sexual size dimorphism in higher vertebrates, both tending to result in larger male size. There is one other major influence on relative sizes which in some circumstances will override the importance of the other two and result in the evolution of larger females.

In some birds the female is larger than the male (see Table 2), a situation known in higher vertebrates as reversed sexual dimorphism. A loading constraint on the female could have resulted in the independent evolution of reversed sexual dimorphism in several unrelated groups of birds. In the period prior to egg laying the female increases in weight due to enlargement of the ovaries, reserves for egg production and incubation and the developing eggs themselves. This breeding increment in body mass constitutes a dead-weight loading which reduces the aerial speed and agility of the female. Although all flying birds will experience this impairment, the problem will be most critical for those with feeding strategies which rely on high flight performance. These include predators such as most raptors and owls, and aerial kleptoparasites like frigate birds and skuas. Selection to limit the reduction in flight performance may have resulted in the evolution of larger females in each of these groups since, within a species, larger individuals will experience a reduced impairment of their flight characteristics while carrying a given breeding increment. In raptors there is the expected relationship between high flight performance and the degree of reversed sexual dimorphism, ranging from ground feeders with little or no dimorphism to species which pursue and catch their prey in flight, in which the female can be almost twice the mass of the male (Wheeler & Greenwood, 1983). For these groups it is probable that the advantages of maintaining female mobility offered by larger body size outweigh the problems of pre-natal hyperthermia and result in the evolution of reversed sexual dimorphism.

The same selection pressures could also explain the occurrence of reversed sexual dimorphism among many species of Microchiropterans which hunt on the wing. These bats not only experience a pre-natal weight loading but the females of some species continue to carry the young after birth on their foraging flights. Interestingly, it has been noted among Vespertilionid species that those which produce larger litters, and consequently have the greatest weight loading, also exhibit the strongest degrees of reversed sexual dimorphism (Myers, 1978). Among the frugivorous Megachiropterans, as expected, males are normally larger than females.

Conclusion

In this chapter we have been concerned with the major trends in sexual size dimorphism throughout the animal kingdom. As a result we do not claim to be able to explain the degree of dimorphism in every group or species since additional selective pressures (ecological, energetic, physiological) may be important in specific cases. We have instead concentrated on those factors which appear to be of importance, particularly in relation to the marked differences between higher vertebrates and other animals.

The relative sizes of the sexes will result from a balance of selection pressures (see Fig. 1) with different pressures predominating in different groups. In the case of birds and mammals we suggest that in addition to sexual selection and loading constraints as influences on size dimorphism,

Fig. 1. A schematic summary of the major factors influencing sexual size dimorphism in bradymetabolic and tachymetabolic animals. Arrow lengths indicate the possible strength of the factors and breadths of their relative expression in the groups. 'Big mother' is relatively more important in bradymetabolic animals which generally produce large numbers of offspring and often have indeterminate growth. The problem of pre-natal hyperthermia is restricted to birds and mammals. Sexual selection will more often produce larger male size in tachymetabolic animals because of the underlying influence of pre-natal hyperthermia and a reduced 'big mother' effect. Sexual selection may produce extreme forms of dimorphism in some birds and mammals but this mechanism may not be relevant to the majority of species where the degree of dimorphism is only slight. Special loading problems include the maintenance of high flight performance in birds of prey (see text) and, in invertebrates and lower vertebrates, the requirement in many species that one sex must be able to carry the other before, during or after copulation (e.g. amphipods, pond skaters, dungflies and toads). The degree of sexual size dimorphism in any particular species will result from the interaction and relative importance of each factor.

Invertebrates and lower vertebrates (bradymetabolic animals)

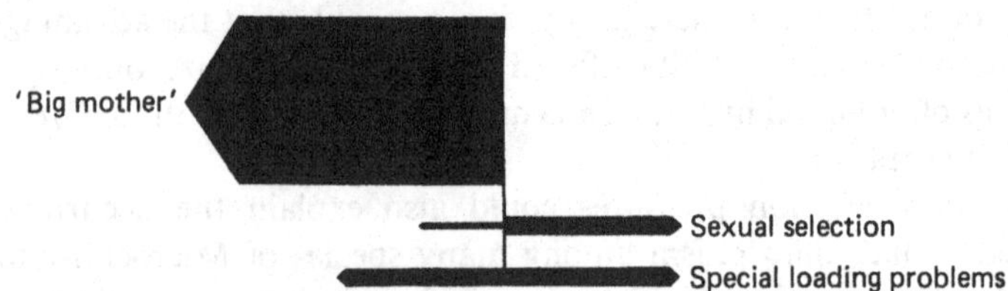

Higher vertebrates (tachymetabolic animals)

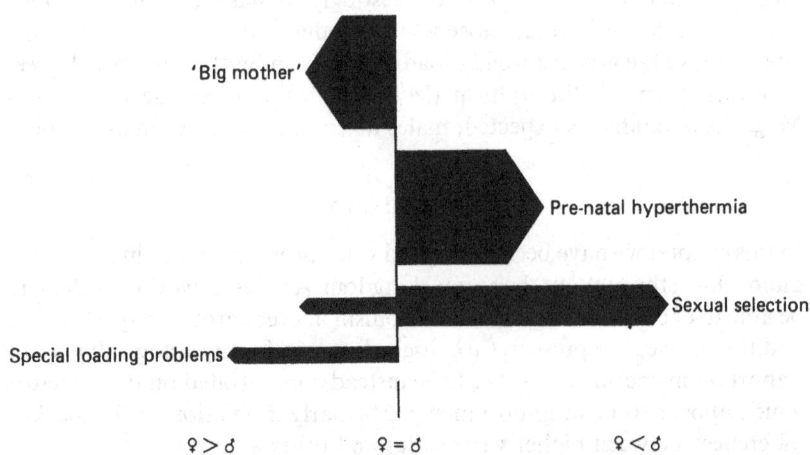

the differential problem of hyperthermia has been a major factor in the evolution of females smaller than males.

References

Adams, J. & Greenwood, P. J. (1983). Why are males bigger than females in pre-copula pairs of *Gammarus pulex*? *Behavioural Ecology and Sociobiology*, **13**, 239–41.

Alexander, R. D., Hoogland, J. L., Howard, R. D., Noonan, M. & Sherman, P. W. (1979). Sexual dimorphisms and breeding systems in pinnipeds, ungulates, primates and humans. In *Evolutionary Biology and Human Social Behaviour*, ed. N. A. Chagnon & W. Iron, pp. 402–35. North Scituate, Massachusetts: Duxbury Press.

Alsop, F. M. (1919). The effect of abnormal temperature upon the developing nervous system in the chick embryos. *Anatomical Record*, **15**, 307–24.

Berry, J. F. & Shine, R. (1980). Sexual size dimorphism and sexual selection in turtles (Order Testudines). *Oecologia (Berl.)*, **44**, 185–91.

Bowler, K. (1967). The effects of repeated temperature applications to testes on fertility in male rats. *Journal of Reproductive Fertility*, **14**, 171–3.

Bradley, S. R. & Deavers, D. R. (1980). A re-examination of the relationship between thermal conductance and body weight in mammals. *Comparative Biochemistry and Physiology*, **65A**, 465–76.

Caputa, M., Kadziela, W. & Narebski, J. (1983). Cerebral temperature regulation in resting and running guinea pigs *Cavia porcellus*. *Journal of Thermal Biology*, **8**, 265–72.

Clutton-Brock, T. H., Albon, S. D. & Harvey, P. H. (1980). Antlers, body size and breeding systems in the Cervidae. *Nature*, **285**, 565–7.

Clutton-Brock, T. H., Harvey, P. H. & Rudder, B. (1977). Sexual dimorphism, socionomic sex ratio and body weight in primates. *Nature*, **269**, 797–800.

Cowles, R. (1965). Hyperthermia, aspermia, mutation rates and evolution. *Quarterly Review of Biology*, **40**, 341–67.

Cowles, R. B. & Nordstrom, A. (1946). A possible avian analogue of the scrotum. *Science*, **104**, 586–7.

Darwin, C. (1871). *The Descent of Man and Selection in Relation to Sex*. London: Murray.

Davies, N. B. & Halliday, T. R. (1979). Competitive mate searching in male common toads, *Bufo bufo*. *Animal Behaviour*, **27**, 1253–67.

Edwards, M. J. (1967). Congenital defects in guinea pigs following induced hyperthermia during gestation. *Archives of Pathology*, **84**, 42–8.

Edwards, M. J. (1968). Congenital malformations in the rat following induced hyperthermia during gestation. *Teratology*, **1**, 173–8.

Edwards, M. J. (1969*a*). Congenital defects in guinea pigs: foetal resorptions, abortions and malformations following induced hyperthermia during early gestation. *Teratology*, **2**, 313–28.

Edwards, M. J. (1969*b*). Congenital defects in guinea pigs: pre-natal retardation of brain growth of guinea pigs following hyperthermia during gestation. *Teratology*, **2**, 329–36.

Edwards, M. J., Penny, R. H. & Zevnick, I. (1971). A brain cell deficit in newborn guinea pigs following prenatal hyperthermia. *Brain Research*, **28**, 341–5.

Fisher, R. A. (1930). *The Genetical Theory of Natural Selection*. Oxford: Clarendon Press.

Ghiselin, M. T. (1974). *The Economy of Nature and the Evolution of Sex*. Berkeley: University of California Press.

Gould, S. J. (1983). Sex and size. *Natural History*, **92**, 24–7.

Harvey, P. H., Kavanagh, M. J. & Clutton-Brock, T. H. (1978). Sexual dimorphism in primate teeth. *Journal of Zoology*, **186**, 475–85.

Herreid, C. F. & Kessel, B. (1967). Thermal conductance in birds and mammals. *Comparative Biochemistry and Physiology*, **21**, 405–14.

Hudson, D. M. & Bernstein, M. H. (1981). Temperature regulation and respiration in flying white-necked swans. *Journal of Experimental Biology*, **90**, 267–82.

Jonson, K. M., Lyle, J. G., Edwards, M. J. & Penny, R. H. C. (1976). Effect of prenatal heat stress on brain growth and serial discrimination reversal learning in the guinea pig. *Brain Research Bulletin*, **1**, 133—50.

MacFarlane, W. V., Pennycuik, P. R. & Thrift, E. (1957). Resorption and loss of foetuses in rats living at 35 °C. *Journal of Physiology*, **135**, 451–9.

Maderson, P. F. A. & Bellairs, A. D'A. (1962). Culture methods as an aid to experiments on reptile embryos. *Nature*, **195**, 401–2.

Moore, C. R. (1926). The biology of the mammalian testis and scrotum. *The Quarterly Review of Biology*, **1**, 4–50.

Moreng, R. E. & Saftner, C. S. (1951). Lethal internal temperature for the chicken from the fertilised egg to the mature bird. *Poultry Science*, **30**, 255–66.

Murton, R. K., Lofts, B. & Orr, A. M. (1970*a*). The significance of circadian based photosensitivity in the house sparrow *Passer domesticus*. *Ibis*, **112**, 448–56.

Murton, R. K., Lofts, B. & Westwood, N. J. (1970*b*). The circadian basis of photoperiodically controlled spermatogenesis in the greenfinch *Chloris chloris*. *Journal of Zoology*, **161**, 125–36.

Myers, P. (1978). Sexual dimorphism in size of Vespertilionid bats. *American Naturalist*, **112**, 701–11.

Patterson, J. W. & Davies, P. M. C. (1978). Preferred body temperatures: seasonal and sexual differences in the lizard *Lacerta vivipara*. *Journal of Thermal Biology*, **3**, 39–41.

Perrins, C. M. (1979). *British Tits*. London: Collins.

Ralls, K. (1976*a*). Mammals in which females are larger than males. *Quarterly Review of Biology*, **51**, 245–76.

Ralls, K. (1976*b*). Extremes of sexual dimorphism in size in birds. *Wilson Bulletin*, **88**, 149–50.

Ralls, K. (1977). Sexual dimorphism in mammals: avian models and unanswered questions. *American Naturalist*, **111**, 917–38.

Riley, G. M. (1937). Experimental studies on spermatogenesis in the house sparrow *Passer domesticus*. *Anatomical Record*, **67**, 327–52.

Riley, G. M. (1940). Diurnal variations in spermatogenic activity in the domestic fowl. *Poultry Science*, **19**, 361 (abstract).

Scholander, P. F., Hock, R., Walters, V., Johnson, F. & Irving, L. (1950*a*). Heat regulation in some arctic and tropical mammals and birds. *Biological Bulletin*, **99**, 237–58.

Scholander, P. F., Hock, R., Walters, V., Johnson, F. & Irving, L. (1950*b*).

Adaptation to cold in arctic and tropical mammals and birds in relation to body temperature, insulation and basal metabolic rate. *Biological Bulletin*, **99**, 259–71.

Scholander, P. F., Walters, V., Hock, R. & Irving, L. (1950*c*). Body insulation of some arctic and tropical mammals and birds. *Biological Bulletin*, **99**, 225–36.

Shine, R. (1978). Sexual size dimorphism and male combat in snakes. *Oecologia (Berl.)*, **33**, 269–77.

Shine, R. (1979). Sexual selection and sexual dimorphism in the Amphibia. *Copeia*, 1979(2), 297–306.

Skreb, N. & Frank, Z. (1963). Developmental abnormalities in the rat induced by heat shock. *Journal of Embryology and Experimental Morphology*, **11**, 445–57.

Sturkie, P. D. (1965). *Avian Physiology*. Ithaca, New York: Comstock.

Taylor, C. R. & Rowntree, V. J. (1973). Temperature regulation and heat balance in running cheetahs: a strategy for sprinters. *American Journal of Physiology*, **224**, 848–51.

Taylor, C. R., Schmidt-Nielsen, K., Dmi'el, R. & Fedak, M. (1971). Effect of hyperthermia on heat balance during running in the African dog. *American Journal of Physiology*, **220**, 823–7.

Trivers, R. L. (1972). Parental investment and sexual selection. In *Sexual Selection and the Descent of Man 1871–1971*, ed. B. Campbell, pp. 136–79. Chicago: Aldine-Atherton.

Wanner, R. A. & Edwards, M. J. (1983). Comparison of the effects of radiation and hyperthermia on pre-natal retardation of brain growth of guinea pigs. *British Journal of Radiology*, **56**, 33–9.

Wheeler, P. & Greenwood, P. J. (1983). The evolution of reversed sexual dimorphism in birds of prey. *Oikos*, **40**, 145–9.

Wiley, R. H. (1974). Evolution of social organisation and life history patterns among grouse. *Quarterly Review of Biology*, **49**, 201–27.

Wolfson, A. (1954). Sperm storage at lower than body temperature outside the body cavity in some passerine birds. *Science*, **120**, 68–71.

20

Leks and the unanimity of female choice

J. W. BRADBURY, S. L. VEHRENCAMP AND R. GIBSON

One of the striking characteristics of lek mating is the large variance in mating success of males assembled on any one arena. On a typical vertebrate lek, as few as 5% of the males may account for as many as 70% of the matings (e.g. Wiley, 1973; Lill, 1974; Bradbury, 1977). This bias exceeds that recorded in nearly all other known mating systems (Payne & Payne, 1977). Because variance in male copulation rates is a major factor driving sexual selection (Wade & Arnold, 1980), it is not surprising that most lek species are highly sexually dimorphic, adult males have bizarre shapes and behaviours, and the sexes show divergent life histories.

Although male interference may play a role in some lek species, it is generally accepted that the major determinant of the distribution of matings among males is female choice (see review by Bradbury & Gibson, 1983). Since the measured distributions of matings by males reflect the summed choices of many females, it follows that the high levels of bias in these distributions must result from considerable unanimity in female choice. How might such unanimity be generated? Congruence in evaluation of males by females is most likely when males are quite different, when females are quite accurate in assessing critical cues, or both. Unfortunately, we have information neither about the distributions of critical cues among males nor about the levels of female acuity in assessing them. This is primarily because no one has yet identified the critical cues. Claims to have identified the cues (such as position on a lek) notwithstanding, it remains the case that very little of the variance in male mating success on leks can be explained by variations in any trait or characteristic which has yet been measured. This presents a paradox: if the cues are so obvious to females that they are nearly unanimous in their selection of mates, why cannot frustrated field workers identify them?

This issue prompted the following computer simulations. Our basic aim was to see whether observed levels of bias in male mating distributions could be generated by independent female assessments of male cues having reasonable frequency distributions. Our models thus focussed on a particular cue distribution, drew a specified number of males from that distribution to form a lek, and then let a sequence of females compare these males and make a choice for mating. We assumed in all runs that females selected mates independently of each other, each female mated only once, and male interference did not bias mate choice. Within any one set of simulations, all females relied on a common psychometric function to discriminate among males. These functions gave the probability that a female would select that male with the larger cue value in any pair-wise contrast. Following standard psychological models of discrimination, the argument of the functions was the *relative* difference in cues values of the two contrasted males, and the functional values varied sigmoidally with the argument from a chance value of 50% when cues were identical to an asymptotic value of 100% when they were very different. Based on our own field observations of hammer-headed bats, manakins and sage grouse, we modelled choice as a series of pair-wise contrasts between successive males. Thus each female picked two males at random, rejected one and retained the other after comparing the psychometric probability of correct choice with a randomly drawn fraction, and then began the next contrast between the retained male and one drawn at random from those as yet unexamined. For a lek of N males, the male retained at the end of the $N-1$ contrasts was assigned to the female as her mate. This process was repeated for the same lek of males and a specified number of females, and the final mating distribution summarized using a single index as explained below. For any set of parameters, a total of 100 leks were drawn from the specified cue distribution and the index of mating bias averaged over this entire sample.

To compare the simulation results with real observations, we required a single statistic which summarized the skewness in mating success among males on a common lek. We utilized for this purpose an index H which is a normalized form of the variance in male success:

$$H = N/(N-1) \sum_{i=1}^{N} (C_i - 1/N)^2,$$

where N was the number of males per lek and C_i was the fraction of matings attributed to the ith male. This index varied from 0 (when the division of matings was perfectly equitable) to 1 (when a single male obtained all the matings). Chance values of this index if females were simply to mate randomly are 0.011, 0.010, 0.007, and 0.007 for leks of 10, 20, 50, and 80 males respectively. As can be seen in Table 1, the observed values of H for single-season data on real vertebrate leks are at least an order of magnitude larger than these chance values.

Since we do not know how accurately lek females can assess male cues, we tried to bracket likely psychometric rules with four discrimination functions ranging from the very accurate to the rather crude. These are

Table 1. *Index of skew in male mating success in several lek-mating species (see text for computation)*

Species	Number of males per lek	H	Reference
Hammer-headed bat	60	0.248	Bradbury, 1977
Hammer-headed bat	25	0.236	Bradbury, 1977
Black-and-white manakin	20	0.154	Lill, 1974
Black-and-white manakin	10	0.515	Lill, 1974
Sage grouse	60	0.384	Wiley, 1973
Sage grouse	30	0.298	Wiley, 1973
Sage grouse	41	0.359	Wiley, 1973
Prairie chicken	9	0.492	Robel & Ballard, 1974
Indigo birds	14	0.276	Payne & Payne, 1977
Indigo birds	14	0.441	Payne & Payne, 1977
Indigo birds	7	0.308	Payne & Payne, 1977

shown in Figs 1 and 2. Function A is the most acute with a 67% chance of correct discrimination when cues differ by 5% and rises to a 95% chance of correct choice for a cue difference of only 10%. Function B is about half as accurate, but is more typical of literature values. Functions C and D extend the 95% correct choice level to cue differences of 53% and 86% respectively. They were used on cues which only took integer values.

Three male-cue distributions were considered. The 'size model' presumed a normally distributed continuous trait. Since known coefficients of variation for continuous anatomical traits such as length or body weight in lek males fall in the range of 1–5% (Lill, 1974; Bradbury, 1977; Stiles & Wolf, 1979), we used a maximal value of 5% in all runs with this model. The 'age model' allowed for a cue which continued to increase in value with age. The growth trajectory was arbitrarily set proportional to the age since sexual maturity, taken to the 1/2 power. Specified parameters were survival rate and percentage increase in cue value during the first breeding period. Because the cue was increasing with age at the same time as cohort size was decreasing due to mortality, the age model generated a distribution with a very few large-cue males, and a large number of smaller-cue males. It was intended to mimic the distribution of body size and voice pitch in frogs, and experience-dependent performance of displays such as has been described for blackcock (Kruijt, de Vos & Bossema, 1972). The final cue distribution modelled the 'centre effect' which has been proposed for a number of lek species (Wiley, 1973; Lill, 1974). We assumed a linear preference by females for males closest to the lek centre. Males were thus settled in annuli around a central male, and each individual was given a cue value equal to its

Fig. 1. Female discrimination functions used in mate choice simulations. Relative cue difference is plotted on abscissa, and probability female will pick male with larger cue value is given on ordinate. Cue range gives relative cue values that will produce 67% and 95% correct choices respectively. All functions based on following formula:

$$P = \tfrac{1}{2}[1 + (1 - e^{-R[(C_1 - C_2)/C_1]})^A]$$

where R and A were varied to change function shapes.

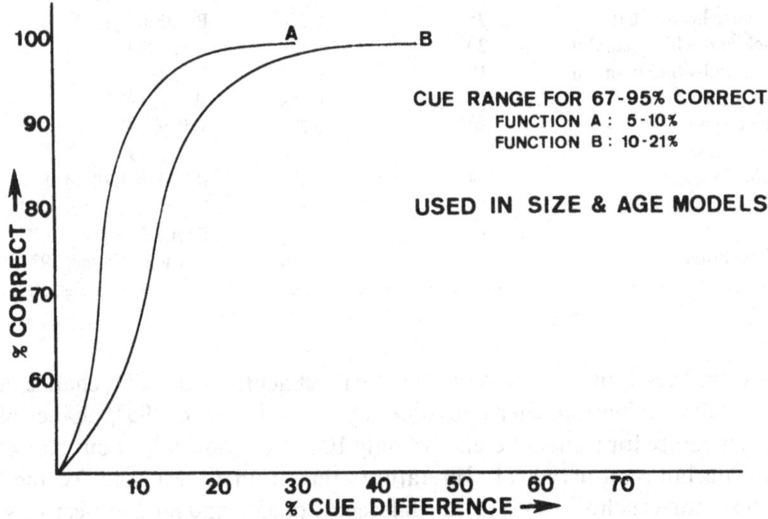

Fig. 2. The three less acute female discriminant functions. The conventions are the same as for Fig. 1.

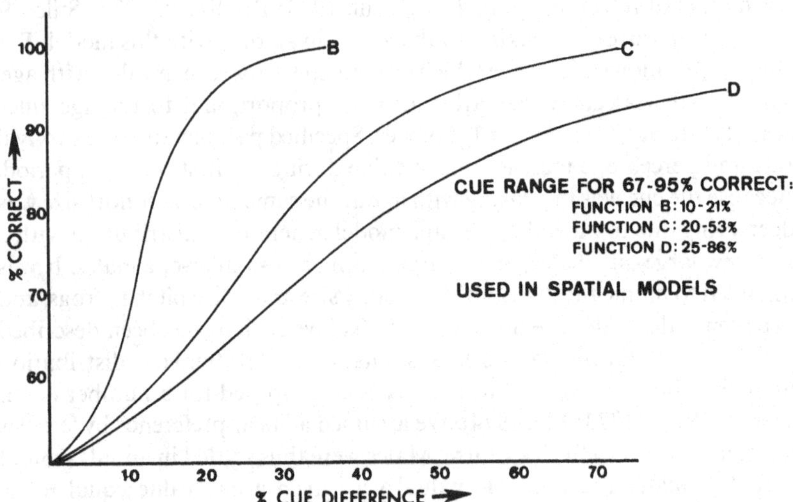

distance (in annuli) from the lek periphery plus one. Three such 'spatial models' were run: (i) a linear lek such as occurs in hammer-headed bats, (ii) a two-dimensional array which was hexagonally packed, and (iii) a rectangularly packed two-dimensional array. Cue values in all spatial models were discrete.

The size model was run with psychometric functions *A* and *B*. Because of the number of possible parameter combinations associated with the age model, only function *B* was utilized for that distribution. Spatial models were run with functions *B*, *C* and *D*. *H* values were found to be most sensitive to small numbers of visiting females, but changed little as female number was increased after 20–30 per lek. Accordingly, all runs reported here used 100 females per lek.

The results of these simulations are summarized in Figs 3–7. Each graph gives the mean and standard deviation of *H* for 100 leks and for four different lek sizes. *H* values for random choice by females were given earlier, and if plotted on these graphs would be indistinguishable from the abscissae. In addition to the simulation predictions, actual *H* values for species appropriate to each model are plotted on each graph. The basic outcomes are described below.

Size model. Except for a few very small leks, the observed *H* values are all much greater than that predicted for independent female choice using physical traits which are continuous. This is true even when the maximal levels of known variation are used, and the most accurate psychometric function is employed (see Fig. 3).

Age model. The fit for age-model distributions varies depending upon the assumed adult survivorship of males. In hammer-headed bats (Bradbury, 1977 and unpublished data), sage grouse (Patterson, 1952), and indigo birds (Payne & Payne, 1977), where adult male survival is 50–60% per year, age-dependent cues would have to increase by 40–50% in the first breeding year to explain the amount of bias seen (Fig. 4). While changes in behavioural performance of this magnitude are possible, it is unlikely that they occur but have been missed in field studies. In contrast, long-lived species such as the manakin (*Manacus manacus*), require only 10–20% increases in cue values during the first year following maturity to explain observed values (Fig. 5). The success of the fit arises because a few older males are so distinct in cue values that they are very easily distinguished from the more numerous younger males by females. Again, the fact that Lill (1974, 1976) was unable to find any clear correlations between a variety of possible cues and mating success in these manakins must raise some doubts about the propriety of this model, despite the fit in *H* values. Since annual survival in frogs varies widely with species, it is possible that there are lek-mating forms in which survival is high enough, and/or growth in the first adult year fast enough, that bias in mate choice may follow strictly from the model. However, even

Fig. 3. Simulation results for size model using trait coefficient of variation of 5% and female choice functions *A* and *B*. Predicted values plotted as means (open circles) and standard deviations of 100 leks (dotted lines). Actual values for relevant lek species plotted as dark circles. Female functions identified by letter and cue range (see Fig. 1).

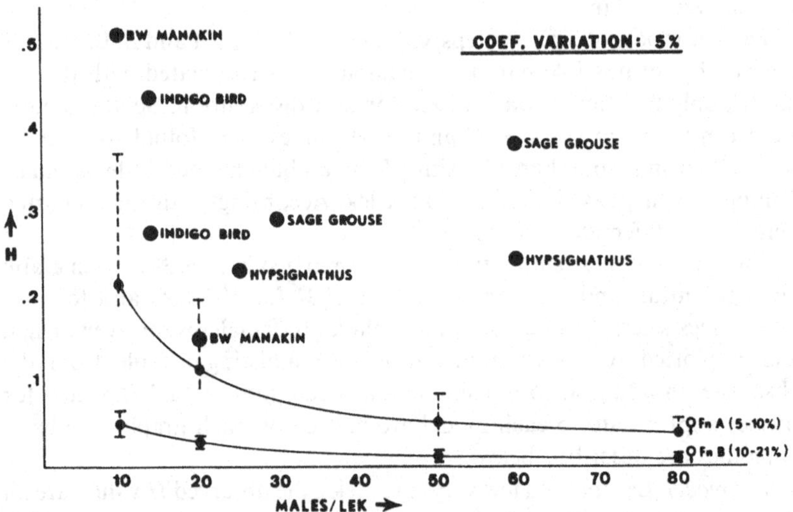

Fig. 4. Simulation results for age model with 50% survival per breeding season and various rates of cue growth in first breeding season. Other notations as in Fig. 3. All runs use female function *B*.

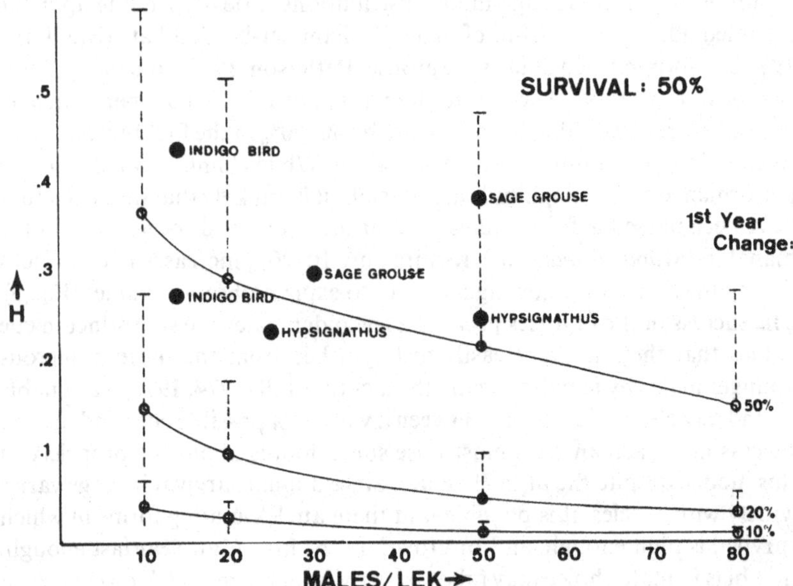

growth changes of 20% in the first year of maturity are high, and we have been unable to find in the literature evidence that any anuran with lek mating fits the specified conditions.

One other comment on the age models is worth making. It is obvious from Figs 4 and 5 that the variances in H for any lek size are much larger than for the other cue distributions. This high variance arises because sampling from a distribution with very few large-cue males and many small-cue ones will sometimes draw older males in the formation of a lek and sometimes not. When only younger males are present, the cue distribution is more similar to that for the size model. If the prevalent cues used by females were age dependent, one would expect to find respectable variations in male mating success bias for leks of the same size in the same populations. In the samples noted here, the indigo bird data do seem to show such variations for similarly sized leks. The two sets of data on manakins also show large variations but they differ in lek sizes, which confounds the interpretation. More extensive within-species data would be of considerable interest in this respect.

Spatial models. All of the spatial-model runs showed poor fits between predicted and observed H values. Figure 6 gives the results for the linear arrays of hammer-headed bats, and Fig. 7 plots the outcomes for

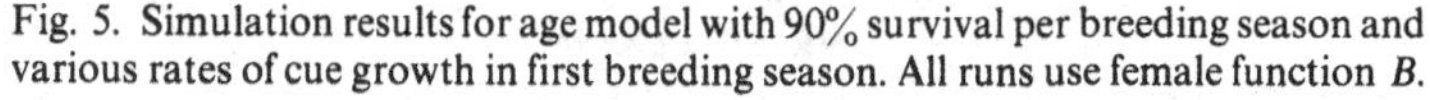
Fig. 5. Simulation results for age model with 90% survival per breeding season and various rates of cue growth in first breeding season. All runs use female function B.

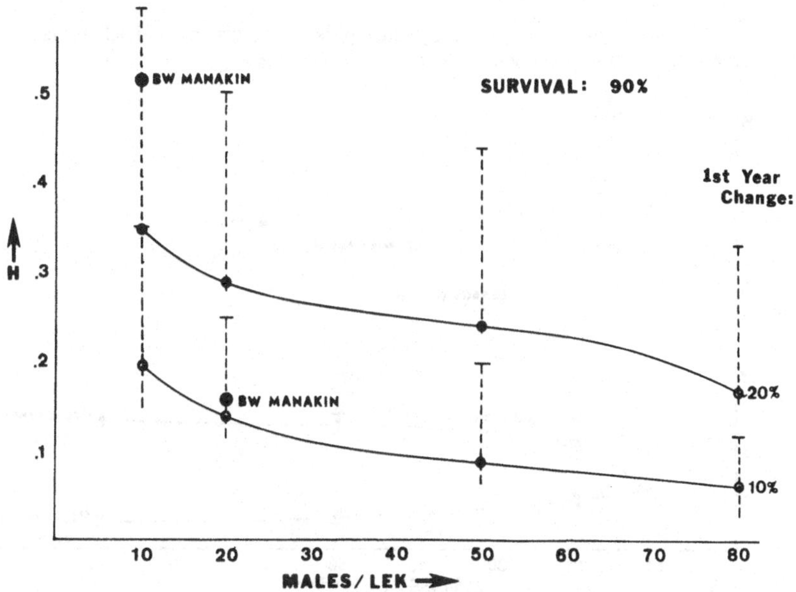

hexagonally packed two-dimensional arrays. The rectangular-packing data are not plotted because they are nearly identical in values to the hexagonal runs. The very low predicted H values for these models all arise because so many males (within a common annulus) have the same cue values.

Fig. 6. Simulation results for linear spatial model and several female discrimination functions.

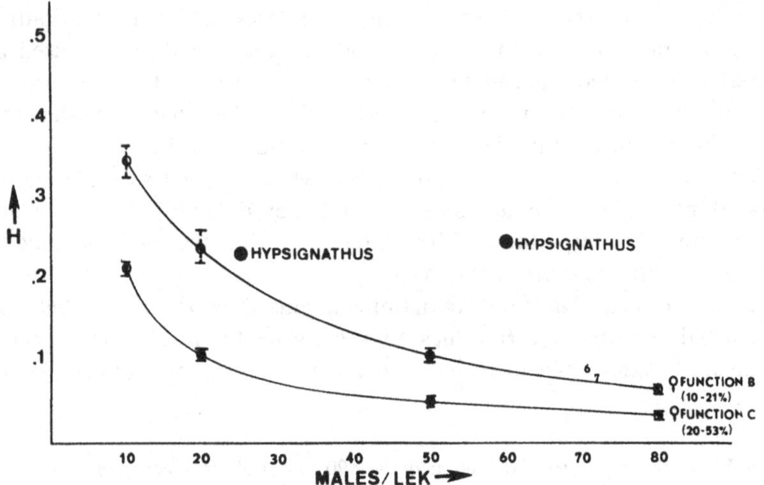

Fig. 7. Simulation results for hexagonally packed two-dimensional lek using several female discrimination functions.

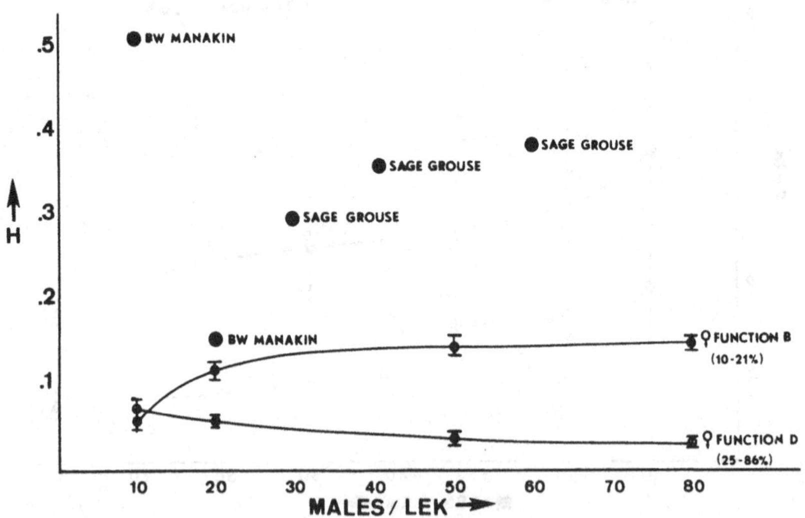

In short, the simple models proposed here for female choice were generally unable to produce the bias in male success seen. The exceptions were small leks in a few cases, and very long-lived species if age-dependent cues are assumed. This leaves a majority of the examples cited unexplainable by the models. As with any simulation model, there are a variety of simplifying assumptions which are called into doubt when the model fails to provide some reasonable fit to observations. We discuss below four assumptions invoked in our models whose failure to be true could both cause the observed mismatch and raise interesting theoretical issues.

1. *The accuracy of female discrimination is higher than assumed here, and/or follows a typical discrimination processes.* It is possible that the correlated selection on female discrimination in these highly sexually selected species has produced animals which are much more accurate at assessing male cues than are other species with other types of discrimination tasks. The best way to examine this possibility is to use standard laboratory discrimination testing on lek-species females. The problem of course is finding the proper cues to be contrasted. A related possibility is that female acuity is not relative but is absolute: our simulations assumed relative acuity (e.g. percent difference between cue values); if female accuracy was fixed for all cue values, this would have made discrimination for larger cue values in the model less accurate than was appropriate. The use of absolute acuity would have little effect on the outcomes of the size models, where males are very similar in cue value, but might enhance the fit for the age-model runs. Again, the patterns of acuity in females of lek species ought to be assessable in laboratory studies.

2. *Females use several cues jointly and multiple-cue assessment is more accurate than single-cue assessment.* Although it is possible that females use a suite of cues in comparing males, it is not obvious that this will increase accuracy. If cues are uncorrelated, one might expect greater 'compound' uniformity among males as viewed by females than would be true for a single cue; the outcome would depend greatly on the rules adopted by females for combining conflicting information. Where cues are correlated, it is not clear that additional information is being provided. We are currently re-running our simulations with several cues and a variety of cue-correlations and decision rules.

3. *Female choice may not be independent.* The possibility that female choice may not be independent has been raised by several authors (e.g. Wiley, 1973; Lill, 1974). In a variety of lek species, females tend to visit males in groups and, in some, they may even move between male-display territories as a unit. Certainly, the potential for female copying is present in a variety of forms. If females copied each other's choices sufficiently frequently, this would generate higher unanimity and thus biasses in male success than would be expected given our models. Assuming females copied

both 'correct' and 'erroneous' choices with equal fidelity, high levels of female copying might also explain why there are such poor correlations observed between the values of likely cues and male mating success. *Why* females might bother to copy is another issue. Where first-year females constitute a sufficiently large fraction of the population, copying of older experienced females by younger ones might be both advantageous and sufficiently prevalent to affect the final mating distributions. Whether females of similar experience ought to copy is an issue that deserves theoretical attention. In any case, our simulations point up the possibility that female copying is occurring and this at least ought to alert field workers to spend more time trying to detect and quantify it.

4. *Male-cue distributions are more variable than the three models used here would predict.* It is unlikely that physical traits of males are more variable than our three models would predict. However, this may not be true of the behavioural and facultative components of male display. Presumably, all behavioural components have some maximal levels of intensity and/or frequency, and these are likely to be distributed in a manner similar to physical traits. If males all displayed at their respective maximal levels, the values of behavioural components would be distributed according to one of our models. However, if males had reasons to individually reduce facultative components *below* maximal levels, a much greater variance in the values of male behavioural components would be possible.

There are imaginable conditions which might promote reductions in display by some males on a lek. The most obvious scenario is as follows: (*a*) display at maximal levels early in a lek season allows males to make estimates of their relative abilities; (*b*) rate or intensity of display is important in female choice, but is positively correlated with the risk of either energetic 'burn-out' or predation; (*c*) males who view themselves as unlikely competitors with the apparent top males reduce their own display to enhance their own chances of remaining on the lek and then acceed to top status when the original top males are killed or exceed their energy reserves. To justify this reduced display, it must be the case that there is a sufficiently high chance of improving status later to justify the loss of matings now.

The conditions favouring this type of display reduction can be exemplified in the following simple example. Consider a two-male lek such as occurs in some manakins (Foster, 1977). We assume that display rate or intensity is important in female choice and is directly correlated with the risk of dying or running out of sufficient energy to remain on the lek. During establishment of the lek, we presume that the two males estimate the relative fractions of matings each would hope to get if it displayed at its maximal level. Suppose the more effective male expects to get a fraction q of the matings, where q is greater than $\frac{1}{2}$, whereas the less-effective male only expects to get $(1-q)$. To simplify the mathematics, we divide the season into two consecutive periods. If either male displays at its maximal level during the first period, it has a probability S_1 of being present on the lek during the second period. If, instead, it reduces its display level to the minimum necessary to hold its territory through the first period, it has an enhanced probability, S_2, of still being on the lek during the second period. If one male reduces its display to the minimum, while the other retains maximal display levels, the second obtains a fraction z of the matings during that period, $(z > q)$, and the reduced displayer obtains only $(1 - z)$ of the matings. If both reduce display to minimal levels, we assume they split matings equally. Should either male die or disappear during the first period, we assume the survivor gets all matings during the second period and that there is no change in the total number of matings due to the absence of one male. (The latter assumption is reasonable if males display where females are likely to go, but not if females are attracted to sites males choose to use for display.)

These conditions generate the following asymmetric game table (all payoffs are entire-season values):

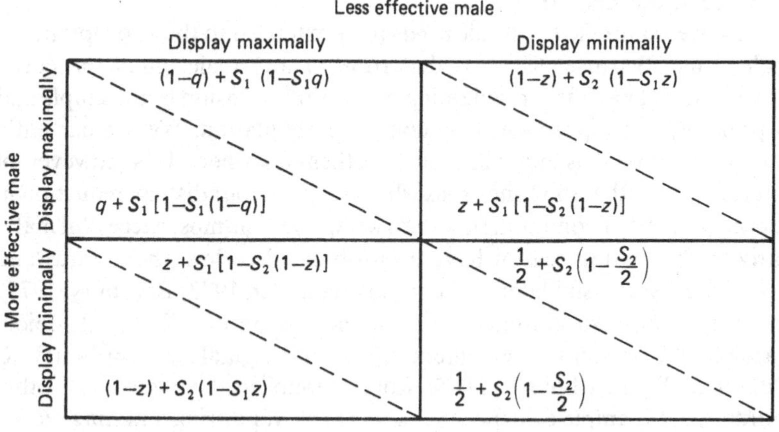

Less effective male
Display maximally
Display minimally
More effective male
Display maximally
Display minimally
$(1-q) + S_1\ (1-S_1q)$
$q + S_1\,[1-S_1(1-q)]$
$(1-z) + S_2\ (1-S_1z)$
$z + S_1\,[1-S_2(1-z)]$
$z + S_1\,[1-S_2(1-z)]$
$(1-z) + S_2(1-S_1z)$
$\frac{1}{2} + S_2\left(1-\frac{S_2}{2}\right)$
$\frac{1}{2} + S_2\left(1-\frac{S_2}{2}\right)$

Rather than solve this table analytically, it is sufficient for our point to see whether reasonable parameter values can make reduced display by the less effective male an evolutionarily stable strategy (ESS). Considering a species with moderately low initial survival, we can set $S_1 = 0.4$, $S_2 = 0.6$, $q = 0.6$, and $z = 0.9$. This is a situation in which the initial bias (set by maximal abilities) is small, and the change in survival gained by display reduction is modest. This set of parameter values results in the single ESS in which both males display maximally throughout the season. However, if the gains in survival generated by display reduction are increased, e.g. by letting $S_1 = 0.2$ and $S_2 = 0.8$ (other parameters remaining unchanged), the single ESS is for the more effective male to display maximally while the less effective male should display minimally. This same ESS can be generated by returning survival values to the initial 0.4 and 0.6, respectively, but increasing the initial skew so that $q=0.8$. In fact, display reduction by the less effective male and maximal display by the more effective male is generally the ESS when $(S_2-S_1)/[(1-q)-(1-z)]$ is sufficiently large.

While it is clear that display reduction *within* a period can increase the variance in male mating success for that time interval, it is not obvious that it might alter the variance for the entire season. Using the second parameter set in our example above, two males which both displayed maximally would produce an H of 0.023; were one to maintain maximal display, and the other to display at reduced levels, the resultant H would be 0.032. The third set of parameter values gives an H of 0.180 if the two males ignore the ESS and both display maximally; the value if they adopt the ESS is a higher value of 0.203. These numerical examples do not prove that display reduction *always* increases the variance in male mating success, but they demonstrate that such reduction can be favoured and increase male variance at the same time.

This over-simplified example needs to be extended to the appropriate N-male game. Because death or departure of any male alters the entire distribution of payoffs for remaining males, this extension is not simple and is probably best approached by computer simulation. We are currently running such models and will report on them elsewhere. It is, however, of interest to note that the conditions which might favour display reduction in leks may not be uncommon. In several lek species, the most successful males early in the mating season have been observed to lose their status and become less successful later in the season (Hartzler, 1972; Bradbury, 1977). In Uganda kob, successful males frequently drop out of the lek, rejoin bachelor herds, and later return to their original territories (H. K. Buechner, W. Leuthold & H. D. Roth, unpublished material). In other words, there is ample evidence that some turnover during a mating season does occur. Because we do not know the critical cues in any species, we do not yet know whether the second condition, that behaviourally variant components of display are important in mate choice, is true or not. It is certainly the case that the studies providing the best evidence for identification of the critical cues implicated behavioural, not physical, attribute in mate choice (Hartzler, 1972; Payne & Payne, 1977). Finally, very few studies have even considered the relationships between likelihood of remaining on a lek throughout a season and the levels of display. As noted, males may disappear from leks either because predators capture them or because they run out of sufficient energy. Predation on lek males is common in the frog *Physalaemus* (Tuttle & Ryan, 1981) and sage grouse (Patterson, 1952; Hartzler, 1972), but no study has looked to see whether risk has any relation to levels of display. The cost of display in lek species, relative to overall energy budgets, has again not been examined in any detail. Bucher, Ryan & Bartholomew (1982) have shown that calling in the frog *Physalaemus* is substantial, but whether this is limited by the entire budget is unknown. The abandonment of display in both sage grouse and hammer-headed bats (J. Bradbury, S. Vehrencamp & R. Gibson, unpublished observations; Bradbury, 1977) after several days of colder than average weather (but not necessarily *on* a cold day), suggests that males of these species may be under some significant energy constraints. Clearly, detailed studies on the relative costs of display in a variety of lek species would be of interest both to physiologists and to those concerned with mating strategies in this type of breeding system.

Many of the computer simulations cited here were undertaken at the University of Sussex during a sabbatical visit by J. Bradbury and S. Vehrencamp. The authors want to thank John Maynard Smith for sponsoring that visit and this work, and Paul Harvey, Brian Charlesworth and Deborah Charlesworth for stimulating discussions and pleasant evenings at various Brighton restaurants and Lewes pubs.

References

Bradbury, J. W. (1977). Lek mating behavior in the hammer-headed bat. *Zeitschrift für Tierpsychologie*, **45**, 225–55.

Bradbury, J. W. & Gibson, R. (1983). Leks and mate choice. In *Mate Choice*, ed. P. P. G. Bateson, pp. 109–38. Cambridge University Press.

Bucher, T. L. Ryan, M. J. & Bartholomew, G. A. (1982). Oxygen consumption during resting, calling, and nest building in the frog *Physalaemus pustulosus*. *Physiological Zoology*, **55**, 10–22.

Buechner, H. K., Leuthold, W. & Roth, H. D. Lek territory occupancy in male Uganda kob antelope (unpublished MS).

Foster, M. S. (1977). Odd couples in manakins: a study of social organization and cooperative breeding in *Chiroxiphia linearis*. *American Naturalist*, **111**, 845–53.

Hartzler, J. E. (1972). An analysis of sage grouse lek behavior. PhD thesis, University of Montana, Missoula, Montana.

Kruijt, J. P., de Vos, G. J. & Bossema, I. (1972). The arena system of black grouse. *Proceedings of the XVth International Ornithological Congress*, pp. 399–423. Leiden: E. J. Brill.

Lill, A. (1974). Social organization and space utilization in the lek-forming white-bearded manakin, *M. manacus trinitatis*. *Zeitschrift für Tierpsychologie*, **36**, 513–30.

Lill, A. (1976). Lek behavior in the golden-headed manakin (*Pipra erythrocephala*) in Trinidad (West Indies). *Fortschrift Verhaltensforschung*, Heft 18, pp. 1–84. Berlin/Hamburg: Verlag Paul Parey.

Patterson, R. L. (1952). *The Sage Grouse in Wyoming*. Denver: Sage Books.

Payne, R. & Payne, K. (1977). Social organization and mating success in local song populations of village indigo birds, *Vidua chalybeata. Zeitschrift für Tierpsychologie*, **45**, 113–73.

Robel, R. J. & Ballard, W. B. (1974). Lek social organization and reproductive success in the Greater Prairie chicken. *American Zoologist*, **14**, 121–8.

Stiles, F. G. & Wolf, L. L. (1979). Ecology and evolution of lek mating behavior in the long-tailed hermit hummingbird. *Ornithological Monographs*, **27**, 1–78. Washington, DC: American Ornithologists' Union.

Tuttle, M. & Ryan, M. J. (1981). Bat predation and the evolution of frog vocalizations in the neotropics. *Science*, **214**, 677–8.

Wade, M. J. & Arnold, S. J. (1980). The intensity of sexual selection in relation to male sexual behavior, female choice, and sperm precedence. *Animal Behaviour*, **28**, 446–61.

Wiley, R. H. (1973). Territoriality and non-random mating in sage grouse, *Centrocercus urophasianus. Animal Behaviour*, **6**, 85–169.

John Maynard Smith, FRS, Publications

1952 J. Maynard Smith – The importance of the nervous system in the evolution of animal flight. *Evolution*, **6**, 127–9.

1953 J. Maynard Smith – Birds as aeroplanes. *New Biology*, **14**, 64–81.

1954 J. Maynard Smith & S. Maynard Smith – Genetics and cytology of *Drosophila subobscura* – VIII. Heterozygosity, viability and rate of development. *Journal of Genetics*, **52**, 152–64.

T. Koske & J. Maynard Smith – Genetics and cytology of *Drosophila subobscura* – X. The fifth linkage group. *Journal of Genetics*, **52**, 521–41.

1955 J. M. Clarke & J. Maynard Smith – The genetics and cytology of *Drosophila subobscura* – XI. Hybrid vigour and longevity. *Journal of Genetics*, **53**, 172–80.

M. J. Hollingsworth & J. Maynard Smith – The effects of inbreeding on rate of development and on fertility in *Drosophila subobscura. Journal of Genetics*, **53**, 295–314.

J. Maynard Smith, J. M. Clarke & M. J. Hollingsworth – The expression of hybrid vigour in *Drosophila subobscura. Proceedings of the Royal Society B*, **144**, 159–71.

1956 J. Maynard Smith – Fertility, mating behaviour and sexual selection in *Drosophila subobscura. Journal of Genetics*, **54**, 261–79.

J. Maynard Smith – Acclimatization to high temperatures in inbred and outbred *Drosophila subobscura. Journal of Genetics*, **54**, 497–505.

J. Maynard Smith & R. J. G. Savage – Some locomotory adaptations in mammals. *Journal of the Linnean Society* (*Zool.*), **42**, 603–22.

1957 J. Maynard Smith – Temperature tolerance and acclimatization in *Drosophila subobscura. Journal of Experimental Biology*, **34**, 85–96.

1958 J. Maynard Smith – Prolongation of the life of *Drosophila subobscura* by a brief exposure of adults to a high temperature. *Nature*, **181**, 496–7.

J. Maynard Smith – The effects of temperature and of egg-laying on the longevity of *Drosophila subobscura. Journal of Experimental Biology*, **35**, 832–42.

J. Maynard Smith – *The Theory of Evolution.* London: Penguin Books.

1959 J. Maynard Smith – Sex-limited inheritance of longevity in *Drosophila subobscura. Journal of Genetics*, **56**, 1–9.

J. Maynard Smith – A theory of ageing. *Nature*, **184**, 956–8.

J. Maynard Smith – The rate of ageing in *Drosophila subobscura.* In *CIBA*

Foundation Symposium on The Lifespan of Animals, 269–81.
J. Maynard Smith & R. J. G. Savage – The mechanics of mammalian jaws. *The School Science Review*, **141**, 289–301.

1960 J. M. Clarke, J. Maynard Smith & P. Travers – Maternal effects on the rate of egg development in *Drosophila subobscura. Genetical Research, Cambridge*, **1**, 375–80.
J. Maynard Smith – Continuous, quantized and modal variation. *Proceedings of the Royal Society B*, **152**, 397–409.
J. Maynard Smith & K. C. Sondhi – The genetics of a pattern. *Genetics*, **45**, 1039–50.

1961 J. M. Clarke & J. Maynard Smith – Independence of temperature of the rate of ageing in *Drosophila subobscura. Nature*, **190**, 1027–8.
J. M. Clarke & J. Maynard Smith – Two phases of ageing in *Drosophila subobscura. Journal of Experimental Biology*, **38**, 679–84.
J. M. Clarke, J. Maynard Smith & K. C. Sondhi – Asymmetrical response to selection for rate of development in *Drosophila subobscura. Genetical Research, Cambridge*, **2**, 70–81.
J. Maynard Smith – Evolution and history. In *Darwinism and the Study of Society*, ed. Michael Banton, pp. 83–93. London: Tavistock Publications.
J. Maynard Smith & K. C. Sondhi – The arrangement of bristles in *Drosophila. Journal of Embryology and Experimental Morphology*, **9**, 661–72.

1962 J. Maynard Smith – *An Agnostic View of Evolution*, pp. 49–73.
J. Maynard Smith – I. The causes of ageing. In *Review Lectures on Senescence. Proceedings of the Royal Society, B*, **157**, 115–27.
J. Maynard Smith – Disruptive selection, polymorphism and sympatric speciation. *Nature*, **195**, 60–2.
J. Maynard Smith – The limitations of molecular evolution. In *The Scientist Speculates*, ed. I. J. Good, pp. 252–6.

1963 J. Maynard Smith – Temperature and the rate of ageing in poikilotherms. *Nature*, **199**, 400–2.
J. Maynard Smith – Letter to the Editor – *The Lancet*, 358.

1964 M. J. Lamb & J. Maynard Smith – Radiation and ageing in insects. *Experimental Gerontology*, **1**, 11–20.
J. Maynard Smith – Group selection and kin selection. *Nature*, **200**, 1145–47.
J. Maynard Smith & D. Michie – Machines that play games. *New Scientist*, **12**, 367–9.

1965 J. Maynard Smith – The spatial and temporal organization of cells. In *Mathematics and Computer Science in Biology and Medicine*, pp. 247–54. London: Medical Research Council.
J. Maynard Smith – An agnostic view of evolution. In *Biology and Personality*, ed. I. Ramsey, pp. 49–73. Oxford: Blackwell.
J. Maynard Smith – Professor J. B. S. Haldane, FRS. *Nature*, **206**, 239–40.
J. Maynard Smith – The evolution of alarm calls. *The American Naturalist*, **94**, 59–64.
J. Maynard Smith – Eugenics and Utopia. *Daedalus*, **94**, 487–505.

1966 J. M. Clark & J. Maynard Smith – Increase in the rate of protein synthesis with age in *Drosophila subobscura. Nature*, **209**, 627–9.
J. Maynard Smith – Theories of ageing. In *Topics in the Biology of Ageing*, ed.

P. L. Krohn, pp. 1–35. New York: Wiley.

J. Maynard Smith – Sympatric speciation. *The American Naturalist*, **100**, 637–50.

S. Maynard Smith & J. Maynard Smith – The preservation of genetic variability. In *Regulation and Control in Living Systems*, ed. H. Kalmus, pp. 328–48. London: Wiley.

1967 J. Maynard Smith – The genetics of ageing. The Woodhull Lecture 3 November 1967. *Proceedings of the Royal Institute*, **42**, 69–74.

1968 F. Dingley & J. Maynard Smith – Temperature acclimatization in the absence of protein synthesis in *Drosophila subobscura*. *Journal of Insect Physiology*, **14**, 1185–94.

J. Maynard Smith – Evolution in sexual and asexual populations. *The American Naturalist*, **102**, 469–73.

J. Maynard Smith – 'Haldane's Dilemma' and the rate of evolution. *Nature*, **219**, 1114–16.

J. Maynard Smith – *Mathematical Ideas in Biology*. Cambridge University Press.

1969 F. Dingley & J. Maynard Smith – Absence of a life-shortening effect of amino-acid analogues on adult *Drosophila*. *Experimental Gerontology*, **4**, 145–9.

M. J. Lamb & J. Maynard Smith – Radiation-induced life-shortening in *Drosophila*. *Radiation Research*, **40**, 450–64.

J. Maynard Smith – Limitations on growth rate. *Symposia of the Society for General Microbiology*, No. 19, 1–13.

J. Maynard Smith – The status of neo-Darwinism. In *Towards a Theoretical Biology*, vol. 2, *Sketches*, ed. C. H. Waddington, pp. 82–9. Edinburgh: Edinburgh University Press.

1970 J. Maynard Smith – Time in the evolutionary process. *Studium Generale*, **23**, 266–72.

J. Maynard Smith – Population size, polymorphism, and the rate of non-Darwinian evolution. *The American Naturalist*, **104**, 231–7.

J. Maynard Smith – Natural selection and the concept of a protein space. *Nature*, **225**, 563–4.

J. Maynard Smith – The causes of polymorphism. *Symposia of the Zoological Society of London*, **26**, 371–83.

J. Maynard Smith – Genetic polymorphism in a varied environment. *The American Naturalist*, **104**, 487–90.

J. Maynard Smith, A. N. Bozcuk & S. Tebbutt – Protein turnover in adult *Drosophila*. *Journal of Insect Physiology*, **16**, 601–3.

1971 J. Maynard Smith – What use is sex? *Journal of Theoretical Biology*, **30**, 319–35.

J. Maynard Smith – Population genetics and molecular evolution. In *De la Physique Théorique à la Biologie*, pp. 230–9. Comptes rendus de la Seconde Conférence Internationale de Physique Théorique et Biologie. Palais des Congrès, Versailles, 30 Juin–5 Juillet 1969. Publiés sous la direction de M. Marois. Editions du Centre National de la Recherche Scientifique 15, quai Anatole-France – Paris VIIe.

1972 J. Haigh & J. Maynard Smith – Population size and protein variation in man. *Genetical Research, Cambridge*, **19**, 73–89.

J. Haigh & J. Maynard Smith – Can there be more predators than prey? *Theoretical Population Biology*, **3**, 290–9.

J. Maynard Smith – *On Evolution.* Edinburgh: Edinburgh University Press.

J. Maynard Smith & M. G. Ridpath – Wife sharing in the Tasmanian native hen (*Tribonyx mortierii*): a case of kin selection? *The American Naturalist*, **106**, 447–52.

1973 J. Maynard Smith & G. R. Price – The logic of animal conflict. *Nature*, **246**, 15–18.

J. Maynard Smith & M. Slatkin – The stability of predator–prey systems. *Ecology*, **54**, 384–91.

1974 J. Maynard Smith – The theory of games and the evolution of animal conflicts. *Journal of Theoretical Biology*, **47**, 209–21.

J. Maynard Smith – *Models in Ecology.* Cambridge University Press.

J. Maynard Smith – Recombination and the rate of evolution. *Genetics*, **78**, 299–304.

J. Maynard Smith & J. Haigh – The hitch-hiking effect of a favourable gene. *Genetical Research, Cambridge*, **23**, 23–35.

1975 J. Baker, J. Maynard Smith & C. Strobeck – Genetic polymorphisms in the bladder campion, *Silene maritima. Biochemical Genetics*, **13**, 393–410.

1976 E. Charnov, J. Maynard Smith & J. J. Bull – Why be an hermaphrodite? *Nature*, **263**, 125–6.

J. Haigh & J. Maynard Smith – The hitch-hiking effect – a reply. *Genetical Research*, **27**, 85–7.

L. Lawlor & J. Maynard Smith – The co-evolution and stability of competing species. *American Naturalist*, **110**, 79–99.

J. Maynard Smith – Sexual selection and the handicap principle. *Journal of Theoretical Biology*, **57**, 239–42.

J. Maynard Smith – A comment on the Red Queen. *American Naturalist*, **110**, 325–30.

J. Maynard Smith – What determines the rate of evolution. *American Naturalist*, **110**, 331–8.

J. Maynard Smith – Evolution and the theory of games. *American Scientist*, **64**, 41–5.

J. Maynard Smith – Group selection. *Quarterly Review of Biology*, **51**, 20–9.

J. Maynard Smith – A short-term advantage for sex and recombination through sib-competition. *Journal of Theoretical Biology*, **63**, 245–8.

J. Maynard Smith & G. A. Parker – The logic of asymmetric contests. *Animal Behaviour*, **24**, 159–75.

C. Strobeck, J. Maynard Smith & B. Charlesworth – The effects of hitch-hiking on a gene for recombination. *Genetics*, **82**, 547–58.

1977 J. Maynard Smith – Parental investment: a prospective analysis. *Animal Behaviour*, **25**, 1–9.

J. Maynard Smith – Why the genome does not congeal. *Nature*, **268**, 693–6.

J. Maynard Smith – The limitations of evolution theory. In *The Encyclopaedia of Ignorance*, ed. R. Duncan & M. Weston-Smith, pp. 235–42. Oxford: Pergamon Press.

J. Maynard Smith – The sex habit in plants and animals. In *Measuring Selection in Natural Populations*, ed. F. B. Christiansen & T. H. Fenchel, pp. 265–73. Berlin: Springer-Verlag.

1978 J. Maynard Smith – The handicap principle – a comment. *Journal of Theoretical Biology*, **70**, 251–2.

J. Maynard Smith – The evolution of behaviour. *Scientific American*, **239**, 176–91.

J. Maynard Smith – Optimization theory in evolution. *Annual Review of Ecology and Systematics*, **9**, 31–56.

J. Maynard Smith – The concept of sociobiology. In *Morality as a Biological Phenomenon*, ed. G. S. Stent, pp. 23–34. Berlin: Verlag-Chemie.

1979 R. Heller & J. Maynard Smith – Does Muller's ratchet work with selfing? *Genetical Research, Cambridge*, **32**, 289–93.

W. G. S. Hines & J. Maynard Smith – Games between relatives. *Journal of Theoretical Biology*, **79**, 19–30.

J. Maynard Smith – Hypercycles and the origin of life. *Nature*, **280**, 445–6.

J. Maynard Smith – The effects of normalizing and disruptive selection on genes for recombination. *Genetical Research*, **33**, 121–8.

J. Maynard Smith – The ecology of sex. In *Behavioural Ecology – An Evolutionary Approach*, ed. J. R. Krebs & N. B. Davies, pp. 159–79. Oxford: Blackwell.

J. Maynard Smith – Game theory and the evolution of behaviour. *Proceedings of the Royal Society of London, B*, **205**, 475–88.

M. Slatkin & J. Maynard Smith – Models of coevolution. *Quarterly Review of Biology*, **54**, 233–63.

1980 J. Maynard Smith – A new theory of sexual investment. *Behavioural Ecology and Sociobiology*, **7**, 247–51.

J. Maynard Smith – Selection for recombination in a polygenic model. *Genetical Research*, **35**, 269–77.

J. Maynard Smith – Models of the evolution of altruism. *Theoretical Population Biology*, 151–9.

J. Maynard Smith & R. Hoekstra – Polymorphism in a varied environment: how robust are the models? *Genetical Research*, **35**, 45–57.

1981 J. Maynard Smith – Macroevolution. *Nature*, **289**, 13–14.

J. Maynard Smith – Will a sexual population evolve to an ESS? *American Naturalist*, **117**, 1015–18.

J. Maynard Smith – Learning the evolutionarily stable strategy. *Journal of Theoretical Biology*, **89**, 611–33.

J. Maynard Smith – Symbolism and chance. *Scientific Philosophy Today*, ed. J. Agassi & R. S. Cohen, pp. 201–6. D. Reidel Publishing Company.

J. Maynard Smith – Overview – unsolved evolutionary problems. In *Genome Evolution*, Genome Conference, Cambridge, June 1981, ed. G. A. Dover & R. B. Flavell, pp. 375–82. London: Academic Press.

1982 J. Maynard Smith – Commentary on Webster and Goodwin's 'The origin of species: a structuralist approach'. *Journal of Social and Biological Structures*, **5**, 49–68.

J. Maynard Smith – The evolution of social behaviour – a classification of models. In *Current Problems in Sociobiology*, ed. Kings College Sociobiology Group, pp. 30–44. Cambridge University Press.

J. Maynard Smith – Do animals convey information about their intentions? *Journal of Theoretical Biology*, **97**, 1–5.

J. Maynard Smith (ed.) – *Evolution Now*. London: Macmillan.

J. Maynard Smith & N. Warren – Models of cultural and genetic change. *Evolution*, **36**, 620–7.

1983 J. Maynard Smith – Game theory and the evolution of cooperation. In *Evolution from Molecules to Men*, ed. D. S. Bendall, pp. 445–56. Cambridge University Press.

J. Maynard Smith – Evolution and development. In *Development and Evolution*, ed. B. C. Goodwin, N. Holder & C. C. Wylie, pp. 33–45. Cambridge University Press.

J. Maynard Smith – Contemporary aspects of evolution. *Experientia*, **39**, 805–6.

J. Maynard Smith – *Evolution and the Theory of Games*. Cambridge University Press.

J. Maynard Smith – Models of evolution. *Proceedings of the Royal Society of London, B*, **219**, 315–25.

J. Maynard Smith – The genetics of stasis and punctuation. *Annual Reviews of Genetics*, **17**, 11–25.

J. Maynard Smith – Adaptation and satisficing. *Behavioral and Brain Sciences*, **6**, 370–1.

J. Maynard Smith – Current controversies in evolutionary biology. In *Dimensions of Darwinism*, ed. M. Greene, pp. 273–86. Cambridge University Press.

1984 J. Maynard Smith – Science and myth. *Natural History*, **93**, 10–24.

J. Maynard Smith – Game theory and the evolution of behaviour. *Behavioural and Brain Sciences*, **7**, 95–125.

J. Maynard Smith – The population as a unit of selection. In *Evolutionary Ecology*, ed. B. Shorrocks, pp. 195–202.

J. Maynard Smith – Matchsticks, brains and curtain rings. *New Scientist*. 16 Feb. pp. 9–10.

J. Maynard Smith & S. E. Riechert – A conflicting-tendency model of spider antagonistic behaviour. *Animal Behaviour*, **32**, 564–78.

1985 R. L. Brown & J. Maynard Smith – Competition and body size. *Theoretical Population Biology*, in press.

J. Maynard Smith – The evolution of intelligence. In *Mind Machines and Evolution*, ed. C. Hookway. Cambridge University Press (in press).

J. Maynard Smith – *The Problems of Biology*. Oxford University Press.

Index